Insect Reproduction

Edited by
Simon R. Leather and Jim Hardie
Imperial College of Science, Technology, and Medicine
University of London
Department of Biology
Imperial College at Silwood Park
Ascot, England

CRC Press
Taylor & Francis Group
Boca Raton London New York

CRC Press is an imprint of the
Taylor & Francis Group, an **informa** business

First published 1995 by CRC Press
Taylor & Francis Group
6000 Broken Sound Parkway NW, Suite 300
Boca Raton, FL 33487-2742

Reissued 2018 by CRC Press

Library of Congress Cataloging-in-Publication Data

Insect reproduction / Simon R. Leather and Jim Hardie, editors.
p. cm.
Includes bibliographical references (p.) and index.
ISBN 0-8493-6695-X (alk. paper)
1. Insects--Reproduction. I. Leather, S. R. (Simon R.)
II. Hardie, Jim.
QL495.I4985 1995
595.7'016--dc20 95- 16294

A Library of Congress record exists under LC control number: 95016294

Publisher's Note
The publisher has gone to great lengths to ensure the quality of this reprint but points out that some imperfections in the original copies may be apparent.

Disclaimer
The publisher has made every effort to trace copyright holders and welcomes correspondence from those they have been unable to contact.

ISBN 13: 978-1-315-89450-8 (hbk)
ISBN 13: 978-1-351-07360-8 (ebk)

Visit the Taylor & Francis Web site at http://www.taylorandfrancis.com and the
CRC Press Web site at http://www.crcpress.com

INTRODUCTION

This book, consisting of ten review chapters contributed by leading workers in their respective fields, from around the world, covers the whole subject of insect reproduction. It begins with the basic physiological questions of insect reproduction, moves on to discuss the new advances seen in the fields of behavioral and ecological mechanisms, and culminates by examining the recent work on evolutionary biology and its application in the field.

Each chapter, although including a brief review of the basic seminal work, focuses mainly on the advances made within the last ten years and highlights those areas in which the respective authors see the greatest scope for further important advances. By allowing each author full rein to explore their chapter subject using their particular "hobby horse," it has been possible to make this not just a book of review chapters, but one in which exciting new ideas have been raised.

This book should be of general interest to all entomologists, whether they are in pure or applied fields, and should also be an important asset to any teaching program where entomology is taught at the undergraduate and post-graduate level.

SRL, JH.

THE EDITORS

Dr. Simon R. Leather is presently Lecturer in Applied Ecology and Pest Management at Imperial College, London. He obtained his B.Sc. from the University of Leeds, England in 1977 with first class honors in Agricultural Zoology. After receiving his Ph.D. in 1980 from the University of East Anglia in Norwich, he embarked on further research in aphid ecology. Then followed a ten year spell with the British Forestry Commission where he worked in the Research Division, primarily on the population biology of forest pests with particular reference to their reproductive behavior. He started his current position in 1992. Dr. Leather is a Fellow of the Royal Entomological Society, a Member of the British Ecological Society, a Member of the Flora and Fauna Preservation Society, and a Member of the Institute of Biology. He sits on the Council of the Royal Entomological Society and edits their journal *Antenna*.

Dr. Jim Hardie is presently a Principal Research Fellow at Imperial College, London. He obtained a Ph.D. degree from the University of Birmingham, England in 1975 and a D.Sc. from London University in 1989. He has worked in the field of aphid physiology for more than twenty years and is regarded as one of the leading figures in this area. He is a Fellow of the Institute of Biology, the Royal Entomological Society, and the Royal Microscopical Society.

CONTRIBUTORS

Roger L. Blackman, B.Sc., Ph.D.
Department of Entomology
The Natural History Museum
London, England

Carol L. Boggs, Ph.D.
Center for Conservation Biology
Department of Biological Sciences
Stanford University
Stanford, California
and Rocky Mountain Biological Laboratory
Crested Butte, Colorado

A. F. G. Dixon, B.Sc., D. Phil.
School of Biological Sciences
University of East Anglia
Norwich, England

Cedric Gillott, B.Sc., Ph.D., D.Sc.
Department of Biology
University of Saskatchewan
Saskatoon, Canada

Jim Hardie, B.Tech., Ph.D., D.Sc.
Department of Biology
Imperial College of Science, Technology, and Medicine
University of London
Silwood Park
Ascot, England

Klaus H. Hoffmann, Prof. Dr.
Department of Animal Ecology I
University of Bayreuth
Bayreuth, Germany

Simon R. Leather, B.Sc., Ph.D.
Department of Biology
Imperial College of Science, Technology, and Medicine
University of London
Silwood Park
Ascot, England

Athol McLachlan, Ph.D., D.Sc.
Department of Agriculture and Environmental Sciences
University of Newcastle-upon-Tyne
Newcastle-upon-Tyne, England

Rachel Neems, B.Sc., Ph.D.
Department of Genetics
University of Leeds
Leeds, England

Richard Wall, B.Sc., Ph.D.
School of Biological Sciences
University of Bristol
Bristol, England

Christer Wiklund, Ph.D.
Department of Zoology
University of Stockholm
Stockholm, Sweden

TABLE OF CONTENTS

DEDICATION

This book is dedicated to our families in recognition of the support given during the somewhat lengthy process that ensued once we had embarked upon this task. So, thank you Gill, Fern, John, James, Thomas, and Matthew from Simon, and thank you Ros, Sally, Nicola, and Robert from Jim.

Chapter 1

OOGENESIS AND THE FEMALE REPRODUCTIVE SYSTEM

Klaus H. Hoffmann

CONTENTS

I. INTRODUCTION

Insect reproduction results from a succession of interdependent steps which are often quite different in nature and take place at various stages of the insect life cycle. The main reproductive events in females are sex determination, gonial mitoses and meioses, differentiation of the

0-8493-6695-X/95/$0.00+$.50

reproductive organs, previtellogenesis and vitellogenesis, accessory gland functioning, sexual behavior, mating, ovulation, and oviposition. It is the function of the female reproductive tract to produce the eggs and to deposit them at an appropriate time and in an appropriate place. In addition, the female reproductive tract must receive the spermatozoa from the male and transport them to the spermatheca where they are stored until they are used to fertilize the eggs as they are oviposited.

Insect reproduction strictly depends on environmental factors. Factors which may affect reproduction include temperature, humidity, photoperiod, nutritive conditions, and a suitable egg-laying substrate.

Regulation of insect reproduction involves numerous sensory receptors, nerve transmission, and integration in the brain, which regulates the synthesis of the two groups of insect developmental hormones, the juvenile hormones and the ecdysteroids, and produces its own neurohormones.[1] The mechanisms which regulate each reproductive step may vary with insect species (see Chapter by Hardie).

Present knowledge of oogenesis has progressed variously, depending on the event considered. The best-known field is that of vitellogenesis and its endocrine regulation, while the early events of ovarian development and its "fine-tuning" control by hormones are less well understood.

Sexual reproduction is the general rule in insects, although many exceptions and modifications are observed (see Chapter by Dixon). In the more highly evolved social insects (Hymenoptera), reproduction is limited to a small number of individuals, often one queen and a small number of males (drones).

II. MORPHOLOGY OF THE FEMALE REPRODUCTIVE SYSTEM

The central site of egg production is the ovary. Ovaries are usually located dorsolateral to the gut and each comprises a number of tubular ovarioles ensheathed by a network of connective tissue (Figure 1). Each ovariole is composed of somatic and germ cell tissues. The number of ovarioles per ovary varies from one (e.g., in some viviparous aphids and in dung beetles) to about 3000 (in some higher termite queens).[2] Three basic types of ovary organization are found in insects (Figure 2). The panoistic ovarian, probably the most primitive one, is found in the oldest families of insects, such as in Archaeognatha, Zygentoma, Odonata, Plecoptera, and in most orthopteroid insects, but also in some Megaloptera (Corydalidae) and most Siphonaptera. In the panoistic ovary, specialized nutritive cells are absent and most of the informational resources of the oocyte are provided by the synthetic activity of the oocyte nucleus itself. The polytrophic meroistic ovary is found in most endopterygotes, and in Dermaptera, Psocoptera, and in Phthiraptera. In polytrophic ovaries, a number of nurse cells are enclosed in each follicle along with an oocyte. In the Hemiptera, polyphage Coleoptera, Megaloptera (Sialidae)/Raphidioptera, and also in the most "primitive" winged insects, the Ephemeroptera,[3] telotrophic meroistic ovaries are found. In this ovarian type, a syncytium of nutritive cells is connected with each oocyte by means of a trophic cord. In the vast majority of insect species, the nutrive, or yolk, contribution is supplied largely by the fat body, but in some cases the follicle cells can also serve as additional source of yolk. Accumulation of yolk (vitellogenesis; Section IV.C) normally occurs only in the terminal oocyte, that is, the oocyte closest to the oviduct. Another follicle cell function is the formation of the protective layers of the egg. These include the vitelline membrane and the chorion (Section IV.D).

Besides the ovary, oviduct-associated secretory cells can have important contributions to egg production (Section II.C). The most significant of such structures are the spermathecal accessory glands and the female accessory glands, such as the colleterial glands. Last, secretory functions of the vagina may play a role in egg production. Like other terrestrial

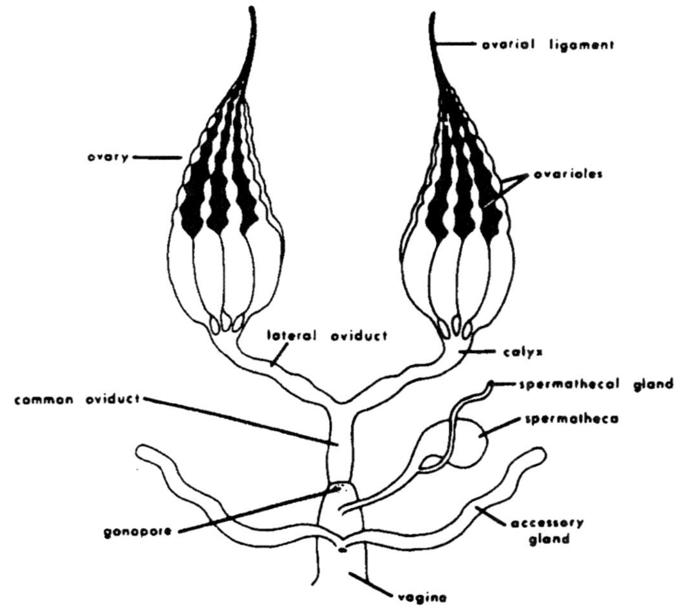

FIGURE 1. Female reproductive system: diagram of common type found in many insects. (From Gillott, C., *Entomology,* Plenum, New York, 1980, chap. 19. With permission.)

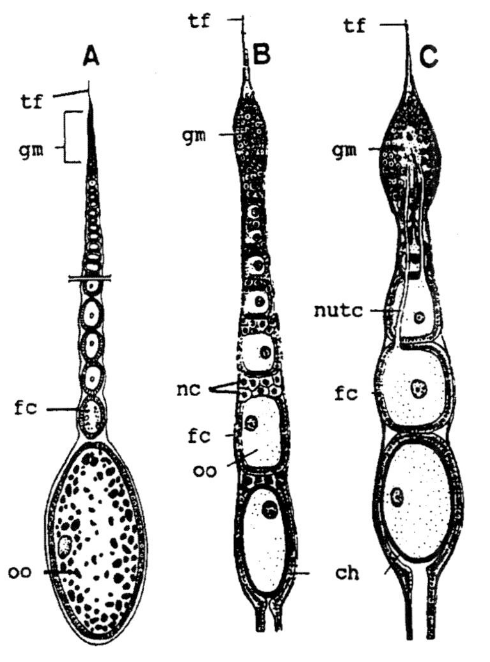

FIGURE 2. Schematic diagrams showing the types of ovarioles. Meroistic ovaries may be organized in the telotrophic and the polytrophic way, respectively.[4,70] ch = chorion; fc = follicle cell; gm = germarium; nc = nurse cells; nutc = nutritive cord; oo = oocyte; tf = terminal filament. (From Gillott,C., *Entomology,* Plenum, New York, 1980, chap. 19. With permission.)

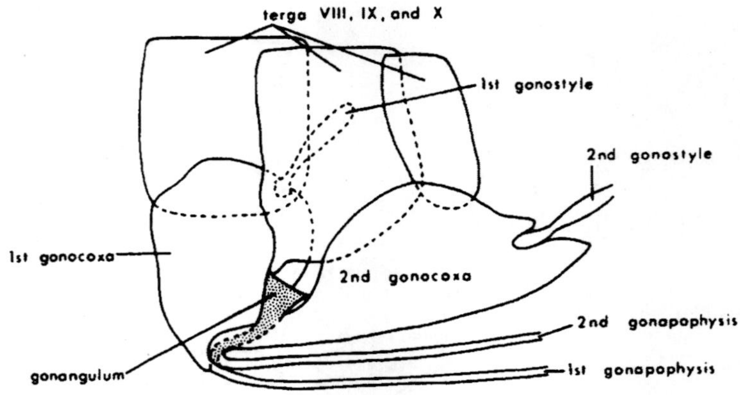

FIGURE 3. The primitive structure of the pterygote ovipositor in the thysanuran *Lepisma*. (From Gillott, C., *Entomology*, Plenum, New York, 1980, chap. 19. With permission.)

animals, insects have had to solve the problem of bringing together sperm and egg in the absence of surrounding water (internal fertilization, see Chapter by Wall). Almost all insects store the spermatozoa received from the male in a specialized organ, the spermatheca, until they are used to fertilize the mature eggs.

A. EXTERNAL GENITALIA

The morphology of the organs specialized for copulation and oviposition is highly varied. In the mayflies, the oviducts open directly to two genital pores behind the seventh abdominal segment. In most insects, the appendages of the genital segments (eighth and ninth abdominal segment) form an ovipositor. In the apterygotes and some of the winged insects, the ovipositor is a simple opening both for copulation and for the deposition of eggs. The primitive structure of the pterygote ovipositor can be seen already in the thysanuran *Lepisma* (Figure 3).[4] Among Pterygota, an ovipositor is found in Notoptera (Grylloblattodea), Dictyoptera, Ensifera, Caelifera, and Hymenoptera, some Odonata and most Hemiptera, Thysanoptera, and Psocoptera. The structure and elaborateness of the ovipositor is determined by the site of egg deposition. The ovipositor of Hymenoptera may be considerably modified for boring, piercing, sawing, and stinging. In the stinging Hymenoptera, such as bees, the eggs are released at the base of the ovipositor, and the ovipositor is modified by the addition of poison glands and reservoirs that evacuate the venom through the hollow sting. The ovipositor of *Drosophila* has sharpened ends that penetrate the surface of fruit, while the ovipositors of some of the predatory wasps are long (up to 15 cm in length) and sharp to penetrate the body of the insect prey.[5]

B. STRUCTURE OF THE OVARY

Insects have become particularly adept in manufacturing large numbers of oocytes within the ovary. The fruit fly, *Drosophila melanogaster*, can produce, during a 10-week reproduction period, a quantity of eggs equivalent to 30 times her body weight.[6] Her newly formed oocytes can undergo a 100,000-fold increase in volume within 3 days. Such reproductive feats are possible because of certain evolutionary adaptations: (1) Oocytes have developed methods for incorporating massive quantities of female-specific proteins (vitellogenins) which are transported in the hemolymph from the fat body to the ovary. (2) Mechanisms have evolved for loading unfertilized eggs with the ribosomes, tRNA, and long-lived mRNA that are required in early embryogenesis. (3) The ovarioles are supplied by tracheae of the aeriferous type, with a great diversity in the methods by which oxygen is delivered to the individual oocytes.[7]

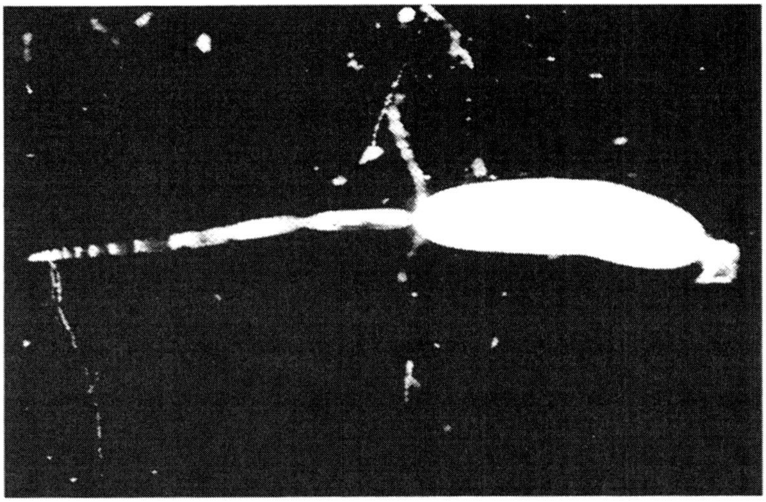

FIGURE 4. Panoistic ovariole of a cricket, *Gryllus bimaculatus* de Geer, with a mature egg in the terminal oocyte. Photograph courtesy of K. H. Hoffmann, Bayreuth.

Panoistic ovarioles can be developed by blocking germ cell cluster divisions totally, as is found in most "primitive" insects, or after germ cell cluster formation by final cleavage of cystocytes, all of which develop as oocytes as found in stone flies or thrips.[8] In the panoistic ovary, each of the ovarioles is composed of a terminal filament, the germarium, a series of oocytes at the previtellogenic phase of development, one or more oocytes in the process of vitellogenesis, and last, the mature egg (Figure 2A; Figure 4). The terminal filament is made up of a group of flattened cells, surrounded by a basement lamina and an ovarian sheath, both of the latter surrounding the entire ovariole. The oogonia are located in the most anterior region of the germarium, followed by a zone of oocytes in the early stages of meiosis. At the posterior end of the germarium, the oocytes are beginning to be surrounded by a monolayer of follicle cells. The size increase of oocytes at the previtellogenic stage is accomplished by an expansion of the cytoplasmic volume. In many cases, a multilayered pad of interfollicular tissue is located between successive oocytes.[9] In many insects with panoistic ovaries, vitellogenesis commences in the penultimate oocyte only after ovulation at the terminal one (e.g., in *Locusta migratoria* and *Schistocerca gregaria* [Caelifera]), or in the case of ovoviviparous cockroaches, after the loss of the egg case. The inhibitory effect is mediated by the interfollicular cells which pass an inhibitory substance from the anterior to the more posterior oocyte. In contrast, secretion from such cells located proximal to the oocyte stimulates vitellogenesis. In other species (e.g., in *Periplaneta americana* [Blattodea] and *Melanoplus sanguinipes* [Caelifera]), two or more oocytes may be vitellogenic at the same time, although at different stages of the yolk deposition cycle. All the above-mentioned insects produce eggs in batches. In the stick insect *Clitumnus extadentatus*, the different ovarioles can mature asynchronously, and the female lays a few eggs per day for several weeks. The number of ovarioles in panoistic ovaries can range from 4–3000 (5 in Acrididae, 15–30 in Tettigonioidea, 150–170 in Gryllidae, about 3000 in Isoptera queens).

A terminal filament and a germarium are also found in polytrophic meroistic ovaries (Figure 2B). In this case, the anterior region of the germarium contains one or more stem-line oogonia and a number of daughter cells or cystoblasts. The cystoblasts divide to give a cluster of cells remaining connected by structures called "ring canals" or "intercellular bridges." The innovation of the polytrophic ovary is the differentiation of only one oocyte, which generates from one, central cell of the cluster, whereas all other siblings are transformed into nurse cells.

In many cases, clusters follow the 2^n-rule (1 oocyte + $2^n - 1$ nurse cells), in which n represents the number of cell cycles. Identical characters in polytrophic meroistic ovaries among hemimetabolous and holometabolous insects indicate a "basic type" of common origin.[8] In *Drosophila*, where the number of cells in the cluster is 16, the clustering of the cells, as well as the formation of the ring canals, is mediated by structures called fusomes. The fusomes contain a random array of membranous vesicles and fibrils,[10,11] but no mitochondria and few ribosomes and microtubules. In the central region of the germarium, prefollicular cells grow around the oocyte-nurse cell cluster, while in the posterior section of the germarium, typical egg chambers are detectable; which means the oocyte-nurse cell complex is completely surrounded by a unilayer of follicle cells. Previtellogenesis includes the enlargement of the oocyte, an increase in the number of follicle cells, and the polyploidization of the follicle cells and nurse cells. An epithelial sheath surrounds each ovariole and consists of a thin, acellular, inner membrane; a median cellular network of muscle and tracheal cells; and an outer epithelial membrane. The number of ovarioles in polytrophic ovaries can be highly variable (usually 4 in the Lepidoptera, 10–30 in *Drosophila*, and 70–100 in *Musca* or *Lucilia* [Diptera]). Commonly, a single oocyte per ovariole will become vitellogenic at one time.

Telotrophic ovaries differ from polytrophic ovaries by retention of all nurse cells in an anterior trophic chamber and by changing oocyte-nurse cell determination. This type of ovary developed independently three times (in Hemiptera, Raphidioptera/Megaloptera [Sialidae], and in polyphagous Coleoptera) from polytrophic ancestors and once directly from panoistic ancestors in mayflies.[8] Despite fundamental differences between the subtypes of telotrophic ovarioles, they share some common characters. As the oocytes move towards the region of follicle formation, they become surrounded by prefollicular cells. The connection to the tropharium is retained via a cytoplasmic strand, the nurse strand or trophic cord. The young oocytes subsequently enlarge by incorporation of nurse cell material, transported through the nutritive cords (previtellogenesis). The vitellogenic growth phase begins when yolk spheres are observed to accumulate in the oocytes. The nutritive cords collapse during vitellogenesis. The follicular cells surround each growing oocyte to form a monolayered epithelium, and this tissue will secrete a vitellin membrane and the chorion. An interesting variant of the telotrophic ovary is found in the polyphagous Coleoptera. The number of ovarioles remains more or less constant in each species, but varies between species from 1 (some Scarabaeinae) to about 1000 in blister beetles (*Meloe proscarabaeus*).[6] In *Creophilus maxillosus* (Staphylinidae), the differentiation of nurse cells and oocytes occurs within linear chains of sibling cells (linear chain model). Only the most basal member of the sibling cluster develops into an oocyte; the others differentiate into nurse cells.[6,9]

C. STRUCTURE OF THE FEMALE ACCESSORY REPRODUCTIVE SYSTEM

The bottom of each ovariole forms a small duct or pedicel (Figure 1). The pedicels of each group unite to form a calyx, and each calyx opens into a lateral oviduct. Usually the oviducts of the two sides join to form a single median oviduct. The common oviduct is ectodermal in origin and typically it is heavily invested with muscles. The lateral oviducts can be of mesodermal or ectodermal origin. The presence or absence of a cuticular intima indicates the origin. In *Oncopeltus fasciatus* (Hemiptera), the lateral oviducts undergo a drastic larval-adult transformation during the last (fifth) larval stadium.[12] The long and thin larval oviducts shorten and become very wide. This transformation is ecdysteroid dependent in a dose-related manner and only takes place in the absence of juvenile hormone. The morphological transformation is accompanied by dramatic cytological changes. Whereas the cells of the anterior part of the oviduct commence with a strong secretory activity, the cells of the caudal part form a bizarre pattern of cell projections which deposit the cuticle. The secretory material of the ductal system may act as a lubricant for egg passage, as protective oothecal coverings, or as glues to attach eggs to various substrates or to hold batches of eggs together.[9] The distal end of the common oviduct is called the gonopore, which serves for the discharge of eggs. In the

Dermaptera, the gonopore is the external opening, located on the posterior part of the seventh abdominal segment. In most insects, however, the gonopore opens into a genital chamber. The opening to the outside is the vulva. The genital chamber can be of variable complexity and is often associated with an ovipositor. Within some orders, an invagination of the primitive genital chamber forms a distinct intermediate structure, the vagina, between the external vulva and the gonopore. Generally, the vagina is not secretory and consists of a single layer of epithelial cells, covered by a cuticular intima and surrounded by muscle. In many species, the genital chamber has become modified to form a bursa copulatrix. An important function for this organ is to receive spermatophores or seminal fluid. In Lepidoptera, the bursa remains in the eighth abdominal segment, whereas the eggs are deposited through a separate opening, the oviporus, on segment IX. In this case, the bursa is connected with the vagina by the seminal duct. In other insects, however, there is no separation between the copulatory and egg-laying apertures, and both of them open as the vulva on segment IX. In the bursa of some butterflies (e.g., genus *Danaus*), tooth-like dentata are present and may function in tearing open the spermatophore. Some secretory activity may also be associated with the bursa copulatrix, since empty spermatophores are digested within the bursa of some insects. In ovoviviparous and viviparous species, a brood pouch or uterus forms as an expansion of the vagina wall.

Two ectodermal glands (ectademia) are connected with the genital chamber or vagina. One is the spermatheca in which spermatophores are stored and that has a gland attached to its duct. The other is a paired structure, the accessory glands or colleterial glands, with functions associated with egg deposition.

1. Spermatheca and Spermathecal Accessory Glands

The morphology and arrangement, as well as the number of spermathecae, vary from species to species. In most insects, the spermatheca is a single organ, spherical or ovoid in form. In some insects, such as cockroaches and mosquitoes, secretory cells are associated with the walls of the storage portion of the spermatheca, whereas in others, such as honeybees and many beetles, the distal end of the spermatheca is specialized for secretion and is referred to as the spermathecal accessory gland. In the genus *Rhodnius* (Heteroptera), the spermathecae are a pair of blind tubules which open into the common oviduct near its junction with the lateral oviducts.

In spite of the morphological diversity, the cellular elements of the secretory part of the spermatheca are rather constant. A generalized diagram of a spermathecal secretory cell is given in Figure 5. The cells form a cup-shaped cavity at their apical surface, and the membrane in this region is thrown into numerous microvilli. The central cavity communicates with the spermathecal lumen via a cuticle-lined ductule. In other forms, however, several ductules may open at a single pore. The cuticle lining the ductule and the spermathecal lumen is produced by separate duct cells which are interposed between the secretory cells and the lumen of the spermatheca. Ultrastructural studies showed that the secretory cells have a phenotype associated with cells which are specialized for export protein synthesis.[9]

The secretion produced by the gland cells is a mucoprotein or mucopolysaccharide and may be used by the spermatozoa for an energy source.[13] Removal of the spermathecal accessory gland from females of *Anthonomus grandis* (Coleoptera) results in a gradual loss of motility among spermatozoa in the spermatheca. The epithelium of the spermatheca, particularly in the storage part, exhibits characteristics of ion-transporting epithelia and thus may be responsible for providing an ionic milieu within the lumen of the spermatheca different from that in the hemolymph. In some insects, the spermathecae may be sufficiently permeable to allow the passage of various male secretory products.

Information on the control of spermatheca differentiation is rather limited. In the beetle *Tenebrio molitor*, the differentiation of the spermatheca occurs in the pupal stage prior to the eclosion to the adult. The differentiation process can be divided into three phases: (1) cellular proliferation, (2) cellular morphogenesis, and (3) cuticulogenesis. From *in vitro* culture

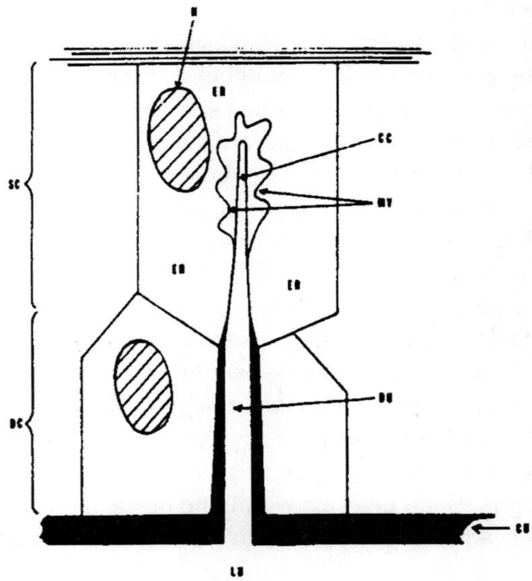

FIGURE 5. Diagrammatic representation of a spermathecal secretory cell and its spatial relationship to the duct cell and basement membrane. CC, central cavity; CU, cuticle; DC, duct or canal cell; DU, ductule; ER, rough endoplasmic reticulum; LU, lumen; MV, microvilli; N, nucleus; SC, secretory cell. (From Kaulenas, M. S., *Insect Accessory Reproductive Structures. Function, Structure, and Development*, Springer-Verlag, Berlin, 1992. With permission.)

experiments, it was concluded that cuticulogenesis is under control of 20-hydroxyecdysone; the hormone is necessary to initiate cuticle deposition.[14] With regard to the control of the acquisition of definitive levels of differentiated functions by the spermatheca, some information is available for a number of Orthopterans. In *Chorthippus curtipennis* (Caelifera, Acridinae), allatectomy leads to degeneration of the spermatheca, while in *Gomphocerus rufus* (Caelifera, Acridinae), allatectomy resulted in the inability of females to dissolve transferred spermatophores, suggesting a failure in the production of spermathecal proteolytic enzymes.[9] The results suggest that juvenile hormone is necessary to initiate and possibly maintain differentiative secretory function in these grasshoppers. On the other hand, allatectomy in *L. migratoria* (Caelifera, Locustinae) does not alter the histology of the spermatheca. In *Rhodnius prolixus* (Heteroptera), removal of the neurosecretory cells in the pars intercerebralis results in a failure of the secretion to appear in the lumen of the gland as a result of the absence of a myotropin from the neurosecretory cells. The myotropin acts on the muscles of the spermatheca to squeeze the secretion from the reservoir beneath the intima into the lumen.[13]

2. Colleterial Glands and Mesodermal Accessory Glands

Many insects produce protective coatings for the newly laid eggs. The Lepidoptera coat individual eggs with a proteinaceous glue that hardens on contact with air and attaches the eggs onto appropriate substrates. Accessory glands in the genus *Musca* (Diptera) contribute in aiding in the fertilization process by providing secretions which assist in liberating sperm acrosomal contents.[15] Aquatic insects often produce egg cases of a gelled substance, but the most complex of these structures may be the tough ootheca of cockroaches and mantids. In mantids, the glands produce a polymerized protein foam. In cockroaches, several layers of tanned protein form a complicated egg chamber with elaborate respiratory apparatus and release valves. The sources of these protective devices are the colleterial glands which branch off the vagina or genital chamber. In most cases, the colleterial glands are a pair of organs composed of a number of multibranched tubules and are formed from invagination of the epidermis (Figure 1). The morphology of the colleterial glands has been well documented. Five cell types compose the left colleterial gland; four types of glandular cells are homologous with the dermal glands of the integument, and a chitinogenic cell type is homologous with the epidermal cell of the integument. The latter cell type is found interspaced between the glandular cells and secretes the protective intima which lines the lumen of the tubule. Each of the gland cell types is distinct from the other, and is located in separate regions of the tubule. Type 1 cells are found in the most posterior region of the tubule and are in a presecretory stage. Type 2 gland cells are found anterior to type 1 cells and lack a well-defined rough endoplasmic reticulum (ER), but are packed with mitochondria. The morphology of the type 3 cells resembles that of both type 2 and type 4, and appears to be a transient intermediate form. The type 4 cells dominate the anterior ends of the tubules and are packed with rough ER and with mitochondria.[16]

The colleterial gland of *P. americana* has been studied most extensively.[9,13] The left side tubules (type 4 cells) secrete the proteins (oothecins) which make up the structure of the ootheca, together with a polyphenol oxidase (type 2 cells) as well as the precursor of the tanning agent, the 4-O-β-glucoside of protocatechine acid. The right colleterial gland secretes the enzyme β-glucosidase. The opening for the duct from the right gland is more anterior than the opening for the duct from the left gland. At the time of ovulation (see Section VI), the chorionated oocytes migrate down the oviduct to the genital atrium for fertilization. After fertilization, the eggs are transported more posteriorly and pass the duct of the right gland where the contents of the gland (β-glucosidase) are secreted onto the fertilized egg. Thereafter, as the egg moves more posteriorly, it encounters the secretions of the left colleterial gland. The subsequent mixing of the secretions from both glands results in the hardened ootheca.[16]

Apparently, juvenile hormone affects the synthesis of oothecins in left colleterial glands.[9,13,16,17] Analyses of the *Periplaneta* oothecin cDNA sequences and of the resultant predicted amino acid sequences have confirmed the existence of 11 major glycine-rich oothecins which represent six size classes with molecular weights of 14.5, 15.5, 17–18.5, 23.5–26, 28.0, and 37–39 kDa, respectively.[18] The oothecin sequences have numerous similarities to silkmoth chorion proteins. In mantids, the chemistry of the ootheca proteins shows some unique features. The glucosides identified in five mantid species are 3-O-β-glucosides of *N*-β-alanyldopamine and *N*-(*N*-malonyl-β-alanyl)dopamine. The light color of the ootheca and the occurrence of phenolic compounds modified at the β position of the parent compounds suggest that β-sclerotization occurs in mantid oothecae.

In *S. gregaria* and other Orthopterans, the foamy ootheca is produced, at least in part, by mesodermal accessory glands (mesadenes) which consist of convoluted blind tubules opening into the proximal end of the lateral oviducts. In spite of the great structural differences between glands responsible for oothecal formation in cockroaches and locusts, similar mechanisms may operate to harden the ootheca.

3. Milk Glands

In tsetse flies, members of the genus *Glossina*, the accessory glands of the female are transformed into milk glands and supply a secretion upon which the developing larva feeds (adenotrophic vivipary). In these flies, the female ovulates a single egg into a uterus, where it hatches. The growth of the larva is rapid under such circumstances. In *Glossina austeni* for example, development from an egg to a fully grown larva weighing some 30 mg requires only 9–10 days. The fully mature larva is then "larviposited." The milk gland in *Glossina* is a branched tubular structure ramifying throughout the abdomen and emptying into the uterus via a single muscular collecting duct which contains two channels.[13] The tubules consist of a single layer of secretory cells similar to those in other accessory glands. The tubules undergo cyclic changes in diameter, largely as a result of changes in the volume of secretory reservoirs. The secretory reservoir is an extracellular structure, formed by a cup-shaped invagination of the apical membrane of each of the secretory cells. In *G. austeni*, the tubular diameter reaches a peak of about 100 μm 3 days before each larviposition, and a diameter of about 30 μm at each larviposition, followed by a resynthesis of the "synthetic machinery." Removal of the corpora allata greatly reduces the production of milk in tsetse flies.

Milk glands also occur in some cockroaches. In the viviparous cockroach *Diploptera punctata*, nutrient is supplied to the developing embryos, which increase in weight by a factor of about 50 before they hatch. The developing embryos receive this nutrition by ingesting a fluid secretion ("milk") in the brood pouch. The brood pouch is a part of the genital chamber. The intima of this cuticle-lined chamber is penetrated by pores, each of which is the opening of a ductule leading from a secretory cell. In *Diploptera*, juvenile hormone is necessary to allow the decline of milk gland synthesis at the termination of pregnancy.[19]

D. ENDOCRINE CONTROL OF DIFFERENTIATION OF ACCESSORY GLANDS AND DUCTS

The dependency on ecdysone and 20-hydroxyecdysone of the organogenesis of accessory glands and genital ducts has been demonstrated in several insect species.[1,20] The development of the colleterial glands of *P. americana*, which occurs at the end of the last larval instar, requires ecdysone. In young females of *L. migratoria* and *S. gregaria*, implantation of additional corpora allata accelerated the development of both the oviducts and colleterial glands. In *T. molitor, Ephestia kühniella* and *Samia cynthia, L. migratoria*, and *O. fasciatus*, duct differentiation only occurs in the presence of ecdysone *in vivo* and *in vitro*. However, other tissues of epidermic origin, such as the spermatheca, escape mitotic stimulation by ecdysone.

Besides the effects exerted by hormones upon the reproductive organs, humoral relationships between gonads and ducts have been demonstrated. In *Drosophila*, female ducts were sometimes found to be attached to the gonads, thus causing degeneration. In the bugs *Dysdercus fasciatus* and *Triatoma infestans*, atrophy of one ovary was observed together with the regression of the upper part of the corresponding lateral oviduct, and the experimental section of the oviduct had the same effect in both species.[1]

III. ORIGIN AND FORMATION OF THE GERM CELLS

Both sperm and eggs are derived from primordial germ cells set aside very early in the development of the embryo. Among the orthopteroid insects, in species where the germ cells appear early in development, such as in the house cricket *Acheta domesticus* and the grasshopper *Melanoplus differentialis*, the germ cells appear to be of ectodermal origin, forming at the posterior pole of the egg at the time of mesoderm segregation.[9] Later in development in *A. domesticus*, they become associated with the mesoderm of the second and third abdominal segments; in *M. differentialis* they migrate into the coelomic cavities of the first to eighth abdominal segments, where they associate with the splanchnic wall and form a genital strand, from which the gonad differentiates later in embryogenesis. In species where the germ cells are first recognizable slightly later in embryonic development, they are associated with the median walls of the dorsal cavities in the abdominal segments (e.g., *L. migratoria, Blattella germanica, P. americana*). In each of these species, a genital ridge containing the germ cells is formed on each side of the embryo. The typical genital rudiment in these insects, during or just after involution of the germ band (anatrepsis), consists of a terminal filament membrane; a mesodermal dorsal cell mass, ventral to the filament membrane; a central cell mass, composed of primordial germ cells and mesodermal cells; a ventral cell strand of mesodermal cells which are the primordia of the gonadal portion of the genital ducts; and a surrounding epithelial membrane which envelops all of the above (Figure 6).

Among the endopterygotes, germ cell formation is well understood in the Diptera. The germ cells are formed from pole cells, which are established very early in development at the posterior pole of the embryo. In *Drosophila*, about 18 energids (cleavage nuclei) enter the posterior pole plasma and are pinched off as pole cells. The pole cells continue to divide, to

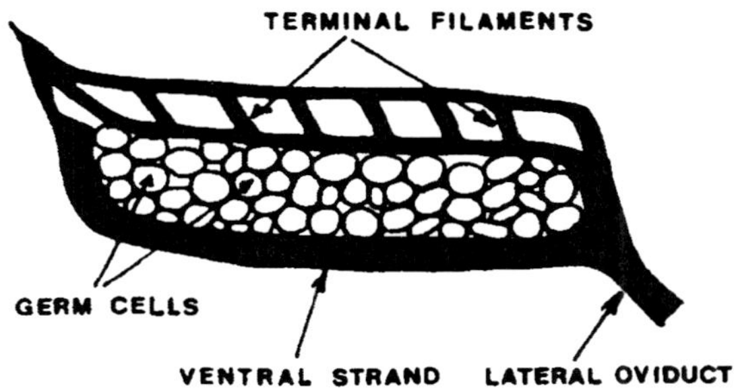

FIGURE 6. Diagrammatic representation of the early development (late embryonic stage) of the female reproductive system in an orthopteroid insect.(From Kaulenas, M. S., *Insect Accessory Reproductive Structures. Function, Structure, and Development*, Springer-Verlag, Berlin, 1992. With permission.)

produce eventually between 37 and 71 cells.[9] Only some of the pole cells migrate to the presumptive gonads, which lie on either side of the gut and are mesodermally derived. The final number of pole cells in the gonad has an upper limit of about 13 pole cells per gonad at about stage 16 of embryogenesis.[21] The pole cells which fail to reach the gonad probably degenerate later on. The determination of the pole cells as presumptive germ cells depends upon the interaction of the entering energids with the cytoplasm of the posterior pole cells (polar plasm). Likely candidates for cytoplasmic elements important for germ cell determination are polar granules, which are concentrated at the posterior pole of late stage oocytes and early embryos.[22] Functionally similar posterior pole plasms, which determine germ cell differentiation, occur in the Coleoptera and the Hymenoptera. In the Coleoptera, germ cells become distinguishable at the time of blastoderm formation, at the posterior end of the egg. In most Hymenoptera, germ cells first become recognizable during gastrulation or later, forming from the mesodermal tube. In many Lepidoptera, germ cells appear at the posterior pole just after blastoderm formation.

Germ cells (oogonia) are the only cells that normally exhibit genetic programs that lead to the construction of eggs.

IV. OOGENESIS

The first events in oogenesis include mitosis, the onset of meiosis, and ovariole differentiation. In the panoistic ovary, all oogonia (except stem line oognoia) are transformed to oocytes, whereas in the meroistic type oogonia generate both oocytes and nurse cells.

A. EARLY EVENTS IN OOGENESIS

Since more information is available concerning oogenesis in *D. melanogaster* than for any other insect, this fruit fly will be used to illustrate the early events of ovarian development that are shared by many evolutionary advanced insects with a polytrophic meroistic ovary.[6] Each *Drosophila* ovariole contains a collection of egg chambers in which each oocyte is one member of a clone of interconnected cells. In brief, four consecutive mitoses of the cystoblast and its daughter cells in region 1 of each germarium give rise to the 16 germ-line cells (cystocytes), which remain connected via intercellular bridges due to incomplete cytokinesis (Figure 7). King et al.[11] suggest that the interactions of the centrioles and fusomes during the cystocyte divisions are responsible for the multiple-branched canal system that results. In region 2, the individual 16-cell clusters become separated by invading somatic follicle cells. One cell with four intercellular bridges (ring canals), the prospective oocyte, moves from a central position in the cell cluster to a posterior location between regions 2 and 3 of the germarium. This spatial reorganization of the 16 germ-line cells results in a polarization of the cluster: the prospective oocyte becomes positioned posterior to the remaining 15 cells, which differentiate into polyploid nurse cells. The forming follicle has established an anteroposterior

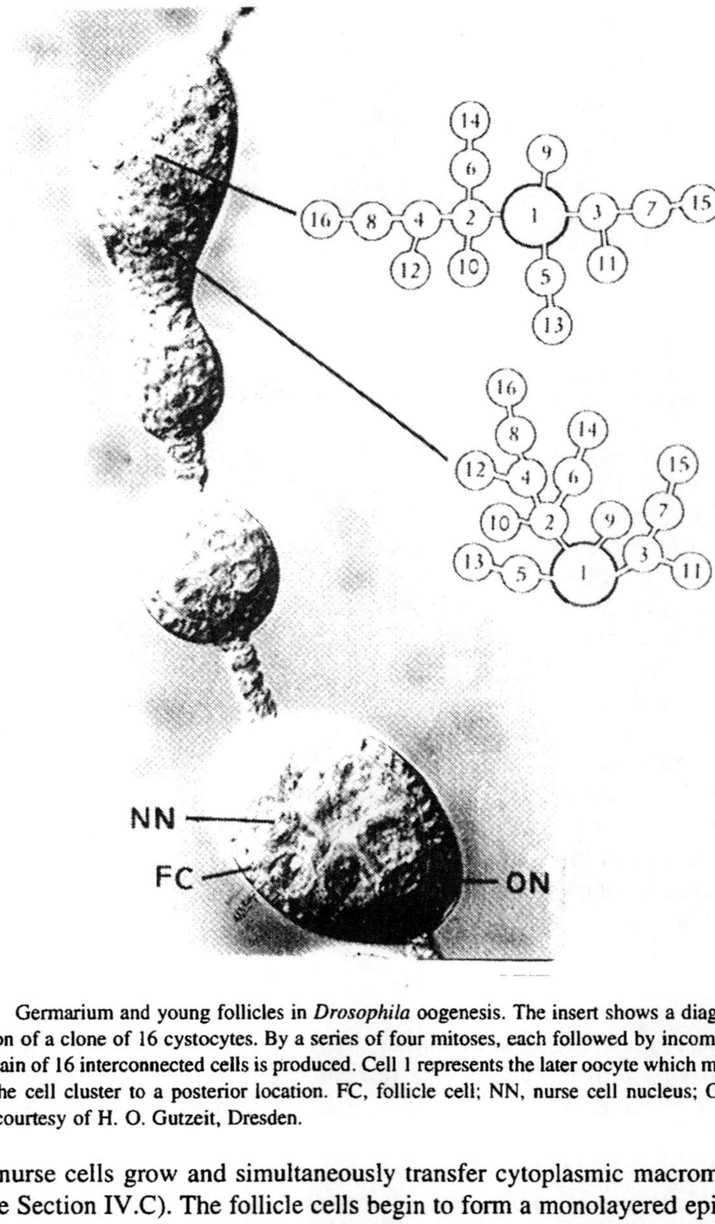

FIGURE 7. Germarium and young follicles in *Drosophila* oogenesis. The insert shows a diagram of the steps in the production of a clone of 16 cystocytes. By a series of four mitoses, each followed by incomplete cytokinesis, a branching chain of 16 interconnected cells is produced. Cell 1 represents the later oocyte which moves from a central position in the cell cluster to a posterior location. FC, follicle cell; NN, nurse cell nucleus; ON oocyte nucleus. Photograph courtesy of H. O. Gutzeit, Dresden.

axis. The nurse cells grow and simultaneously transfer cytoplasmic macromolecules to the oocyte (see Section IV.C). The follicle cells begin to form a monolayered epithelium around the germ-line cells in region 3 of the germarium. This process starts at the posterior end of the follicle. At the anterior end, a special group of follicle cells forms a stalk ("stalk cells"), thus separating the follicle from the germarium and releasing it into the vitellarium.[23]

In the silkmoth *Bombyx mori*, there are only three cystocytes divisions resulting into eight germ-line cells ($n = 2^3$). In the braconid wasp *Habrobracon juglandis*, the final number of cystocytes per cluster is 32; in other wasps, the number is not fixed but varies from 20–80.[24] Fleas with polytrophic meroistic ovarioles (some species of the Hystrichopsylloidea) have germ cell clusters consisting of 32 cells (2^5) which are generated by five mitotic cycles during the pupal stage. One of the cells containing five intercellular bridges becomes the oocyte; the others serve as nurse cells. However, nurse cells remain small and show the same ultrastructural characters as the oocyte.[25] Certain species of lacewings do not obey the $N = 2^n$ rule.[6] For example, in *Chrysopa perla*, egg chambers contain 12–14 cystocytes. In this case, first- and second-generation cystocytes divide in synchrony, whereas at M_4 (see Figure 7), cells 3, 4, 7, and 8 divide; the rest do not. In the earwigs (Dermaptera), each follicle in the vitellarium

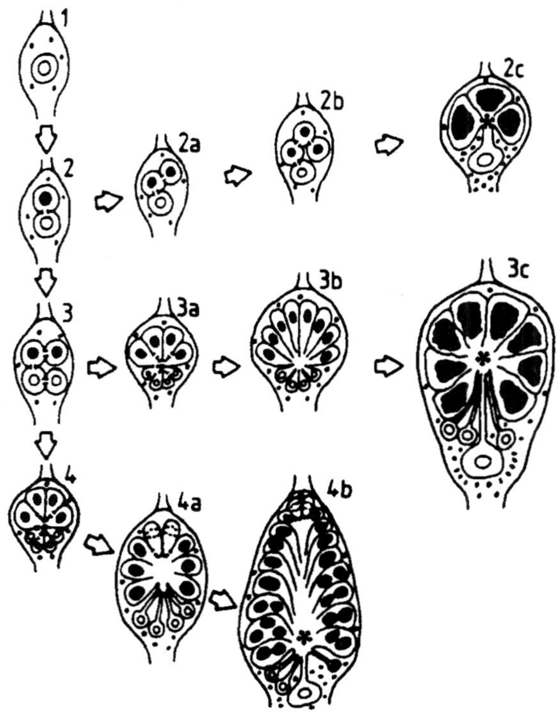

FIGURE 8. Hypothetical diagrammatic representation of germ cell cluster formation in subgroups of Hemiptera: (1) development begins in one persisting germ cell; (2) the germ cell divides by mitosis, followed by incomplete cytokinesis. One of the daughter cells is determined as a presumptive nurse cell (black nucleus), the other will be a presumptive oocyte (white nucleus); (2 a–c) germ cell division in scale insects (Coccina); (3) in other groups, the presumptive nurse cell and the presumptive oocyte divide, giving rise to a cluster of four germ cells arranged in a rosette configuration; (3a–c) germ cell division in aphids (Aphidina); (4) further divisions of presumptive nurse cells and oocytes; (4 a–b) germ cell division in bugs (Heterotera), where the oocyte subclone has its divisions limited. Asterisk = region of microtubule matter. For further details see text. (Reprinted from King, R. C. and Büning J., *Comprehensive Insect Physiology, Biochemistry, and Pharmacology,* Vol. 1, Kerkut, G. A. and Gilbert, L. I., Eds., Copyright 1985, ch. 3. With permission from Pergamon Press Ltd., Headington Hill Hall, Oxford OX3. OBW, UK.)

consists of one oocyte and one nurse cell, surrounded by a single layer of follicle cells. Formerly, the polytrophic meroistic ovary of earwigs has been looked at as a parallel development, but new findings indicate only one origin of the polytrophic meroistic ovary (see Section II.B).[8]

In all hemipteran species with telotrophic ovarioles, the germ cells are also clearly separated into nurse cells and oocytes. The oocytes are found at the base of the tropharium (see Figure 2C), whereas the nurse cells occupy its upper parts. Two models of germ cell cluster formation have been proposed. Both assume that independent stem cells in the apical region of the tropharium fuse to form germ cell clusters in which basally located germ cells are subsequently determined as oocytes. Based on data derived from several groups of Hemiptera, a new model has been advanced which assumes that cluster formation begins with a single cell. The first division is a differential mitosis, leading to a presumptive oocyte and a presumptive nurse cell. In scale insects (Coccinea), the presumptive nurse cell divides and its apical descendant will divide again. The final configuration is a rosette of four cells in which the intercellular bridges stay close together. During rosette formation, the intercellular bridges then vanish, and the trophic core forms (Figure 8). The last step in cluster formation is the polyploidization of nurse cells. In other groups, the presumptive nurse cell and the presumptive oocyte divide, giving rise to a cluster of four germ cells arranged in a rosette configuration. In aphids, germ cells continue to divide, and subclones of 2^n oocytes and 2^n nurse cells arise. Rosette formation, formation of the trophic core, and polyploidization of nurse cell nuclei are the same as proposed for scale insects. Further divisions of presumptive nurse cells and oocytes lead to the situation in bugs (Heteroptera) with a constant number of oocytes, but an increasing number of nurse cells.

In snakeflies (Raphidioptera) and alderflies (Megaloptera, Sialidae), the same type of telotrophic ovary occurs. Cluster formation starts with germ cell migration into the ovariole anlage. The number of germ cells increases as more germ cells enter the anlage and as those already there divide. Dividing stem cells undergo complete cytokinesis next to the terminal filament (apical region), whereas germ cell clusters arise by incomplete cytokinesis more basally. The clusters are disc-shaped and oriented at right angles to the long axis of the ovariole. Each cluster presumably contains 2^n cells, with five as the maximum number of divisions.[6] Prospective oocytes do not differ from nurse cells in their ultrastructure, except for one fact: nurse cells lose their cell membranes totally to form a syncytium, whereas prospective oocytes remain in their original status, which they acquired at the end of cluster divisions.[8]

In polyphage Coleoptera, germ cells migrate in the ovariole anlagen, and cluster formation is started by mitosis of each germ cell. Each germ cell can undergo only a limited number of mitoses, each followed by incomplete cytokinesis. Later mitoses are highly synchronized within each cluster, but not between different clusters. In most species, the clusters are oriented parallel to the long axis of the tropharium. A three-dimensional network of interstitial cells keeps the nurse cell nuclei in place, when nurse cell-nurse cell membranes are reduced.[6] Oocytes develop at the base of the tropharium, primarily connected to nurse cells by an intercellular bridge. However, recent investigations have shown that cluster formation is more complex than has been assumed before, and that ramifications of clusters occur, even between oocytes.

1. Oocyte Differentiation

Immediately after the cluster of 16 cystocytes is found in a *Drosophila* germarium, both four-canal cells (cells 1 and 2) form synaptonemal complexes in their nuclei during the time they pass through germarial region 2. The synaptonemal complexes form during zygonema, are completed during pachynema, and are responsible for the synapsis of homologous chromosomes during meiotic prophase.[6] Since cells 1 and 2 start meiosis, they are called pro-oocytes. Sometimes cells 3 and 4 also form synaptonemal complexes and enter meiosis, whereas cells 5–16 fail to enter meiotic prophase. In the posterior region 2, one of the two four-canal cells loses its synaptonemal complexes and enters the cycle of endomitosis characteristic of nurse cells. The other cell continues to develop as an oocyte and retains its synaptonemal complexes during the previtellogenic stage of oogenesis in the vitellarium (see Section IV.B). The divergence of the two pro-oocytes takes place in the region where follicle cells first surround the 16-cell clone. It is suggested that the first pro-oocyte to come into contact with a follicle cell is the one that receives the critical stimulus that causes it to continue on the oocyte developmental pathway. Such cellular interactions between the germ-line cells and the somatic follicle cells have been studied for a long time.[26,27] More recent data show that the correct cellular organization and determination of cells in the germarium seem to depend on the activity of several genes. For example, in *Drosphila*, genes like *egaliterian* and *Bicaudal-D* are apparently involved in cystocyte diversification, since in mutant follicles a nurse cell differentiates instead of the oocyte. The genes *dicephalic* and *spindle-C* are required for the correct spatial arrangement of the cystocytes. Moreover, the mutant *dicephalic* illustrates the importance of early cellular interactions between somatic follicle cells and germ-line cells.[28] However, it remains to be analyzed which kind of specific signals the pro-oocytes receive in the germarium.

In some carabid beetles, all cystocytes of a clone form synaptonemal complexes, and consequently all nurse cells enter meiotic prophase together with the presumptive oocyte. In some telotrophic ovaries, the cell which differentiates as an oocyte also seems to depend upon its position relative to certain somatic tissues. These facts again are best explained by involving signals passed to the cystocyte by adjacent somatic cells.

In all insects with panoistic ovaries, fully activated oocyte chromosomes are required. In these cases, the oocyte chromosomes enter a lampbrush state in the diplotene stage of meiotic prophase. Lampbrush chromosomes also occur in many insects possessing telotrophic meroistic ovaries.

2. Endocrine Control of Early Oogenesis

In the blood-sucking bug *Panstrongylus megistus*, determination of mitosis, the onset of meiosis, and ovariole differentiation take place during the last larval instar and begin 24–48 h after a blood meal.[1] Observation of the neurosecretory (ns) cells of the pars intercerebralis during the last larval instar from the time of the blood meal until imaginal moult showed the presence of four distinct ns cell types with different patterns of activity. In type A cells, the ns material is released before the mitotic crisis, whereas in type A′ cells it is discharged before the onset of meiosis. Hence the ns cells appear to trigger both mitosis and meiosis. Electrocoagulation of pars intercerebralis prevented ovarian development and caused its degeneration. Destroying only type A ns cells resulted in an inhibition of the mitotic crisis, but the onset of meiosis was unaffected. Removal of the prothoracic glands showed that ecdysteroids do not trigger the onset of mitosis, but are necessary for that of meiosis. Hemolymph ecdysteroid determinations demonstrated two peak values; the first coincided with the onset of meiosis, the second preceded moulting. Juvenile hormone (JH) does not seem to intervene, either in early gonial mitosis or in the initiation of meiosis. In *L. migratoria, P. americana,* and *Gryllus bimaculatus* (Ensifera, Gryllidae), it was shown that the meiosis reinitiation in the vitellogenic oocyte (see Section IV.C) is preceded by an increase of free ecdysone of the oocytes. *In vitro*, the reinitiation of meiosis can be determined by the addition of ecdysone.[29]

Not only the germ cells but also the mesodermal elements surrounding them have to go through certain steps in order to develop adult structures, and these events are controlled by hormones. It was shown both *in vivo* and *in vitro* that growth and differentiation of nurse cells and follicle cells require the presence of 20-hydroxyecdysone, whereas JH may suppress the effects of 20-hydroxyecdysone. In conclusion, 20-hydroxyecdysone and JH are both necessary for ovariole differentiation and function, but the two hormones must act separately and successively.

3. Follicle Cell Differentiation

During oogenesis, the follicle cells follow a characteristic differentiation program. Three processes have been studied particularly thoroughly:

1. The production of localized developmental signals which bind to their respective receptors in the oocyte membrane. These signals play a role in the axial determination of the embryo.[30]
2. The contribution of the follicle cells to the synthesis and uptake of yolk proteins (see Section IV.C.3).
3. The formation of the egg shell and its genetic control (see Section IV.D).

General follicle cell morphology has been well described for the cockroach *P. americana*. The earliest follicle cells are of the squamous type, with the apical ends of the cells applied closely to the oolemma. As the oocyte grows, the follicle cells rapidly increase in number and gradually become cuboidal. The follicle cells send out processes which interdigitate with the microvilli of the oocyte. The cuboidal shape is maintained until prior to vitellogenesis when the follicular epithelium becomes columnar. Similar arrangements have been found also in other species (e.g., *L. migratoria, Galloisiana nipponensis* [Grylloblattodea], and ants of the genus *Formica* [Hymenoptera]). At the start of the vitellogenic phase, the cell shape changes

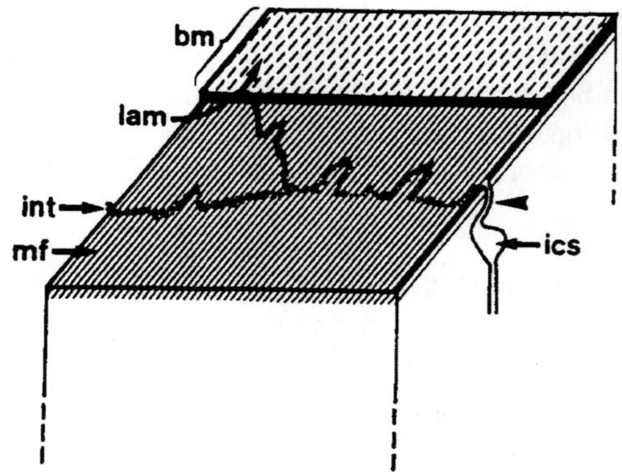

FIGURE 9. Diagrammatic representation of the follicular epithelium from a *Drosophila* ovary (vitellogenic follicle). The basement membrane (bm) with laminin (lam) in circular orientation is partly removed to expose the basal face of the follicle cells. Int, high concentrations of PSβ integrin at the contact site of the cells; mf, parallel microfilament bundles below the cell membrane which extend in the same circular direction as laminin; ics, intercellular space. Figure courtesy of H. O. Gutzeit, Dresden.

again to assume a somewhat spherical (*P. americana*) or flattened character (*L. migratoria*).[9] Ultrastructural studies showed that the cytoplasm of the follicle cells at the late previtellogenic and vitellogenic stages contain large numbers of mitochondria, multivesiculated bodies, and Golgi complexes, which are characteristics of a highly active tissue. The columnar cells are well supplied with rough ER and large amounts of ribonucleoproteins. In some dipteran and hymenopteran insects, in addition to septate desmosomes and adhesion plaques, intercellular bridges interconnect adjacent follicle cells. The bridges appear to result from incomplete cytokinesis and may serve to synchronize differentiation and function of the follicular epithelium. Several authors have described the occurrence of gap junctions between adjacent follicle cells (*L. migratoria*), as well as between follicle cells and oocyte (*R. prolixus, Tribolium destructor* [Coleoptera, Tenebrionidae], *D. melanogaster*).[9] The junctions disappear during the chorion formation phase and during atresia.

The typical transition of follicle cell morphology from cuboidal to columnar and flattened, with large intercellular spaces, suggests that cytoskeletal changes (microtubuli and microfilaments) are responsible for the cell shape transformations. The maintenance of the columnar shape is associated with a well-organized, cylindrical orientation of the microtubular cytoskeleton (*P. americana*). A random distribution of the microtubules might facilitate the transition to a more flattened morphology. Alterations in microtubular association seem to be juvenile hormone dependent. In *Drosophila*, parallel microfilament bundles were shown to be present at the basal side of the vitellogenic follicle cells facing the basement membrane (Figure 9). The density of the microfilament bundles increases during the course of oogenesis. Indirect evidence from a variety of experiments using proteolytic digestion of collagen and inhibitor studies suggests that the microfilaments are required for the adhesion of the follicle cells to the basement membrane. In the absence of parallel microfilaments, the cells lose their epithelial character and round up.[23] Components of the extracellular matrix may affect the organization of the cytoskeleton. Consequently, the parallel microfilaments, together with the extracellular matrix glycoprotein laminin (and possibly additional components of the basement membrane), act in concert in shaping the follicle.

The total number of follicle cells associated with a single oocyte varies with the developmental stage and with the size of the oocyte. A vitellogenic oocyte of *Leucophaea* has approximately 27,000 investing follicle cells. *Drosophila* follicles when first formed have 80 cells, increasing to about 1200 in mature follicles.[9]

In *Drosophila*, a rather interesting additional feature is found. Once the maximum follicle cell number is reached, some of the follicle cells undergo a series of migrations (Figure 10).

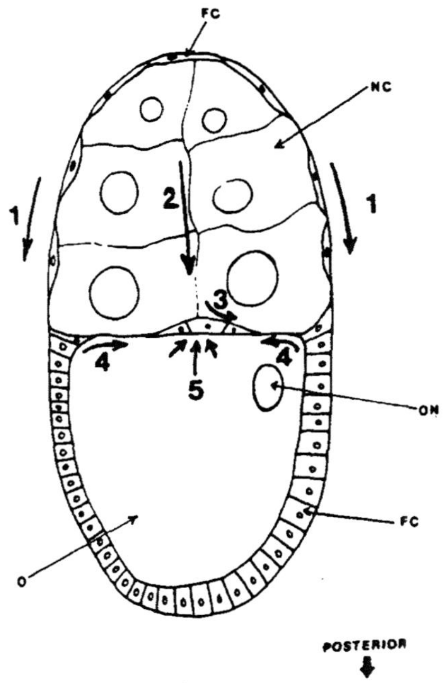

FIGURE 10. Diagrammatic representation of *Drosophila* midvitellogenic follicle, indicating follicle cell migratory pathways (1–5). FC, follicle cell; NC, nurse cell; O, oocyte; ON, oocyte nucleus. (From Kaulenas, M. S., *Insect Accessory Reproductive Structures. Function, Structure, and Development,* Springer-Verlag, Berlin, 1992. With permission.)

The various cell migrations are microtubule dependent.[31] In addition to the involvement of microtubules, the large, steady electric currents which have been shown to traverse the follicles are proposed to direct the follicle cell migration.[32]

In biological systems, electric current is carried by ions, not by electrons. Ion asymmetries within the follicle, which may be associated with the electrical phenomena, have been confirmed, e.g., in *Drosophila* and *Hyalophora cecropia* (Lepidoptera) meroistic ovaries. In the Cecropia moth, a potential difference of about 6 mV is observed over the 30-μm-wide open cytoplasmic bridge which interconnects the oocyte with a trophocyte. The oocyte is at positive potential as related to the tropharium.[33] Most studies of currents associated with the vitellogenic phase of an oocyte have reported an anterior inward current and a posterior outward directed current, rotationally symmetric about the oocyte's long axis. In polytrophic ovaries, the anterior to posterior currents are suggested to provide an electrophoretic force for distributing negatively charged nurse cell products to the vitellogenic oocyte. In panoistic ovaries, currents in this type of oocyte might play a role in the intracellular distribution of cell organelles or products throughout the oocyte (see Section IV.C).

Before or during the vitellogenic stage, the follicular epithelial cells become polyploid. In panoistic ovaries, an exact doubling of DNA during polyploidization is typical, whereas in meroistic ovaries, polyploidization is generally not exact. For example, in *Drosophila hydei*, DNA sequences for the ribosomal RNAs are severely under-replicated, and it has been suggested that polyploidization increases the concentration of those genes which play important roles in follicle cell functions.[9] All the major functions of the follicle cells seem to be juvenile hormone dependent.

4. Trophic Function of Nurse Cells

In both the polytrophic and the telotrophic ovaries, the nurse cell-oocyte syncytium is a polarized structure, with the site of "nutrient" formation distinct and separate from the recipient cell, the oocyte. The most significant difference between the two ovary types is the greater separation between donor and recipient poles in the telotrophic ovary, making it more difficult to use diffusion as a transport mechanism for nutrients.

The trophic function of the nurse cell is enhanced by endoreplication of the DNA. In *Drosophila*, the nurse cells begin their cycle of endomitotic DNA replication in region 3 of the germarium. In the vitellarium, each nurse cell undergoes another seven replications. The maximum level of polyploidy reached by *Drosophila* nurse cells is 2^{10}. In the giant moth *Antheraea polyphemus*, each of the seven nurse cells reaches ploidy levels approaching 2^{16}.[34] In most polytrophic meroistic ovaries, even in *Drosophila*, an apical-basal gradient in polyploidization exists among nurse cells. Highest ploidy levels are found in basal nurse cells.[35] Whereas in panoistic ovaries an exact doubling of DNA content during polyploidization seems to be typical, in meroistic ovaries, polyploidization is generally not exact, but some sequences (e.g., ribosomal RNA genes, histone genes, telomere sequences, satellite sequences) can be under-replicated.[36] Very young nurse cell nuclei have distinct polytenic (giant) chromosomes, but chromosomes cannot be distinguished in large nuclei, as different sections are replicated to varying degrees. In the last endoreplication cycle, all sequences, including the previously under-replicated, replicate fully. In telotrophic ovaries, trophocyte nuclear DNA also undergoes multiple rounds of duplication — a total of seven in *Dysdercus intermedius* (Heteroptera).[37]

In the case of polytrophic ovaries, all or most of the nurse cell cytoplasm is transferred to the oocyte towards the end of oogenesis, whereas for telotrophic ovaries the process could be more selective. Among the major products accumulated by oocytes are large quantities of mitochondria. Also produced are massive stores of ribosomes, which are used by embryonic cells during the early periods of development, when little or no rRNA synthesis takes place. In most meroistic ovaries, the expanded ribosomal gene numbers in the polyploid nurse cells provide sufficient templates for the massive rRNA synthesis. In a few cases, however, additional extrachromosomal rDNA amplification is encountered, as in water beetles.[38] A large variety of nonribosomal transcripts is also synthesized in nurse cells and transferred to the oocyte. Among the most interesting and important gene transcripts synthesized in nurse cells and stored in oocytes are those which specify embryo polarity.[39,40] Other mRNAs which are transcribed in the nurse cells and later transferred to the oocyte include those for heat-shock proteins.[41]

Despite a substantial amount of work, the mechanisms of transport of macromolecules from the trophic cells to the oocyte remain to be totally defined. In polytrophic ovaries, the total nurse cell cytoplasm flows through the ring canals into the oocyte during the final phase of vitellogenesis. Electrophysiological studies have shown that electrophoresis may regulate the distribution of charged molecules between the nurse cells and oocyte. The electrophoretic current from nurse cells to oocyte is driven by the voltage gradient produced by an egg chamber (see Section VI.A.3). Since the equilibrium potential of the nurse cells is several mV more negative than that of the oocyte, macromolecules carrying a net negative charge may be carried by electrophoresis to the oocyte. However, some authors have failed to demonstrate intercellular electrophoresis in movement of materials in polytrophic ovaries. The potential difference between nurse cells and oocyte, therefore, may serve primarily as a regulatory gate effect rather than providing the principal force for macromolecule transport.[9] In a variety of polytrophic ovaries, the nurse cell cytoplasmic streaming can be reversibly inhibited by cytochalasins, so it is likely that microfilament contraction plays some role in the cytoplasmic streaming phenomenon, possibly by squeezing the nurse cell contents into the oocyte.[42] However, mechanisms for capturing specific regionalized compounds in the oocyte must also exist.

In telotrophic ovarioles, the distance between the place of synthesis of macromolecules (the trophic cells in the tropharium) and the place of deposition (the growing oocytes in the vitellarium) can be enormous (see Figure 2). Therefore, most workers have assumed that molecules are actively transported to the oocyte. After the discovery of a system of parallel microtubules in the nutritive cords of heteropterans, suggestions were advanced that these organelles may play a role in the active transport.[6] However, various species of polyphage Coleoptera have nutritive cords that lack microtubules, and in other insect species, the microtubules in the nutritive cords are randomly oriented. Another favored mechanism for the transport of molecules involves electrophoresis and a flow of material assisted by differences in hydrostatic pressure between the trophic area and the oocyte, which may be created by ionic current asymmetries around the ovarioles.[43] Besides active transport, peristaltic movement of the musculature of the epithelial sheath which surrounds the ovariole may cause some cytoplasm to flow from the tropharium to the oocytes. Additional work is required to resolve all the components acting during macromolecule transport in polytrophic as well as in telotrophic ovaries.

B. PREVITELLOGENESIS

The period when young oocytes enlarge by incorporation of nurse cell material (see Section IV.A.4) is called the previtellogenic growth phase, or just previtellogenesis. Previtellogenesis is difficult to investigate because it takes place chiefly in the penultimate oocyte during vitellogenesis of the terminal oocyte and is thus simultaneously subject to its own control and the control exerted on the terminal oocyte. Generally, previtellogenesis begins in young adults, late in pupal development, in nymph, or last instar larvae. However, in insects with panoistic ovarioles, previtellogenic growth begins during earlier nymphal stages. In those insect species which hibernate as adults, growth of the oocytes may be stopped at the beginning of previtellogenesis. The arrest and the onset of previtellogenesis are part of the adult diapause and may be under hormonal control[1] (see Chapter by Hardie).

C. VITELLOGENESIS

Vitellogenesis is the most important metabolic event in the adult life of the female insect. The vitellogenic growth phase begins when yolk spheres are first observed to accumulate in the oocytes. Vitellogenesis often occurs in the adult insect but also may take place earlier. For example, in *Sialis flavilatera* (Megaloptera), vitellogenesis begins early in pupal life. Such a shift into preadult stages will become necessary when the adult lives only a few days, as in species with polytrophic meroistic ovaries that do not feed as adults. In many insects, however, vitellogenesis and egg production is dependent on food availability. An extreme example of a cyclic yolk production with feeding as an initial trigger for vitellogenesis is found, e.g., in *Aedes, Phormia,* or *Rhodnius* (anautogenous insects).

Vitellogenesis involves the production of female-specific proteins termed vitellogenins (vg) and their entry into the oocyte. When vitellogenin is taken into the oocyte, it is processed to vitellin (vn). The vitellogenins are mostly produced in the fat body but may be also produced in the ovary. They are transported by the hemolymph, in which their titer is high during vitellogenesis, and accumulate in the oocyte against a concentration gradient 20–100 times their concentration in the hemolymph. In most species, vitellins comprise 60–90% of the total soluble egg yolk protein. As noted above, vitellogenesis occurs in the terminal oocyte within an ovariole, yet in many species the process is highly synchronized among ovarioles and between ovaries. The synchronization results in a production of egg batches. In some females, vitellogenesis in the penultimate oocyte appears to be inhibited even after the terminal oocyte has completed its yolk deposition and has become chorinated, provided that the mature egg is not laid.

1. Vitellogenin and Vitellin Chemistry

Data on yolk protein chemistry have expanded since new techniques of DNA analysis are providing cloned cDNA and genomic-DNA with which the relationships between vitellogenin genes and the final secretion products can be studied (see Section IV.C.2).

In all insect species investigated, vgs and vns are immunochemically identical. They are glycolipophosphoproteins of native molecular masses ranging between 190 and 650 kDa and often are composed of several polypeptides of variable sizes. In some species, endogenous proteolytic cleavage changes the pattern of vn peptides compared to vgs. In *Leucophaea maderae, B. germanica,* and *L. migratoria,* there seem to be large precursor molecules that are proteolytically cleaved either in the fat body, hemolymph, or oocyte itself.[44] Also, subtle differences in vg and vn lipid and carbohydrate moieties may exist. However, the differences between vg and vn are small and, therefore, the chemistry of both polypeptides will be discussed together.

Cloning the vitellogenin genes enables the putative amino acid sequences of the primary products to be determined; but only complex protein chemistry will unravel the processing and modification of these molecules during their secretion, transport, uptake, and deposition.[45] Harnish and White[46] have characterized the vitellins of a number of insect species and report the existence of three definable groups. The largest group (group I) comprises insects from several orders, including Ephemeroptera, Orthoptera, Dictyoptera, Hemiptera, Dermaptera, Coleoptera, and Lepidoptera. The native proteins are between 380 and 470 kDa, and upon denaturation two distinctly different size classes of subunits are released. The high molecular mass group ranges from 100–180 kDa, the low mass group from 43–86 kDa. The simplest patterns exhibit one polypeptide in each size class, but several in each class is very common. Group II vitellins occur in the orders Hymenoptera (e.g., *Apis mellifera*) and from the more ancient dipterans, the suborder Tipolomorpha (*Aedes aegypti*). In denaturing SDS gels, group II vitellins release only high molecular weight polypeptides. The third type (group III) is found only in higher Diptera (e.g., *Drosophila, Calliphora, Lucilia*). These proteins appear to have molecular masses of about 200 kDa and are composed entirely of small polypeptides of about 50 kDa. The evidence for this schema, however, is not entirely convincing.

Carbohydrate has been found covalently linked to purified vgs and vns in every instance it has been sought. The average carbohydrate content is 1–11%. In many cases, mannose and glucosamine were the only identified sugars involved. In group I proteins, the oligosaccharides are attached only to the heavy subunit. Lipids are also integral to all vgs and vns characterized (7–15%). Phospholipids, diacylglycerides, and cholesterol comprise the bulk lipid components. Vns appear to contain less lipids than vgs. In locust eggs, conjugated ecdysteroids have been found noncovalently bound to vitellin.

Esterified phosphate is another integral part of some vgs and vns. In *L. maderae* vitellogenin, the covalently attached phosphorus is distributed in an uneven fashion among the five subunits. Phosphorylation of vitellogenin occurs posttranslationally in the fat body endoplasmic reticulum.

The fat body of vitellogenic mosquitoes was found to synthesize and secrete another protein, in addition to vitellogenins, that accumulated in developing oocytes. This 53-kDa protein is glycosylated, and immunoblot analysis demonstrated the immunological identity of the 53-kDa polypeptides from the fat body and the ovary.[47] In eggs of some lepidopteran insects, vitellin comprises only half of the total yolk proteins, and the yolk contains significant amounts of other kinds of proteins. Silkworm (*B. mori*) eggs contain a vitellin (M_r 420 kDa) belonging to group I which consists of two heavy subunits (178 kDa) and two light subunits (42 kDa). The second major yolk protein group is composed of non-sex-linked serum proteins with a molecular mass of 30 kDa. They are also produced in the fat body, released into the hemolymph, and finally sequestered into developing oocytes. *Bombyx* 30-kDa proteins are a mixture of three monomers (29.5 to 32 kDa) and contain various lipids and carbohydrates. The third main protein of silkworm eggs is the so-called egg-specific protein, which is produced by the ovary itself and accumulates especially in developing oocytes. The egg-specific protein is a trimer (225 kDa) of two heavy subunits (72 kDa) and one light subunit (64 kDa).[48]

2. Vitellogenin Genes

A large amount of information has been developed in the last decade on vitellogenin gene sequences. Most of the information is on the genes of *Drosophila*. In *Drosophila*, three distinct genes coding for yolk proteins are present, namely, YP1, YP2, and YP3. The genes were shown to be single copy, with the YP1 and YP 2 closely spaced and YP3 approximately 1000 kilobases (kb) distant on the X chromosome. YP1 and YP3 each have a single species of transcript, of about 1.6 and 1.54 kb, respectively. YP 2 produces transcripts of two sizes, 1.59 and 1.67 kb. All three genes have been sequenced.[49] With the availability of base sequences for the yolk protein genes, more recent work has been concentrating on their regulation[40] (see also Section IV.C.3).

A. aegypti and *L. migratoria* are two other insects in which the yolk protein genes have been cloned. In both, the primary transcripts are very large, over 6000 nucleotides in length. The results on *A. aegypti* suggest that there may be a total of five different vitellogenin genes.[50] Sequence information and analysis of any gene control region are not yet available. In *L. migratoria*, two genes, VgA and VgB, coding for vitellogenins have been identified, with only little homology between them (at the 5′ ends).[51] Both are located in the X chromosome. The homologous regions in their 5′ flanking sequences may be important for their control by juvenile hormone.

3. Vitellogenin Synthesis

The fat body is the major, and in many cases, the only site of vitellogenin synthesis. Among some of the Holometabola (Diptera, Lepidoptera, Coleoptera), however, the ovarian follicle cells are also involved in yolk protein production. In the Diptera, the same structural genes, synthesizing identical proteins, are active in both the fat body and the follicle cells.

Three cell types are commonly found in the insect fat body — trophocytes, urocytes, and mycetocytes. The trophocyte is the principal cell type and it functions in a metabolic and storage capacity. The cells are characterized by the presence of lipid droplets, protein spheres, and glycogen granules in a metabolically active cytoplasm.[16] The majority of the trophocytes are found at the periphery of the fat body lobe. This distribution allows the trophocytes to absorb or release products efficiently into the hemolymph.

In the majority of insects, juvenile hormone appears to be the key element in the control of yolk protein production (see Chapter by Hardie).[52] The primary mode of action for this hormone is at the fat body by initiating vitellogenin synthesis, with a secondary function in the regulation of yolk uptake by the ovary (Section IV.C.5). The induction of vg synthesis by JH in the fat body provides a system of hormonal control of gene expression. In response to JH stimulation, the nucleus of the fat body trophocytes enlarges, while the cytoplasm develops extensive rough ER and Golgi complexes. At the macromolecular level, the cells undergo rapid synthesis of DNA, RNA, and protein in response to JH. In *L. migratoria*, a primary stimulation by JH or JH mimics results in a rapid synthesis of rRNA, while the accumulation of vitellogenin mRNA can be detected only after a lag phase. A second dose of JH leads to a more rapid accumulation and translation of vg mRNA, but lowers the production of rRNA. The picture of JH action obtained in insect fat body parallels the finding for steroid hormone-stimulated vitellogenin synthesis in vertebrate systems.[52]

Since JH is a terpenoid and therefore different in chemical structure from that of a steroid hormone, it is of interest to know whether a specific JH receptor is present in the cytosol and nuclei of the target tissues. Cytosolic receptors may function in the translocation of the hormone to nuclear acceptor molecules, the latter being essential for the initiation of gene transcription.[53] Cytosolic fat body preparations of *L. maderae* adults contained a population of JH binding compounds with a high affinity (K_d ca. 1 n*M*), which could not be found in nymphal tissues. A JH binding compound with similar affinity was extracted from nuclei of vitellogenic fat body cells of *L. maderae*. Putative juvenile hormone receptors have also been identified in locust (*L. migratoria*) fat bodies. In the absence of JH, the adipokinetic hormone (AKH-I), which is involved in mobilizing diglycerides, may inhibit vitellogenin gene expression in the locust fat body.[54] Signals from the ovary are supposed to terminate vitellogenin synthesis in the fat body since ovariectomized females continue to produce vitellogenin, which accumulates in the hemolymph. These signals may operate via modulating the activity of the corpora allata.[55]

The majority of insects conform more or less to the regulating scheme described above, but some display significant variations. For example, among Coleoptera, while JH is necessary to set off the initial vitellogenic response, continued yolk production then becomes autonomous. In many Lepidoptera, vitellogenesis appears to be a part of a programmed developmental response to metamorphosis. Among the Hymenoptera, honeybee queens show no dependence on JH or ecdysone for the production of vitellogenins. In some Hemiptera, ecdysone seems to be responsible for triggering elevated levels of yolk protein production.

The induction of vitellogenin synthesis is normally limited to adult females. It is possible, however, to induce vitellogenin synthesis in adult males and in nymphs with large doses of juvenile hormone or JH analogues. This has been demonstrated for *L. migratoria* and several species of Dictyoptera.[56] In male fat bodies of some Diptera, vitellogenin synthesis could be induced by 20-hydroxyecdysone, and not by JH.[57] Female- and male-produced vitellogenins may be different in their polypeptide compositions.

Details of ovarian yolk protein production are best understood for *Drosophila*. Using a radioactive labeled probe containing the coding regions of yolk protein genes and *in situ* hybridization techniques, it has been shown that the follicular epithelium is the specific site of vitellogenin synthesis. The maximum level of yolk protein synthesis by the follicle cells occurs in early vitellogenic stages. The follicular epithelium contributes ca. 35% of the yolk proteins 1 and 2 to the total oocyte content, but only about 10% of the yolk protein 3 polypeptide. Since all three yolk proteins are transcribed at similar rates, yolk protein 3 mRNA seems to be destabilized in the ovarian follicle cells, accounting for the reduction in its steady state level. In the housefly *Musca domestica*, at the start of vitellogenesis the fat body appears to be the main site of vitellogenin synthesis; later, the dominant role is taken over by the ovaries. Overall, the follicle cell contribution of yolk proteins to the oocyte exceeds that of the fat body.

As in the Diptera, the yolk proteins of the Lepidoptera are synthesized in the fat body and the ovarian follicle cells. In Lepidoptera, however, the follicle cell vitellogenins are the product of genes different from those responsible for vg synthesis in the fat body. In a moth, *Plodia interpunctella*, fat body and follicle cell yolk proteins show no immunological cross-reactivity, either as native proteins or as individual subunits.

Diptera also handle the hormonal regulation of vg synthesis somewhat differently from most other insects (see above). 20-Hydroxyecdysone appears to be the main hormonal trigger in the activation of the vitellogenin genes, but JH is involved in facilitating 20-hydroxyecdysone action in the fat body and in the regulation of yolk protein uptake by the ovary[58] (see Chapter by Hardie). In mosquitoes, an oostatic hormone may act to inhibit vitellogenin production.[59]

4. Vitellogenin Secretion

The mechanism for the export of the yolk proteins both from the fat body and the follicle cells involves the usual route through the Golgi and exocytosis at the plasma membrane. The carbohydrate moieties may confer a certain degree of stability to the protein subunits, ensuring proper assembly or preventing aggregation prior to secretion.[60] In a cockroach, *B. germanica*, the vg precursor accumulates but is not secreted when the animal is treated with tunicamycin. Similar observations were done on the export of fat body proteins in *Galleria mellonella* (Lepidoptera). In dipteran follicle cells, the export mechanism can be disrupted by colchicine and other microtubule inhibitors, suggesting an important role of these cytoskeletal compounds.

Yolk proteins excreted from the follicle cells are not normally liberated into the hemolymph at large, but possibly are presented directly to the oocyte surface.

5. Uptake of Vitellogenin by the Ovary

Because of the heterosynthetic nature of vitellogenins in most insects, the oocytes are highly specialized for the specific accumulation of the fat body vitellogenins. The pathway for internalization of vitellogenins and other yolk protein precursors is similar in all types of ovarioles. The onset of vitellogenic uptake is characterized by the formation of gaps and spaces between the follicle cells (Figure 11). Many researchers have utilized various dyes and tracers to demonstrate that this interfollicular route of uptake is universal among insects. Vitellogenins appear to concentrate in the perioocytic space and from there are later taken up by micropinocytosis of the oocyte membrane. The enlargement of the interfollicular channels occurs at the onset of vitellogenesis and has been termed patency. In *R. prolixus*, this condition has been shown to be a development in response to juvenile hormone. In other insects, JH also seems to be involved in yolk protein uptake, but in these cases the precise role of the hormone is not entirely clear (see Chapter by Hardie). The follicle cells lose patency at the time of cessation of vitellogenesis and the deposition of the chorion.

Current evidence clearly points to yolk protein uptake being a receptor-mediated process. The uptake is highly selective for vgs, saturable, species specific, energy dependent, and sensitive to conditions of pH, temperature, and divalent cation concentrations (Ca^{2+}). Studies on binding of vg to membranes isolated from follicles have been made on only a few insects. Binding under equilibrium conditions demonstrated the existence of a saturable, single class of binding sites of follicle or ovary membranes (except for the cockroach *Nauphoeta cinerea*, where two separate binding sites for vg have been suggested). The dissociation constant (K_d) varied between 100 n*M* (for *L. migratoria*) to 13 n*M* (for *Manduca sexta*). In *L. migratoria*, the vitellogenin receptor protein has been isolated and purified by immunaffinity chromatography.[61] It is an acidic, negatively charged (pI = 3.5) 180 kDa glycoprotein (nonreducing conditions) with large amounts of N- and O-linked oligosaccharides, among them neuraminic acid, which has been found to be essential for receptor function. The vg receptor was localized in oocyte membranes and in endocytotic vesicles. So far, vg receptors of two other insects have been visualized using the ligand blotting technique, from *N. cinerea* (M_r = 200 kDa) and *A. aegypti* (M_r = 205 kDa).[62] The fact that the receptors bind vg only when separated under nonreducing conditions suggests that disulfide bonds within the receptor molecule are neces-sary for its biological activity.

Receptor-bound vitellogenins accumulate in specialized regions of the surface membrane, the coated pits, and are then internalized by the formation of coated vesicles (Figure 12). Coated vesicles are suggested to be universal organelles for a specific macromolecular transport in eukaryotic cells. The characteristic feature of the coated vesicles is an outer proteinaceous polyhedral cage enclosing its membrane. The major protein of this cage (50% of the total protein) is "clathrin." Three molecules of clathrin heavy chain (M_r 180 kDa) together with three molecules of clathrin light chain (33 and 36 kDa) form a subunit of the coat called triskelion. The triskelions are self-assembled into the polyhedral cage on the membrane of the coated pits. Once the coated vesicle is formed and pinched off the plasma membrane, rapid uncoating follows, releasing the vesicle and allowing the coat components to recycle to the plasma membrane.[62] The released vesicles fuse with endosomes. Within the endosomal compartment, the adsorbed yolk precursors dissociate from the membrane to become homogeneously distributed in the lumen. As a final transformation, the transitional yolk body changes into a mature yolk body with the crystallization of the vitellins. Mature yolk bodies will store the yolk proteins until initiation of embryonic development, during which they are utilized.

The developmental processes of vitellogenin synthesis and vg uptake by the maturing ovaries occur independently of the presence of conspecific males. However, in *Drosophila*,

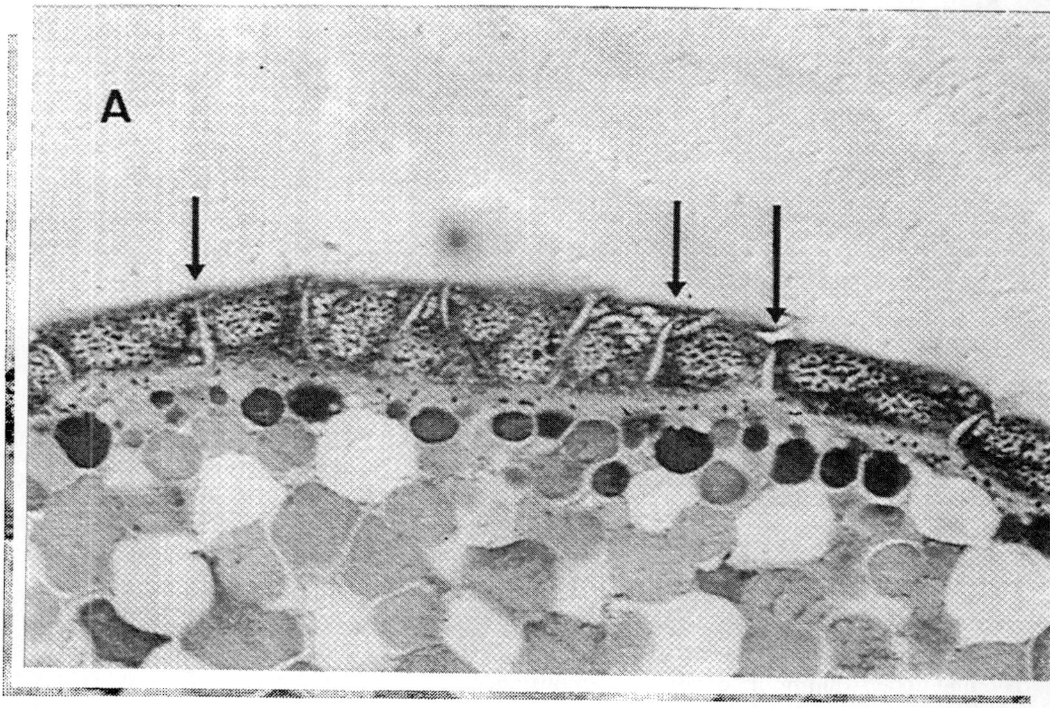

FIGURE 11A

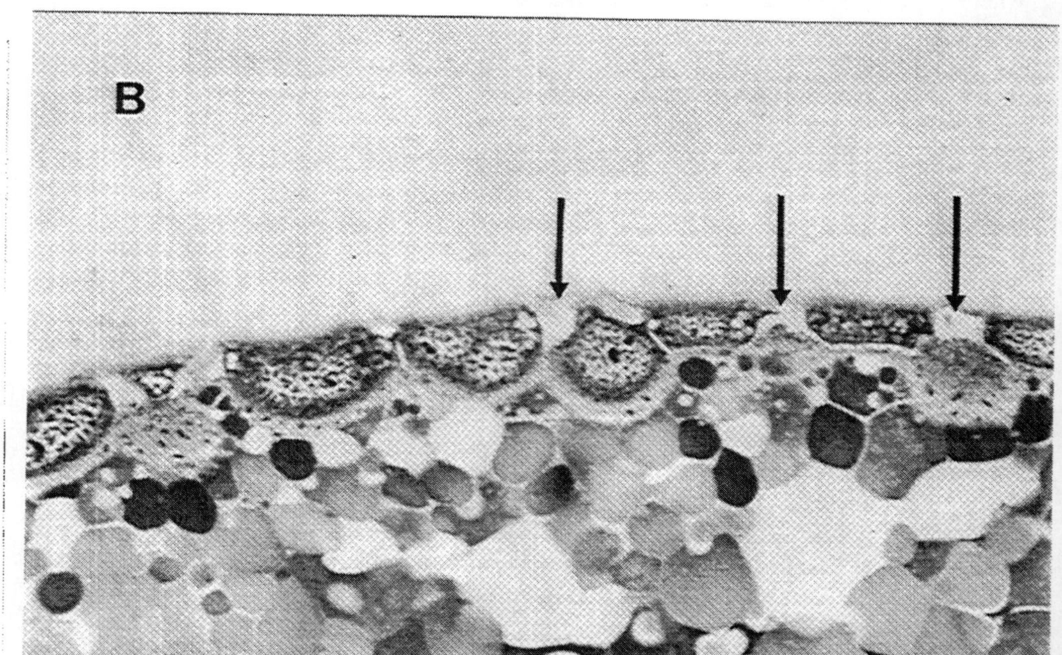

FIGURE 11B

FIGURES 11A and B. Follicular epithelium in *Blattella germanica* oocytes. Sections at the equatorial zone in oocytes from 3-day-old (A) and 5-day-old (B) females with large intercellular spaces (arrows) in (B). Photographs courtesy of X. Bellés, Barcelona.

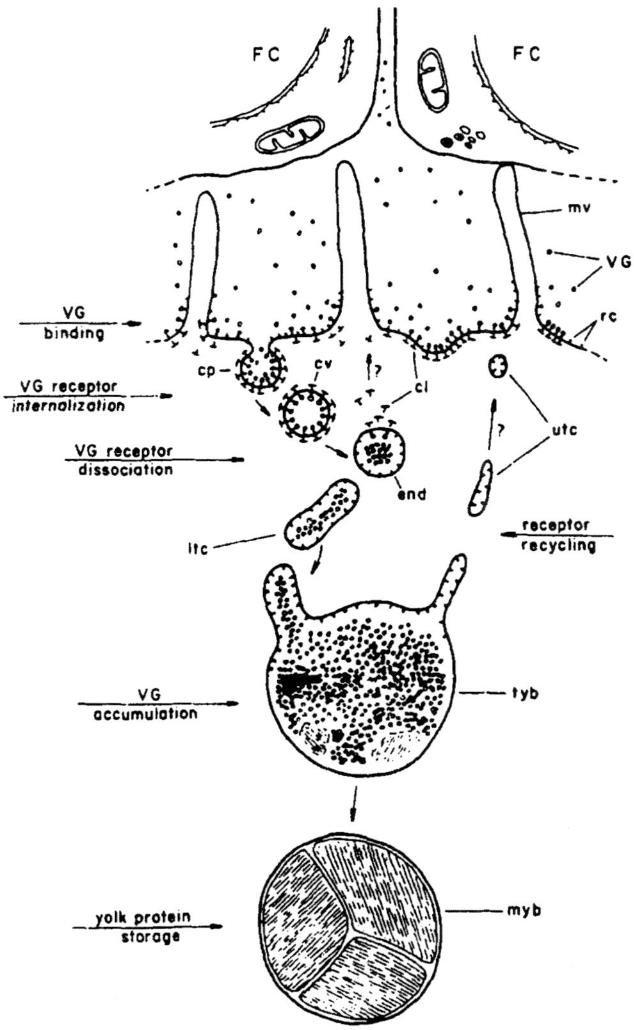

FIGURE 12. Schematic representation of the vitellogenin internalization pathway in mosquito oocyte: cl, clathrin; cp, coated pit; cv, coated vesicle; end, vesicular endosome; FC, follicle cell; ltc, tubular endosome labeled positively with anti-vg antibodies; mv, microvilli; myb, mature yolk body; rc, receptor; tyb, transitional yolk body; utc, tubular compartment labeled negatively with anti-vg antibodies; vg, vitellogenin. (Reproduced, with permission, from the *Annual Review of Entomology*, Vol. 37, p. 217, ©1992 by Annual Reviews Inc.)

it has recently been shown that mature males significantly accelerate the onset of vitellogenesis, and thus ovarian maturation overall, by about 4 days. Although the proximate stimulus is not known, it is conceivable that social signals from the males (during courtship attempts) elicit changes in the hormonal levels of females, thereby initiating the events leading to vitellogenesis. Again, this effect of male behavior on female reproductive biology is similar to that reported in vertebrates.[63]

D. CHORIONIZATION

When vitellogenesis is completed, the vitelline membrane and, later, the chorion (eggshell) are formed (Figure 13).

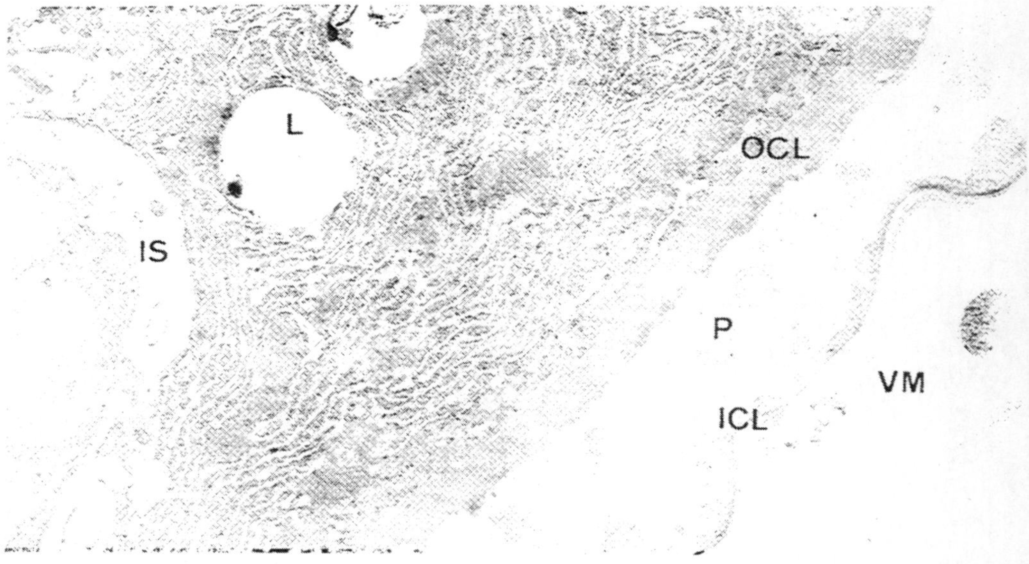

FIGURE 13. Follicle cells from a basal oocyte of *Blattella germanica* at late chorion formation. The perioocyte space contains the vitelline membrane (VM), the inner chorion layer (ICL) and the complex outer chorion layer (OCL) showing columnar projections (P). IS, intercellular space; L, lipid droplet. Photograph courtesy of X. Bellés, Barcelona.

1. The Vitelline Membrane

Early observations suggested that the vitelline membrane was produced by the oocyte itself. This may be the case in some species, for example, in grasshoppers and Lepidoptera.[4] However, more recent studies have demonstrated that the follicle cells secrete material which forms the vitelline membrane in a large number of insects. Although differences in details of this process have been encountered in all of the systems, vitelline membrane production is preceded by an extensive hypertrophy of rough ER and formation of Golgi complexes, followed by an accumulation of secretory granules (vitelline membrane body precursors) in the apical zones of the follicle cells. In *Simulium vittatum* and other Diptera, the build-up of the ER and Golgi vesicles occurs at midvitellogenesis. By the time that vitellogenesis is almost complete, the Golgi complexes, containing both dense and fibrous materials, become even more prominent, and the previtelline membrane secretory substance begins to be deposited between the follicle cells and the oocyte. The secreted material then begins to coalesce and gradually forms the vitelline membrane (see Figure 13). Most of the ultrastructural observations of vitelline membrane formation in exopterygotes support the above outline for the process.

Electron microscopical studies on vitelline membrane formation in *L. migratoria* showed that this structure is composed of two ultrastructurally distinguishable components, the vitelline membrane bodies (VMB) and a fine granular material which eventually cements the VMBs together to give the complete vitelline membrane. The observation that, in *L. migratoria*, the first condensation of VMBs occurs in the vicinity of the oocyte membrane may indicate that the oocyte is the source of VMB secretion, whereas the second vitelline membrane component is clearly the product of the follicle cells. Other support for a dual origin of the vitelline membrane has come from studies on Lepidoptera. In the butterfly *Calpodes ethlius*, a distinct vitelline membrane is detectable at the end of the yolk uptake phase. It has an electron-dense layer away from the oocyte and an electron-lucent layer apposing the oocyte. Both the oocyte and the follicle cells contain coated vesicles, which appear to be in the process of exocytosis. The electron-dense material seems to be exocytosed from the follicle cells, whereas coated vesicles at the oocyte surface exocytose the electron-lucent granular material.

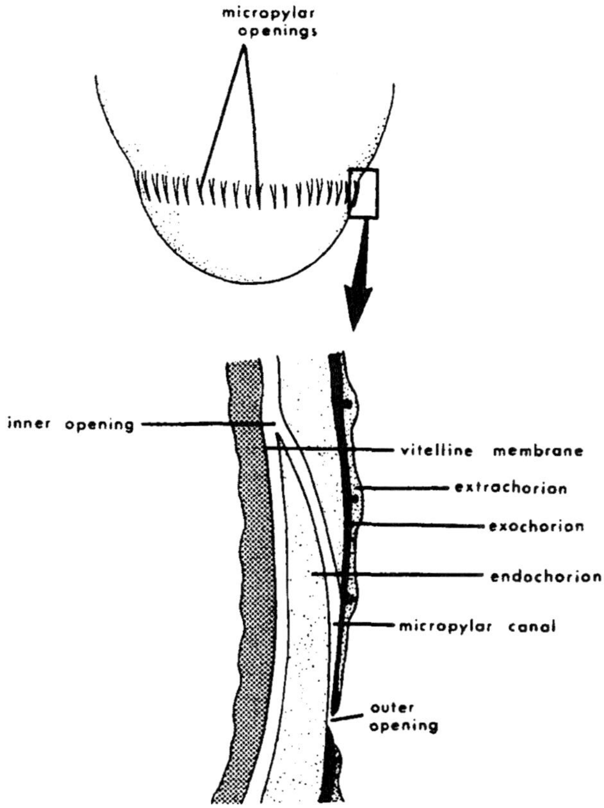

FIGURE 14. Egg of *Locusta migratoria* with a section through the chorion along the micropylar axis. (From Gillott, C., *Entomology*, Plenum, New York, 1980, chap 19. With permission.)

Once fully formed, the vitelline membrane is completely electron dense.[64] A definitive decision on the source of the various contributors to the vitelline membrane cannot be made until it can be demonstrated, with molecular biological techniques now becoming available, that definitive vitelline membrane proteins are synthesized by the oocyte.

2. Chorion Formation

The chorion is usually secreted entirely by the follicle cells and can be seen to comprise two main layers, an endochorion adjacent to the vitelline membrane and an exochorion (Figures 13 and 14). In some insects, e.g., Acrididae, the shell takes on a third layer, the extrachorion, as an oocyte moves through the common oviduct. Although the follicle cells are mesodermal derivatives, the chorion is cuticle-like in nature and contains layers of proteins and lipoproteins, some of which are tanned by polyphenolic substances released by the cells. With chorion secretion, the follicle cells complete their duties and then die.

At the physiological level the chorion functions in protecting the oocyte from mechanical stresses, such as from predators, as well as environmental stresses, such as dessication and drowning, while at the same time permitting gas exchange and sperm penetration.[65] In some species, a wax layer is formed immediately above the vitelline membrane by the coalescence of oil droplets secreted by the follicle cells which renders the chorion waterproof. Viewed from the perspective of the oocyte, the next structure is the basal or inner chorion layer (crystalline chorion layer). The crystalline chorion layer, although flexible, puts a limit to the volume that the oocyte can achieve. Speculations on other functions include a role in the confinement of the wax layer, as well as allowing for gas exchange, through plastron respiration or directly. Distal to the basal chorion is the trabeculate layer which is considered to be a part of the endochorion. The trabeculate layer is characterized by the presence of cavities and pores. The small pores may be formed by the withdrawal of follicle cell processes during and/or after the deposition of this layer. The cavities may interconnect and form extensive channels. The channels may serve as air spaces and open to the exterior via aeropyles (gas exchange). In other cases, the cavities are filled with a mucus-like substance which serves as a reserve to surface-localized adhesive material used to attach eggs to the substrate. The outermost layer of the endochorion, if present, is characterized by the presence of lamellae, based on the helicoidal arrangement of stacks or fibrils. The lamellar layer may be traversed by pores to the exterior. In many Lepidoptera, the lamella layer serves as the outer portion of the shell. In most other insects, however, an exochorion is present. The exochorion consists largely of mucoprotein and contains polysaccharides. In some insects, a ring of follicle cells near the anterior end of the oocyte secretes no exochorion, so that a line of weakness is created at this point, which facilitates hatching. Also, certain follicle cells appear to have larger than normal microvilli which, when withdrawn after chorion formation, leave channels to permit entry of sperm (micropyles; see Figure 14).

The molecular analysis of chorion formation has advanced rapidly. Recent reviews on the biochemistry of chorion proteins, chorion gene structure, and chorion gene expression were presented by Regier and Kafatos[65] and by Kaulenas.[9]

In general, it is assumed that choriogenesis is independent of hormonal control and is initiated at an appropriate late stage of vitellogenesis in response to local signals. However, some information is available that shell formation might depend on brain neurohormones, juvenile hormone, and 20-hydroxyecdysone.[1,66,67] In *L. migratoria*, ecdysteroids are synthesized by the epithelium of the follicle cells at around the time of chorion formation during a short period of 8–12 h. However, ecdysteroid synthesis in follicle cells can also occur earlier, particularly in ovoviviparous or viviparous species, such as in cockroaches and in *Glossina*. Besides helping to regulate chorionization, at least part of these ecdysteroids enter the oocyte, where they are mostly present as polar or apolar conjugates and seem to control the first events in embryogenesis. In a cricket, *G. bimaculatus*, coincident changes in ecdysteroid production for both the ovaries and the abdominal integument were observed.[68] The role of epidermal ecdysteroids during oogenesis is not yet clear.

V. OVULATION AND OVIPOSITION

Ovulation includes the opening of both the oocyte follicle and interfollicular tissue, and involves contractions of ovarioles, pedicels, and oviducts resulting in the expulsion of the egg from the ovary into the oviducts. The follicle then collapses and the follicular cells undergo autolysis. The degenerating follicular cells are generally referred to as a "corpus luteum", and structures of this type have been described in several insect species.[69] Ovulation may occur just before oviposition, or a larger period of time may elapse between the two processes. Several species, including cockroaches and flies, are ovoviviparous and retain their eggs in the genital pouch without supplying them with anything but water. Viviparous insects, on the other hand, supply food to their progeny, either from modified follicle cells or from the accessory glands.

Although in many insects ovulation is rapidly followed by oviposition, there are indications that the two processes are separately controlled. In *R. prolixus*, ovulation is a result of ovarian motility induced by a neurosecretory myotropin released from the corpus cardiacum. The hormone is released in mated females when mature eggs are present in the ovary, as signaled by an increase in ecdysteroid titer in the hemolymph. In other species, ovulation seems to be controlled in a similar way.

In some insects, oocyte resorption, termed oosorption, may take place under various unfavorable environmental conditions. Starvation or the lack of food, mating, or a suitable place to oviposit are frequent causes, but factors such as temperature and change in photoperiod, social life, or maternal care also induce oosorption. In general, oosorption occurs when the external factors do not allow either the survival of eggs and larvae or egg deposition.[20] Oosorption may occur either in young previtellogenic oocytes or in vitellogenic oocytes, and even in chorionated eggs. Oocytes may grow to a certain size and then stop, while the follicle cells begin to change from cuboidal to irregular shapes. The transport of material from the interfollicular spaces to the oocyte ceases, owing to the breakdown of the microvilli of the oocytes and follicle cells. Hydrolytic enzymes produced in the follicle cells cause oocyte lysis, breaking down first the protein and then the lipid yolk globules, penetrating the oocyte and finally themselves degenerating. Often it can be observed that only certain follicles degenerate, while others continue to develop. The appearance of oosorption processes is apparently caused by a decrease in the activity of the corpora allata, but the brain certainly intervenes too, by regulating corpora allata functioning and/or acting humorally on the ovary.

VI. CONCLUDING REMARKS

The development of the female reproductive system is reasonably well understood at the structural or morphological level across a wide range of insect species. Details of the molecular mechanisms involved in the development processes, and especially of the genetic control of gene expression, have been explored in a much lower number of insects. Insect oogenesis comprises many stages which are regulated by certain humoral factors, including neurohormones, juvenile hormone(s), and ecdysteroids, whose importance and modes of intervention vary depending on the species. Detailed knowledge on female reproductive biochemistry, physiology, and endocrinology in a great many insects will be necessary, e.g., to provide a basis for using insect hormones, hormone agonists, or hormone antagonists as "third generation pesticides" in insect pest control. Beyond that, ovarian development and oogenesis are affected by environmental factors, including temperature, humidity, photoperiod, or the finding of food. Therefore, it seems impossible to study the physiology and endocrinology of insect reproduction without taking into account its ecological conditions.

REFERENCES

1. **Raabe, M.,** *Recent Developments in Insect Neurohormones,* Plenum, New York, 1989, chap. 6.
2. **Truckenbrodt, W.,** Über die imaginale Ovarvergrößerung im Zusammenhang mit der Physogastrie bei *Odontotermes badius* Haviland (Insecta, Isoptera), *Insectes Soc.,* 20, 21, 1973.
3. **Gottanka, J. and Büning, J.,** Mayflies (Ephemeroptera), the most "primitive" winged insects, have telotrophic meroistic ovaries, *Wilhelm Roux Arch. Dev. Biol.,* 203, 18, 1993.
4. **Gillott, C.,** *Entomology,* Plenum, New York, 1980, chap. 19.
5. **Berry, S. J.,** Reproductive systems, in *Fundamentals of Insect Physiology,* Blum, M. S., Ed., John Wiley & Sons, New York, 1985, chap. 11.
6. **King, R. C. and Büning, J.,** The origin and functioning of insect oocytes and nurse cells, in *Comprehensive Insect Physiology, Biochemistry and Pharmacology,* Vol. 1, Kerkut, G. A. and Gilbert, L. I., Eds., Pergamon, Oxford, 1985, chap. 3.
7. **Wigglesworth, V. B.,** The distribution of aeriferous tracheae for the ovaries of insects, *Tissue Cell,* 23, 57, 1991.
8. **Büning, J.,** Germ cell cluster formation in insect ovaries, *Int. J. Insect Morphol. Embryol.,* 22, 237, 1993.
9. **Kaulenas, M. S.,** *Insect Accessory Reproductive Structures. Function, Structure, and Development,* Springer, Berlin, 1992.

10. **Storto, P. D. and King, R. C.,** The role of polyfusomes in generating branched chains of cystocytes during *Drosophila* oogenesis, *Dev. Genet.*, 10, 70, 1989.
11. **King, R. C., Cassidy, J. D., and Rousset, A.,** The formation of clones of interconnected cells during gametogenesis in insects, in *Insect Ultrastructure*, Vol. 1, King, R. C. and Akai, H., Eds., Plenum, New York, 1982, 3.
12. **Dorn, A., Rademacher, J. M., and Sehn, E.,** Ecdysteroid-dependent development of the oviduct in last-instar larvae of *Oncopeltus fasciatus*, *J. Insect Physiol.*, 32, 643, 1986.
13. **Davey, K. G.,** The female reproductive tract, in *Comprehensive Insect Physiology, Biochemistry and Pharmacology*, Vol. 1, Kerkut, G. A. and Gilbert, L. I., Eds., Pergamon, Oxford, 1985, chap. 2.
14. **Szopa, T. M. and Happ, G. M.,** Cytodifferentiation of the accessory glands of *Tenebrio molitor*. IX. Differentiation of the spermathecal accessory gland *in vitro*, *Cell Tissue Res.*, 222, 269, 1982.
15. **Degrugillier, M. E.,** *In vitro* release of house fly, *Musca domestica* L. (Diptera: Muscidae), acrosomal material after treatment with secretion of female accessory gland and micropyle cap substance, *Int. J. Insect Morphol. Embryol.*, 14, 381, 1985.
16. **Koeppe, J. K., Fuchs, M., Chen, T. T., Hunt, L.-M., Kovalick, G. E., and Briers, T.,** The role of juvenile hormone in reproduction, in *Comprehensive Insect Physiology, Biochemistry and Pharmacology*, Vol. 8, Kerkut, G. A. and Gilbert, L. I., Eds., Pergamon, Oxford, 1985, chap. 6.
17. **Burns, E. L., Chiang, A.-S., Gadot, M., and Schal, C.,** Juvenile hormone regulation of the left colleterial gland in intact and ovariectomized *Blattella germanica* L. (Dictyoptera: Blattellidae), *J. Insect Physiol.*, 37, 401, 1991.
18. **Pau, R. N., Weaver, R. J., and Edwards-Jones, K.,** The regulation of cockroach oothecin synthesis by juvenile hormone, *Arch. Insect Biochem. Physiol. (Suppl.)*, 1, 59, 1986.
19. **Ter Wee, T. J. and Stay, B.,** Termination of pregnancy and milk synthesis in the viviparous cockroach *Diploptera punctata*: a role of juvenile hormone, *Int. J. Invertebr. Reprod.*, 11, 59, 1987.
20. **Raabe, M.,** Insect Reproduction: Regulation of successive steps, in *Advances in Insect Physiology*, Vol. 19, Evans, P. D. and Wigglesworth, V. B., Eds., Academic Press, London, 1986, 29.
21. **Technau, G. M. and Campos-Ortega, J. A.,** Lineage analysis of transplanted individual cells in embryos of *Drosophila melanogaster*. III. Commitment and proliferative capabilities of pole cells and midgut progenitors, *Wilhelm Roux Arch. Dev. Biol.*, 195, 489, 1986.
22. **Hay, B., Jan, L. Y., and Jan, Y. N.,** Localization of *vasa*, a component of *Drosophila* polar granules, in maternal-effect mutants that alter embryonic anteroposterior polarity, *Development*, 109, 425, 1990.
23. **Gutzeit, H. O., von Seydlitz-Kurzbach, E., and Neuschröer, R.,** How *Drosophila* follicles become spatially organized and obtain their ovoid shape, *Int. J. Insect Morphol. Embryol.* 22, 335, 1993.
24. **Hatakeyama, M., Sawa, M., and Oishi, K.,** Ovarian development and vitellogenesis in the sawfly, *Athalia rosae ruficornis* Jakovlev (Hymenoptera, Tenthridinidae), *Invertebr. Reprod. Dev.*, 17, 237, 1990.
25. **Büning, J. and Sohst, S.,** The flea ovary: ultrastructure and analysis of cell clusters, *Tissue Cell*, 20, 783, 1988.
26. **Telfer, W. H.,** Development and physiology of the oocyte-nurse cell syncytium, *Adv. Insect Physiol.*, 11, 223, 1975.
27. **Gutzeit, H. O.,** Die Entwicklung der Eizelle bei Insekten. Ein Modellsystem für die Entstehung von Zellpolarität, *Biol. Zeit*, 20, 33, 1990.
28. **Bohrmann, J., Frey, A., and Gutzeit, H. O.,** Observations on the polarity of mutant *Drosophila* follicles lacking the oocyte, *Wilhelm Roux Arch. Dev. Biol.*, 201, 268, 1992.
29. **Lanot, R., Roussel, J.-P., and Thiebold, J. J.,** Ecdysteroids and meiotic reinitiation in oocytes of *Periplaneta americana* (Dictyoptera) and *Gryllus bimaculatus* (Orthoptera), *Invertebr. Reprod. Dev.*, 15, 69, 1989.
30. **St. Johnston, D. and Nüsslein-Volhard, C.,** The origin of pattern and polarity in the *Drosophila* embryo, Cell, 68, 201, 1992.
31. **Gutzeit, H. O.,** Transport of molecules and organelles in meroistic ovarioles of insects, *Differentiation*, 31, 155, 1986.
32. **Overall, R. and Jaffe, L. F.,** Patterns of ionic current through *Drosophila* follicles and eggs, *Dev. Biol.*, 108, 102, 1986.
33. **De Loof, A.,** The meroistic insect ovary as a miniature electrophoresis chamber, *Comp. Biochem. Physiol. Sect. A*, 74, 3, 1983.
34. **Berry, Y. S.,** Maternal direction of oogenesis and early embryogenesis in insects, *Annu. Rev. Entomol.*, 27, 205, 1982.
35. **Cardoen, J., Schoofs, L., Broekaert, D., van Mellaert, H., Verachtert, B., and de Loof, A.,** Polyploidization and localisation of poly $(A)^+$ RNA in the different cell types of the vitellogenic meroistic ovary of the fleshfly, *Sarcophaga bullata*, *Histochemistry*, 85, 305, 1986.
36. **Hammond, M. P. and Laird, C. D.,** Chromosome structure and DNA replication in nurse and follicle cells of *Drosophila melanogaster*, *Chromosoma*, 91, 267, 1985.
37. **Dittmann, F., Hörner, R., and Engels, W.,** Endoploidization of tropharium nuclei during larval development and the first gonocycle in *Dysdercus intermedius* (Heteroptera), *Int. J. Invertebr. Reprod.*, 7, 279, 1984.

38. **Dittmann, F., Steinbrück, G., and Münz, A.,** Amplification of tropharium rDNA in the telotrophic ovariole of the bug, *Dysdercus intermedius, Invertebr. Reprod. Dev.*, 17, 9, 1990.
39. **Sander, K. L.,** Specification of the basic body pattern in insect embryogenesis, *Adv. Insect Physiol.*, 12, 125, 1976.
40. **Frohnhöfer, H. G. and Nüsslein-Volhard, C.,** Organization of anterior pattern in the *Drosophila* embryo by the maternal gene *bicoid, Nature (Lond.)*, 324, 120, 1986.
41. **Ambrosio, L. and Schedl, P.,** Gene expression during *Drosophila melanogaster* oogenesis: analysis by *in situ* hybridization to tissue selections, *Dev. Biol.*, 105, 80, 1984.
42. **Gutzeit, H. O.,** The role of microfilaments in cytoplasmic streaming in *Drosophila* follicles, *J. Cell Sci.*, 80, 156, 1986.
43. **Diehl-Jones, W. and Huebner, E.,** Pattern and composition of ionic currents around ovarioles of the hemipteran, *Rhodnius prolixus* (Stahl), *Biol. Bull.*, 176(S), 86, 1989.
44. **Della-Cioppa, G. and Engelmann, F.,** The vitellogenin of *Leucophaea maderae.* Synthesis as a large phosphorylated precursor, *Insect Biochem.*, 17, 401, 1987.
45. **Bownes, M.,** Expression of the genes coding for vitellogenin (yolk protein), *Annu. Rev. Entomol.*, 31, 507, 1986.
46. **Harnish, D. G. and White, B. N.,** Insect vitellins: identification, purification, and characterization from eight orders, *J. Exp. Zool.*, 220, 1, 1982.
47. **Hays, A. R. and Raikhel, A. S.,** A novel protein produced by the vitellogenic fat body and accumulated in mosquito oocytes, *Wilhelm Roux Arch. Dev. Biol.*, 199, 114, 1990.
48. **Zhu, J., Indrasith, L. S., and Yamashita, O.,** Characterization of vitellin, egg-specific protein and 30 kDa protein from *Bombyx* eggs, and their fates during oogenesis and embryogenesis, *Biochem. Biophys. Acta*, 882, 427, 1986.
49. **Yan, Y. L., Kunert, L. J., and Postlethwait, J. H.,** Sequence homologies among the three yolk polypeptides (Yp) genes in *Drosophila melanogaster, Nucleic Acids. Res.*, 15, 67, 1987.
50. **Gemmill, R. M., Hamblin, M., Glaser, R. L., Racioppi, J. V., Marx, J. L., White, B. N., Calvo, J. M., Wolfner, M. F., and Hagedorn, H. H.,** Isolation of mosquito vitellogenin genes and induction of expression by 20-hydroxyecdysones, *Insect Biochem.*, 16, 761, 1986.
51. **Locke, J., White, B. N., and Wyatt, G. R.,** Cloning and 5′ end nucleotide sequences of two juvenile hormone-inducible vitellogenin genes of the African migratory locust, *DNA*, 6, 331, 1987.
52. **Hagedorn, H. H. and Kunkel, J. G.,** Vitellogenin and vitellin in insects, *Annu. Rev. Entomol.*, 24, 475, 1979.
53. **Engelmann, F.,** Regulation of vitellogenesis in cockroaches, in *Cockroaches as Models for Neurobiology: Applications in Biomedical Research*, Vol. 2, Huber, I., Masler, E. P., and Rao, B. R., Eds., CRC, Boca Raton, 1990, 159.
54. **Moshitzky, P. and Applebaum, S. W.,** The role of adipokinetic hormone in the control of vitellogenesis in locusts, *Insect Biochem.*, 20, 319, 1990.
55. **Kempa-Tomm, S., Hoffmann, K. H., and Engelmann, F.,** Vitellogenins and vitellins of the Mediterranean field cricket, *Gryllus bimaculatus:* isolation, characterization and quantification, *Physiol. Entomol.*, 15, 167, 1990.
56. **Lamy, M.,** Vitellogenesis, vitellogenin and vitellin in the males of insects: a review, *Int. J. Invertebr. Reprod. Dev.*, 7, 311, 1984.
57. **Huybrechts, R. and de Loof, A.,** Similarities in vitellogenins and control of vitellogenin synthesis within the genera *Sarcophaga, Calliphora, Phormia* and *Lucilia* (Diptera), *Comp. Biochem. Physiol.*, 72B, 339, 1982.
58. **Kelly, T. J., Adams, T. S., Schwartz, M. B., Birnbaum, M. J., Rubenstein, E. C., and Imberski, R. B.,** Juvenile hormone and ovarian maturation in the Diptera: a review of recent results, *Insect Biochem.*, 17, 1089, 1987.
59. **Borovsky, D., Thomas, B. R., Carlson, D. A., Whisenton, L. R., and Fuchs, M. S.,** Juvenile hormone and 20-hydroxyecdysone as primary and secondary stimuli of vitellogenesis in *Aedes aegypti, Arch. Insect Biochem. Physiol.*, 2, 75, 1985.
60. **Kunke, J. G. and Nordin, J. H.,** Yolk proteins, in *Comprehensive Insect Physiology, Biochemistry and Pharmacology*, Vol. 1, Kerkut, G. A. and Gilbert, L. I., Eds., Pergamon, Oxford, 1985, chap. 4.
61. **Ferenz, H.-J.,** The locust oocyte vitellogenin receptor-function and characteristics, in *Advances in Invertebrate Reproduction*, Vol. 5, Hoshi, M. and Yamashita, O., Eds., Elsevier, Amsterdam, 1990, 103.
62. **Raikhel, A. S. and Dhadialla, T. S.,** Accumulation of yolk proteins in insect oocytes, *Annu. Rev. Entomol.*, 37, 217, 1992.
63. **Craddock, E. M. and Boake, C. R. B.,** Onset of vitellogenesis in female *Drosophila silvestris* is accelerated in the presence of sexually mature males, *J. Insect Physiol.*, 38, 643, 1992.
64. **Griffith, C. M. and Lai-Fook, J.,** Structure and formation of the chorion in the butterfly, *Calpodes, Tissue Cell*, 18, 589, 1986.
65. **Regier, J. C. and Kafatos, F. C.,** Molecular aspects of chorion formation, in *Comprehensive Insect Physiology, Biochemistry and Pharmacology*, Vol. 1, Kerkut, G. A. and Gilbert, L. I., Eds., Pergamon, Oxford, 1985, chap. 5.

66. **Pascual, N., Cerdá, X., Benito, B., Tomás, J., Piulachs, M. D., and Bellés, X.**, Ovarian ecdysteroid levels and basal oöcyte development during maturation in the cockroach *Blattella germanica* (L.), *J. Insect Physiol.*, 38, 339, 1990.
67. **Bellés, X., Cassier, P., Cerdá, X., Pascual, N., André, M., Rosso, Y., and Piulachs, M. D.**, Induction of choriogenesis by 20-hydroxyecdysone in the German cockroach, *Tissue Cell*, 25, 195, 1993.
68. **Weidner, K., Claß, M., Rieck, H., and Hoffmann, K. H.**, Developmental changes in ecdysteroid biosynthesis *in vitro* during adult life and embryogenesis in a cricket, *Gryllus bimaculatus* de Geer, *Invertebr. Reprod. Dev.*, 21, 129, 1992.
69. **Griffith, C. M. and Lai-Fook, J.**, Corpus luteum formation and ovulation in the butterfly Calpodes, *Tissue Cell*, 18, 783, 1986.
70. **Wigglesworth, V. B.**, *The Principles of Insect Physiology*, Chapman and Hall, London, 1972, chap. 15.

Chapter 2

INSECT MALE MATING SYSTEMS

Cedric Gillott

CONTENTS

I. INTRODUCTION

Among the problems that required solution in the evolution of the Insecta as a predominantly terrestrial group was the bringing together of sperm and egg in the absence of an aquatic environment. The solution came through the formation, in most insects, of a spermatophore within which the sperm could be safely transferred to the female reproductive tract, avoiding the risk of both desiccation and predation. In relatively few species, spermatophores are not found and sperm transfer is achieved by means of an intromittent organ. Though sperm production and transfer are the primary functions of the male reproductive system, a number

0-8493-6695-X/95/$0.00+$.50

of secondary functions have evolved, including sperm storage, the generation of "signals," either physical or chemical, that induce significant changes in the female's fecundity and/or receptivity, and the transfer of nutrients to the female.

This chapter will provide the structural and functional background for ensuing chapters that deal with more specific aspects of male insect reproductive biology. In addition, it will focus on selected aspects of the functions of the male reproductive system that are not covered elsewhere but are of particular interest to the author. In keeping with the extreme morphological diversity of the Insecta, the structural, physiological, and biochemical nature of the male system is widely varied. It is not the purpose of this chapter to encyclopedically describe this plethora of detail, but rather to note the generalities that have been established and to point out areas where understanding is still weak.

II. STRUCTURE

A. INTERNAL ORGANS

The male internal organs comprise paired testes (fused to form a single median structure in Lepidoptera), paired vasa deferentia and seminal vesicles, a median ejaculatory duct, and in most species, accessory glands of varied origin and complexity (Figure 1).

Within each testis is a varied number of tubular follicles bound together by a connective tissue sheath. Each follicle connects with a short vas efferens, the vasa efferentia from each gonad opening either confluently or in a linear sequence into the vas deferens. Within each follicle, groups of germ cells in various stages of spermatogenesis (see Section IV.A) may be seen in final instar larvae or adults. As well, glandular cells may occur which, on the basis of their staining properties, have been proposed to produce the nourishment necessary for both maturation of the sperm and their maintenance (storage) within the seminal vesicle. The glandular cells commonly take the form of an ensheathing epithelium around the developing germ cells, the whole structure being known as a "cyst." Though the cyst wall usually breaks open in the final stages of spermatogenesis, the sperm within may remain as a bundle even after insemination. In Acrididae (Orthoptera), one cyst cell differentiates as a "nurse cell" during the spermatid stage.[1] The spermatids in each bundle then become oriented and embed their heads into the cell, which produces a large quantity of mucoprotein, the sperm and mucoprotein cap constituting a spermatodesm. Cantacuzène[1] speculates that the cap does not serve directly as a source of energy for the sperm while they are stored in the seminal vesicle; rather, the enzymes which, she suggests, reside within it degrade the nutrients released by the epithelium which lines the seminal vesicle. The literature also contains a number of reports that implicate the testes as the site of production of chemicals that modify female fecundity and/or receptivity, but these are more appropriately dealt with in Section IV.E.

The vasa deferentia are typically assumed to be merely tubes for conducting the sperm from the testes to the seminal vesicles; indeed, the latter are frequently dilations of the vasa deferentia. However, light microscopical[2-7] and a few electron microscopical[8-12] studies suggest that these tubes have important glandular and phagocytic functions in some species. Rojas-Rousse[2] and Gerber et al.[5] propose that the secretion is used to nourish the sperm within the male reproductive tract, while the presence of numerous lysosomes in the vas deferens cells of *Drosophila melanogaster* may indicate a role in digestion of aged or degenerate sperm.[11] This function is also proposed for the phagocytic cells in the upper vas deferens of the lepidopterans *Anagasta kühniella*[8] and *Calpodes ethlius*.[12] Other proposed functions of the vas deferens secretion include involvement in spermatophore production in Trichoptera[13] and the blister beetle, *Lytta nuttalli*,[14] and production of the spermatodesm in Tettigoniidae (Orthoptera).[2]

In the great majority of insects, the ejaculatory duct is of ectodermal origin and, as such, has a cuticular intima. However, in all Lepidoptera, *Diadromus pulchellus* (Hymenoptera), *Plecia nearctica* (Diptera), and *Nezara viridula* (Hemiptera) an anterior mesodermal component

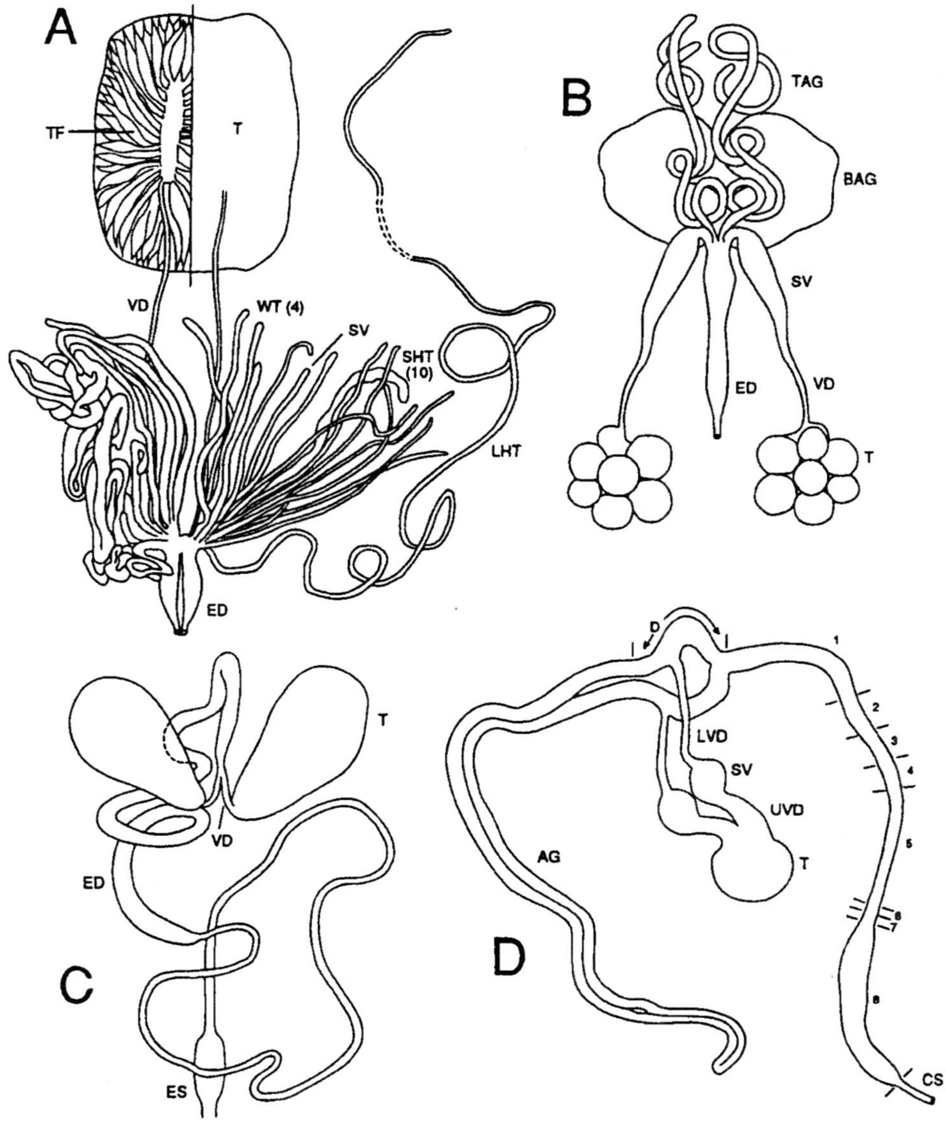

FIGURE 1. Representative male reproductive systems (not to scale). A. *Melanoplus sanguinipes* (Orthoptera); B. *Tenebrio molitor* (Coleoptera); C. *Musca domestica* (Diptera); D. *Anagasta kühniella* (Lepidoptera). *Abbreviations:* AG, accessory glands; BAG, bean-shaped accessory gland; CS, cuticular simplex; D, duplex; ED, ejaculatory duct; ES, ejaculatory sac; LHT, long hyaline tubule; LVD, lower vas deferens; SHT, short hyaline tubules; SV, seminal vesicle; T, testis; TF, testis follicles; TAG, tubular accessory gland; UVD, upper vas deferens; VD, vas deferens; WT, white tubules; 1–8, Eight regions of the noncuticular simplex. (A, original; B, redrawn from Dailey, P.J. et al., *Journal of Morphology,* Vol. 166. Copyright © 1980 John Wiley & Sons, Inc. Reprinted by permission of John Wiley & Sons, Inc. C, redrawn from Luther S. West: The *Housefly: Its Natural History, Medical Importance, and Control.* Copyright © 1950 Comstock Publishing Co., Inc. Used by permission of the publisher, Cornell University Press; D, redrawn from diagram supplied by Dr. J.G. Riemann.)

occurs.[15] The term "simplex" is traditionally used for the ejaculatory duct of Lepidoptera, with the corresponding sections being referred to as the anterior "noncuticular" and posterior "cuticular" simplex regions; further, the noncuticular simplex is subdivisible into as many as seven or eight distinct segments (Figure 1D), each with its own tinctorial and ultrastructural characteristics (see, for example, References 12 and 16).

Secretory cells of both ectodermal and mesodermal origin have been described from the ejaculatory duct, and a variety of functions have been proposed for their products. In Lepidoptera and Coleoptera, components of the spermatophore are derived from the secretions. The noncuticular simplex of some Lepidoptera also produces a sperm activator (see Section IV.F).[17-19] In the midge *Chironomus plumosus*, which lacks accessory glands, seminal fluid is produced by the ejaculatory duct cells.[20] An enzyme, esterase 6, produced in the anterior ejaculatory duct of *D. melanogaster*[21] may have diverse roles in the species' reproductive biology[22] (see Section IV.F). Receptivity-inhibiting substances (see Section IV.E) have also been reported from the ejaculatory duct of *Musca domestica*[23] and *Stomoxys calcitrans*.[24]

The forms and functions of the accessory glands (= collateral glands), which parallel the wide diversity of external form and habits of the Insecta, have been considered in detail by Gillott.[15] In the great majority of insects, the glands are mesodermal and are named "mesadenia". This is clearly the primitive arrangement, though, in a few groups, substitution of ectodermal for mesodermal components in the reproductive system generally results in the formation of ectodermal glands (ectadenia) (see Section III.A). Accessory glands are primitively absent in Thysanura, Ephemeroptera, Plecoptera, Dermaptera, and most Odonata, and have been secondarily lost in many Diptera.

The accessory glands occur in most species as a single pair of tubular structures, though in Coleoptera there are commonly two or three pairs, and in Thysanoptera and Acrididae (Orthoptera) multiple pairs of tubules occur. In the latter arrangement, the cytology of a tubule is generally uniform throughout its length, whereas in species with a single pair of tubules there are often regional or intercellular differences. The secretion of the accessory glands, indeed of individual tubules, is a complex mixture, and this is reflected in the variety of functions that have been ascribed to it, including spermatophore, mating plug and seminal fluid formation; fecundity enhancement and receptivity inhibition; sperm activation; and supply of nutrients to the female. In view of the central importance of the accessory gland secretion to the reproductive biology of the male, its nature and some of its major functions are discussed as distinct topics in Section IV.

B. EXTERNAL GENITALIA

The male external genitalia include two components, the basic structures common to all species and derived from the primary phallic lobes of the embryonic tenth (larval ninth) abdominal segment[25] and secondary structures unique to groups or species formed on adjacent segments. Because of the enormous diversity in form of the external genitalia among different insect groups, only the basic plan will be considered here. For information on specific groups, the reader is referred to the work of Matsuda.[26]

The phallic lobes arise as paired ectodermal outgrowths of the ventral surface of the segment, though only in Ephemeroptera do they remain separate to form the paired penes seen in the adult. In Thysanura the lobes meet in the midline to form a short tubular "penis," a misnomer because the structure is not an intromittent organ. In Odonata, the genitalia on the tenth segment are greatly reduced and, instead, secondary structures develop on the second and third abdominal segments (see Section II.C). In most other insects, each phallic lobe is divided to form a median mesomere and a lateral paramere. Between the fused mesomeres is the gonopore from which the ejaculatory duct arises. Elongation of the mesomeres produces an intromittent organ, the aedeagus, whose opening is the phallotreme and inner channel the endophallus. In some insects, the endophallus is eversible through the phallotreme so that, in effect, the gonopore becomes the external opening. Normally, the parameres differentiate into elongate clasping organs. However, in some cases they fuse with the mesomeres to form the phallobase. This can become an elongate structure, the phallotheca, that encloses the now eversible aedeagus. This telescopic arrangement (i.e., phallotheca, aedeagus, and endophallus), when everted by hemolymph pressure during copulation, may form a very elongate penis that deposits sperm deep within the female system.

C. SPECIALIZED MATING SYSTEMS

According to Matsuda,[26] in Ephemeroptera the primary phallic lobes of the juvenile stage are replaced by adult structures that develop within or in association with those of the nymph; that is, phallic lobe development is biphasic in this order, in contrast to that in all other orders in which continuous development towards the adult form occurs. Further, in this order each imaginal phallic lobe remains separate and does not subdivide into mesomere and paramere. Rather, each lobe develops as a penis, and from each gonopore an ejaculatory duct grows anteriorly to fuse with the vas deferens on each side.

In Odonata, the phallic lobes do not develop to any significant degree so that males have no aedeagus and only an inconspicuous ejaculatory duct. Thus, a unique method of sperm transfer has evolved in this order.[27] Prior to mating, a male coils his abdomen forwards so that the gonopore comes into contact with the intromittent organ on the third abdominal sternum, and a spermatophore is transferred. During copulation, the male grasps the female's head (Anisoptera) or prothorax (Zygoptera) by means of his terminal appendages (modified cerci). The interlocking of the male and female parts may be extremely precise, thus playing an important role in species isolation. The female then rotates her abdomen under that of the male until her genital opening contacts the intromittent organ when the spermatophore is transferred.

Sperm transfer in Strepsiptera, many Cimicoidea, and at least one species of Anthocoridae (Hemiptera) is achieved by hemocoelic insemination.[28-30] Though the details vary, in this most unusual form of insemination the penis penetrates the integument or vagina wall so that sperm is deposited in the hemocoel, sometimes into a special structure, the spermalege. Eventually, some of the sperm makes its way to the conceptacula seminales for storage, the rest being phagocytosed either by hemocytes or by cells of the spermalege. Hinton[28] suggested that hemocoelic insemination may serve to provide females with nutrients to enable them to survive for longer periods in the absence of suitable food.

III. DEVELOPMENT

A. EMBRYONIC DEVELOPMENT

The origin and development of the germ cells and gonads is varied, though two distinct trends can be noted, namely, earlier segregation of the primordial germ cells and the restriction of these cells to fewer abdominal segments.[31,32] In some Thysanura and Orthoptera, the germ cells do not become distinguishable until they appear in the splanchnic walls of several abdominal segments. In *Locusta migratoria*, for example, they arise in the second through tenth pairs of abdominal somites although they persist only in pairs three to six.[33] These segmental groups of cells proliferate and fuse longitudinally to form the gonad on each side. In contrast, in Dermaptera, Psocoptera, Homoptera, and many endopterygotes the germ cells become obvious during blastoderm formation as roundish cells at the posterior end of the egg. In due course, in exopterygotes the cells move anteriorly through the yolk to become enclosed within the splanchnic mesoderm of the third/fourth abdominal segments. The germ cells subsequently separate into left and right halves, from which the testes develop. In endopterygotes, the picture is similar except that division of the germ cells into two groups occurs prior to migration through the yolk.[31,32]

In exopterygotes, the vasa deferentia develop from one or more posterior pairs of abdominal somites (e.g., those of segments nine and ten in phasmids[34]) while the ejaculatory duct is an ectodermal invagination, usually arising behind the ninth sternum. In lower endopterygotes, for example, *Tenebrio molitor*,[6] embryonic development of the gonoducts is similar to that of exopterygotes, and all the elements of the reproductive system can be identified at hatching. In Diptera and Lepidoptera, the ductal components develop postembronically from a single midventral or a bilateral pair, respectively, of genital imaginal discs that arise late in embryogenesis.[35-37] In all insects except, apparently, Acrididae (Orthoptera) the formation of the primary phallic lobes and subsequent development of the external genitalia occur during the larval stage.[26]

B. POSTEMBRYONIC DEVELOPMENT

1. General

In exopterygotes, the reproductive system, like other organ systems, grows steadily during the larval period; differentiation of the various components does not occur, however, until the final juvenile stadium. In contrast, in endopterygotes very little growth occurs through the larval period; rather, growth and differentiation occur simultaneously and are compressed into the pupal stadium. "Differentiation," including in its broadest sense both spermatogenesis and the organogenesis of the tubular components, is regulated hormonally as well as being affected by various exogenous factors. The endocrine control of spermatogenesis is discussed by Hardie in Chapter 5.

2. Endocrine Regulation

As with the differentiation of other tissues, it is changes in the hormone balance within the insect that induce development of the reproductive tract. Though I am not aware of studies on the phenomenon, it seems reasonable to propose, in light of what is known regarding hormonal events during metamorphosis, that changes in hormone titers in earlier instars permit (and possibly promote) the mitotic division of the undifferentiated reproductive tract cells and the slight growth of the system seen especially in exopterygotes.

In the final juvenile instar, there occurs a marked decline in the level of circulating juvenile hormone (JH), as well as one or more surges in the level of circulating ecdysteroid.[38,39] This change in the ratio of the two principal "players" permits the expression of adult characters, though the specific roles of the two hormones are only slowly becoming clear. JH appears to influence metamorphosis in two ways: first, by inhibiting the secretion of prothoracicotropic hormone, ecdysteroid production and release is prevented; and second, though this is more speculative, it may directly prevent the action of ecdysteroid at the organ/tissue level.[38,40]

Our rather limited understanding of the involvement of hormones in preimaginal growth and differentiation of the male reproductive system has recently been reviewed by Happ.[41] Almost nothing is known about the site and mode of action of JH. In the locusts *L. migratoria* and *Schistocerca gregaria*, implantation of two corpora allata into the abdomen of male fourth (penultimate) instar nymphs leads to sterility.[1] However, this treatment or injection of synthetic JH does not exert its effect on spermatogenesis, which proceeds normally, but prevents differentiation of the vas deferens, which remains a thin, solid cord of cells.[42] These treatments also prevent differentiation of the accessory glands,[1,42] whereas the reverse (i.e., allatectomy) results in precocious metamorphosis of these structures.[43] Perhaps more important, Cantacuzène's[1] data show that the timing of the treatment is critical, the greatest effect being achieved when insects are treated midway through the stadium. The accessory gland transplantation experiments of Gallois[44,45] support this point. Implantation of these glands from final instar nymphs of varied ages into adult hosts shows that only those glands from nymphs 2, 3, or more days into the instar are competent to differentiate. This critical period (days 2 and 3) coincides with a marked decrease in JH titer (and a small increase in the level of ecdysteroid).[46] As all these investigations utilized whole insects, it is unclear whether JH is acting directly or indirectly on the tissues whose development it is influencing.

Mere absence of JH is insufficient for development to proceed; ecdysteroid (probably in the form of ecdysterone) must be present both for growth and for differentiation of the reproductive tract. The involvement of ecdysteroid has been investigated in several endopterygotes, most notably *Ephestia kühniella*,[47] *Samia cynthia*,[10] *Heliothis virescens*,[37,48,49] *Bombyx mori*,[50] and *T. molitor*.[51-54] In Lepidoptera, both the sperm ducts and the genital discs (which form the seminal vesicles [= upper vasa deferentia of some authors] and the remaining components of the tract, respectively) grow and differentiate under the influence of ecdysteroid. Apparently, ecdysteroid exerts both direct and indirect effects. In *S. cynthia*[10] and *B. mori*,[50] ecdysteroid is able to directly promote development of the tract *in vitro*, whereas in *H. virescens*, ecdysterone alone has no effect.[37,48,49] Development of the sperm ducts or genital discs does occur when fat body or testis sheath is present in the culture medium in addition to the hormone, or when the tissues are cultured with aqueous extract of fat body previously exposed to ecdysteroid for 24 hours.

Only in *T. molitor* and *H. virescens* does the site and mode of action of the ecdysteroid appear to have been studied.[49,51-54] In the bean-shaped glands (BAGs) of *Tenebrio* (Figure 1B) there are two bouts of mitosis as growth occurs, the second of which coincides with the peak of ecdysteroid in the pupal stage. *In vitro* studies showed that the second burst of mitosis requires ecdysteroid for its occurrence,[51] the hormone promoting the flow of cells from the G_2 into the G_1 and S phases.[52] Though Grimnes and Happ[53] showed that the same concentration of ecdysterone which produced maximum rates of mitosis also promotes formation of characteristic secretions and adult-specific antigens in the BAGs, the latter activity *in vivo* is not immediately induced by the ecdysterone but begins at the end of pupation when levels of the hormone are low.[54] In *H. virescens* genital discs, ecdysterone in combination with fat body or testis sheath stimulates tritiated thymidine incorporation into DNA, suggestive of mitotic division.[49]

IV. FUNCTIONAL ASPECTS

A. SPERMATOGENESIS AND SPERM STORAGE

1. Spermatogenesis

In insects, as in other animals, spermatogenesis comprises three phases: (1) the multiplication phase (spermatocytogenesis); (2) the meiotic phase; and (3) the maturation phase (spermiogenesis).[55] In many insects, this temporal sequence of events is "fixed" spatially; that is, the phases occur in the distal, middle, and proximal regions, respectively, of each testis follicle. In the multiplication phase, the spermatogonia (cells derived from germ cells) undergo a species-specific number (usually, five to eight) of mitoses. For most insects (but not, apparently, Diptera), these divisions (and hence the daughter spermatogonia) occur within a "cyst," that is, a capsule of somatic cells formed initially around either a single spermatogonium or a clone of daughter spermatogonia after one or two divisions. When mitosis is complete, the spermatocytes, as the cells are now known, undergo two meiotic (maturation) divisions. Generally, four spermatids result from each spermatocyte, each having a haploid chromosome complement. However, especially among Diptera, variants of the meiotic process occur, so that only one or two spermatids are formed from each spermatocyte. During spermiogenesis, the spermatids differentiate into flagellated spermatozoa. At this time, the enveloping cyst cells are rich in glycogen and may be supplying nutrients; in addition, they phagocytose cytoplasmic remnants sloughed by the spermatids as they transform into spermatozoa. In Acrididae (Orthoptera)[1,56] and Coleoptera,[57] a mucopolysaccharide "cap," the spermatodesm, comes to envelop the sperm heads near the end of spermiogenesis, though it is by no means clear that this is a product of the cyst cells. According to Cantacuzène,[1] the cap is not a direct source of nutrients; rather, it contains enzymes that degrade nutritive molecules secreted by the seminal vesicles into which the mature sperm are now transferred and stored. In Tettigoniidae (Orthoptera), the spermatodesm is formed from secretion of the intratesticular region of the vas deferens.[1]

In Lepidoptera, two types of sperm are produced, eupyrene (nucleate), which fertilize the eggs, and apyrene (anucleate), whose largely speculative functions may include facilitating eupyrene sperm movement from the testes to the vas deferens, assisting the eupyrene sperm in their migration within the female reproductive tract, providing nourishment for the eupyrene sperm, the female herself, or the zygote, and playing a role in sperm competition either by eliminating sperm deposited in previous matings or by preventing further matings[58] (and see also Section IV.F). The two types can be distinguished as early as the secondary spermatocyte stage, following which their changing morphology follows very different paths.[59] Eupyrene sperm formation begins somewhat earlier than that of apyrene sperm, the former being initiated in the last or penultimate larval instar, the latter in the last larval or pupal instar, depending on the species. The eupyrene sperm, like the sperm of other types of insect, remain in bundles until they have been transferred to the female tract; however, the apyrene sperm bundles dissociate, releasing individual sperm, as they move out of the testis.

For most insects, formation of spermatogonia and spermatocytes is reported to occur in the final nymphal or pupal instar, so that the testes of the adult contain only spermatids and spermatozoa or only spermatozoa in those forms with a very short adult life. These observations led, more than 40 years ago, to the proposal that spermatogenesis was regulated by the morphogenetic hormones, ecdysone and JH, the former promoting and the latter inhibiting the process. However, in representatives from diverse orders, meiotic and even premeiotic phases of spermatogenesis can be found in the adult testis.[55,60] Dumser and Davey[60-62a] suggested that the hormones affected only the rate of spermatogonial mitosis, not differentiation per se. However, the subsequent discovery of ecdysteroids in adult males of several species and of other effects of both ecdysteroids[62b] and JH[63] on germ cell maturation and viability has reopened the question of the endocrine control of sperm production. See Chapter 5 for a full discussion of this topic.

2. Sperm Release and Storage

In the majority of insects, the bundles of mature sperm are moved by peristalsis to the seminal vesicle where they are stored until insemination. However, except in Lepidoptera, information on this process is lacking. For several species of Lepidoptera, it has been observed that release of sperm (both types) from the testis and its subsequent movement through the upper vas deferens and seminal vesicle to the ductus ejaculatorius duplex (Figure 1D) shows a daily rhythm (see Reference 64). This rhythm has since been shown to be circadian in nature,[65] with the clock and its associated photoreceptor located in the testis–vas deferens complex.[66] Correlated with the rhythm of sperm movement is cyclical secretory activity by the epithelium of the upper vas deferens, though the purpose of this carbohydrate-rich secretion remains unclear.[64] However, in their passage toward the ductus ejaculatorius duplex, the sperm undergo complex morphological changes, especially with respect to the fibrous sheath that surrounds individual eupyrene sperm.[67] The significance of these changes remains unclear, but it may be speculated that they relate somehow to the ability of the sperm to fertilize. The overall significance of this periodic sperm release and movement is related to the fact that males normally mate but once a day, and at this time they ejaculate all the sperm stored in the duplex. The timing of sperm release and movement (which is species specific) is such that the sperm produced each day move into the duplex shortly after the end of the male's daily period of receptivity to female sex pheromone. Thus, even if a male does mate, a substantial amount of new sperm will be available for insemination the next day.[65] Though the usefulness of this type of a system is readily apparent for species that have diurnal mating rhythms, it remains to be determined how sperm release and storage is regulated in species whose males are more opportunistic with respect to mating.

B. ACCESSORY GLAND ACTIVITY

As noted earlier, the accessory glands take on a vast array of forms, and their secretion a spectrum of functions. Both these aspects have been extensively studied over the past two decades, and the major conclusions are presented in reviews by Hinton,[68] Leopold,[69] Chen,[70] Happ,[71] and Gillott.[15] Also, the endocrine control of accessory gland activity has recently been reviewed.[72]

1. Nature and Formation of Secretion

Numerous histochemical studies have shown the secretion to be a mixture or chemical complex of protein, carbohydrate, and lipid, and electron microscopy has confirmed that the glandular epithelium is designed for production of these materials. Also noteworthy is that some of these early studies have demonstrated the regional variation that occurs both within and between accessory gland tubules with respect to both amount and nature of secretory components. Electrophoretic studies, in some cases combined with histochemistry, have revealed that the secretion contains many distinct proteins (e.g., >40 in the BAGs of *Tenebrio*[73]), including glycoproteins and lipoproteins. Some of these proteins are enzymic in nature while others, which may be quantitatively dominant, play a structural role (see below for details). Free amino acids and small peptides have also been identified. Among the carbohydrates not complexed with protein are glycogen, mucopolysaccharides (acidic, basic and neutral), glucose, and inositol. Neutral lipids and phospholipids have been reported for some species. Guanosine 3′, 5′-cyclic monophosphate occurs in large amounts in the glands of *Acheta domesticus* and several other Gryllinae (Orthoptera), where it may have a role in sperm metabolism and activity.[74,75] Uniquely, the glands of male *Hyalophora cecropia* store massive quantities of JH-I and JH-II which are transferred to the female during copulation.[76,77] Also, the accessory glands of male Blaberoidea (Dictyoptera) accumulate uric acid (up to about 90% of the dry weight in *Blattella germanica*[78]), which is then secreted as the outermost layer of the spermatophore, possibly to dissuade the female from eating the spermatophore before sperm evacuation has occurred.[79,80a] However, the observation by Mullins and Keil[80b] that radiolabeled urate produced by males and ingested by females (as part of the spermatophore) appears in the ootheca suggests that it may represent an important nitrogen resource for reproductive success (and see Chapter 10).

2. Endocrine Regulation and Effect of Mating

Almost all investigations of the hormonal regulation of accessory gland activity have focused on the role of JH. This is quite understandable in view of the demonstration, more than 50 years ago,[81] of the importance of the corpora allata in male reproductive activity and the long-standing presumption that ecdysteroids would not occur in adults following the degeneration of the moult glands at metamorphosis. However, several demonstrations of ecdysteroids in adult males,[82-84] combined with the realization that, in insects which are sexually mature at eclosion, the accessory glands acquire their secretion during the pupal instar when the corpora allata are inactive, raise the possibility that in some species ecdysteroids may be involved.

In males of most species studied, allatectomy prevents or greatly retards the accumulation of secretion by the accessory glands, an effect that can be reversed by application of JH. Without exception, the effect of JH is at the level of protein synthesis, and in the relatively few investigations where it has been examined, JH has been shown to regulate the synthesis of specific proteins. For example, in allatectomized *Melanoplus sanguinipes* the reduced ability of the long hyaline tubule (Figure 1A) to accumulate secretion is due to the failure of the tubule to synthesize a glycoprotein, LHPI, which comprises 51% of the total protein in the tubule of mature virgin males.[85,86] Parallel ultrastructural studies show that allatectomy causes changes in the protein-synthesizing machinery of the gland cells and, in some cases, in the appearance of the secretion.[87-89]

It must be emphasized that the majority of studies, in which whole insects have been used, do not clarify whether the action of JH on the accessory glands is direct or indirect. A few authors have used decapitation,[90] or have removed both the corpora allata and the cerebral neurosecretory system,[91-93] followed by JH application, or have used an *in vitro* system[72,94-96] to confirm that JH acts directly on the glands. How and where JH acts in the accessory glands is, at this time, a matter of conjecture, though Yamamoto et al.[95] have obtained evidence suggesting the existence of a membrane receptor protein for JH in *D. melanogaster*.

Early reports that ecdysterone promoted protein-synthetic activity in accessory glands must be treated cautiously because pharmacological, rather than physiological, doses of hormone were employed.[97,98] The stimulation of RNA and protein synthesis by ecdysterone has been observed under both *in vivo* and *in vitro* conditions in pharate and newly emerged adults of *Chilo partellus* and *Spodoptera litura*.[99-101] These are interesting observations, because in other Lepidoptera ecdysteroid titers are known to be very low just prior to eclosion. Further, as noted earlier (Section III.B.1), though ecdysteroid is required in *T. molitor* and *B. mori* to render the accessory glands competent to produce secretion, when production begins at the end of pupation, ecdysteroid levels are low; in other words, in these species protein synthesis is not directly regulated by ecdysteroids.[50,54]

Only a handful of reports indicate that neurosecretory factors directly affect accessory gland protein synthesis. For example, in *Rhodnius prolixus*, removal of the median neurosecretory cells reduces protein accumulation by the transparent accessory glands, an effect which can be only partially overcome even by multiple doses of JH-I.[93] *In vitro* studies subsequently confirmed that a polypeptide from the brain stimulated protein synthesis in the transparent accessory glands.[102]

An aspect of endocrine control that merits further examination is the movement of proteins between the accessory glands and the hemolymph. For example, in *M. sanguinipes* there are immunologically similar proteins in the accessory glands, fat body, and hemolymph.[103] The accessory glands can accumulate these proteins from the hemolymph in normal, but not in allatectomized, males. In *C. partellus*, *in vitro* co-culture of accessory glands from adult males with fat body and hemolymph from larvae injected several hours previously with [^{35}S]-methionine has demonstrated that the larval proteins are synthesized in the fat body, released into the hemolymph, and are then accumulated by the accessory glands. Accumulation is enhanced by ecdysterone or its agonist RH 5849 but inhibited by JH-I.[104a] Conversely, the transparent accessory gland of *Rhodnius* synthesizes and releases into the hemolymph a 170-kDa polypeptide,[104b] and Sevala and Davey[104b] suggest that the apparently redundant control of protein accumulation in the transparent accessory gland by both JH[96] and neurosecretion[102] may ultimately be explained by assigning a specific function, synthesis, or release, to each of the hormonal factors.

That mating leads to enhanced synthesis of accessory gland secretion has been shown for various species.[93,105-107] However, to date there is no clear understanding of the control pathway by which the mating effect is exerted. Baumann's study[105] suggested that in *Drosophila funebris* mechanical emptying of the gland was not the stimulus for renewed secretory activity, and this author speculated that there might be "a neurohormonal influence." For *D. melanogaster*, also, copulation-enhanced protein synthesis has been suggested to involve neurohormonal factors based on the parallel effects of mating and JH application.[106] In contrast, though JH is essential for the normal expression of protein synthesis in the accessory glands of *M. sanguinipes*,[107] it does not mediate the copulation effect. This is readily seen in mated but allatectomized *M. sanguinipes* which still exhibit a three-fold increase in accessory gland protein synthetic activity compared with unmated allatectomized controls.[107]

C. SPERMATOPHORE FORMATION

Spermatophores are usually presumed to have evolved in association with the taking up of terrestrial life by insect ancestors and have as their primary function the delivery of sperm to the female reproductive tract. However, as Davey[108] pointed out, some marine Crustacea and many aquatic Annelida produce a spermatophore, so that, for Insecta, this should not be thought of as a new structure but rather as an existing structure that has taken on a new function. Though detailed descriptions of the mechanical aspects of spermatophore formation are available for many species, aspects of the process such as the nature of the stimuli that initiate it, its control and coordination, and its biochemical nature are relatively unexplored. This is unfortunate, given the centrality of this method of sperm transfer in the evolutionary success of insects and, from the pest management point of view, its importance in the life history of species.

Generally, spermatophores have their most complex form in primitive groups; in advanced endopterygotes they may be relatively simple or have been secondarily lost. Paralleling this trend is a change in the site of formation of the spermatophore. Gerber[109] recognized four categories of spermatophore formation based on spermatophore complexity and site of formation. In the most primitive, the first male-determined method, the complex spermatophore is formed in the ejaculatory duct or copulatory organ of the male. Such a method is typical of orthopteroid insects. In the second male-determined method, which characterizes some Hemiptera, Coleoptera, and a few Diptera, the spermatophore is again formed in the copulatory sac, but the latter is everted into the bursa copulatrix of the female. After copulation, the sac is withdrawn, leaving the usually less complex spermatophore in the bursa. In Trichoptera, Lepidoptera, some Coleoptera, and a few Diptera, the spermatophore forms directly in the female tract which thus determines its shape (first female-determined method); the spermatophore is relatively simple but nonetheless, like those formed by the male-determined methods, still encloses the sperm. In the second female-determined method, the accessory gland secretions do not enclose the sperm; rather, they follow them into the female genital tract where they may temporarily harden to form, for example, the mating plug of mosquitoes and the honeybee and the sphragis of some Lepidoptera. It is speculated that these "barriers" may prevent loss of semen or further transfer of sperm in a subsequent mating.

In most early analyses of spermatophore formation, only histological methods were employed, so that the origin and nature of spermatophore components remained unclear. A few authors have used histochemistry or surgical removal of certain accessory gland structures to determine the source and broad chemical nature of particular components (see Gillott[15] for references). However, it is only relatively recently that information on the biochemical nature of individual spermatophore components and their mode of formation has become available. For example, in *T. molitor* the use of monoclonal antibody techniques has permitted Happ's group to trace three structural proteins (spermatophorins) from their site of production, distinct cell types in the BAGs, to specific layers within the spermatophore.[110-112] Amino acid analysis of one of these spermatophorins showed that >25% of its residues were proline, which is in keeping with the large amounts of this amino acid reported for other insect structural proteins (in cuticle, egg shell, and ootheca) and for collagen.[112] Though having immunological identicality, it is clear that the nature of the precursor accessory gland secretion may differ from that of the spermatophorin. This was first noted by Grimnes and Happ[110] as a change in solubility and electrophoretic mobility, leading these authors to speculate that proteolytic cleavage of the precursor may have occurred. A proteolytic enzyme probably involved in the formation of a spermatophorin has been identified in accessory gland secretions of *M. sanguinipes*.[113] A precursor protein of mol wt 85 kDa, which appears to be produced primarily in short hyaline tubule 3 (Figure 1A), is cleaved by a trypsin-like enzyme to form the spermatophorin SP62 (mol wt 62 kDa), a major component of the outer layer of the spermatophore. The major products of the white gland tubules of *Melanoplus* (Figure 1A) are also spermatophorins, but it is not known whether they, too, are derived by proteolysis of higher molecular weight precursors. The function of an aminopeptidase also identified in *Melanoplus* accessory gland secretion remains unknown.[113]

Except in Gryllidae (Orthoptera), where a spermatophore is preformed by the male and carried until copulation takes place, spermatophore formation begins only after copulation is initiated, that is, the male has achieved the correct mating position. Sensory input, especially tactile but for some species chemical and visual, triggers the process, which is under motor neuronal control. For most insects studied, the central nervous system must be intact in order for spermatophore formation to start. For example, in *Locusta* the head ganglia are necessary at this stage to interpret input from the cerci, though decapitation 15 min or more after initiation does not disrupt spermatophore formation.[114] Control of the subsequent stages of spermatophore formation resides in the terminal abdominal ganglion, removal of which causes abrupt termination of the process. The presence of the female is necessary only at the start, presumably related to the need for correct sensory input via the cerci. Separation of copulating pairs beyond 15 min does not disrupt spermatophore formation though the structures are malformed because, Gregory[114] suggested, the spermatophore desiccates and the spermathecal duct serves as a mold for the spermatophore tube in this species.

In *Teleogryllus commodus* and perhaps other species of crickets where the spermatophore is produced prior to mating, formation of the spermatophore is under circadian control; that is, a new spermatophore is produced during each 24-h period.[115] According to Loher,[115] who employed a variety of surgical procedures, the pars intercerebralis region of the brain may serve a dual function in regulating spermatophore formation. First, it may hormonally regulate the synthesis of the raw materials in the accessory glands and, second, it may be the coordinating center for release of the materials from the glands as the spermatophore is being formed. However, Loher's work[115] does not clarify whether the pars intercerebralis serves only as a trigger, the rest of the process being regulated within the terminal abdominal ganglion as in other insects, or whether it controls the entire event. The site of the circadian control center is also undetermined but may be within the optic lobes, as their destruction leads to random generation of spermatophores throughout the 24-h period.

D. SEMINAL FLUID

At the outset, it is important to clarify the term "insect seminal fluid" because, in contrast to the mammalian situation in which seminal fluid is a secretion of the seminal vesicles, in many insects the seminal vesicles are not glandular in nature. Potentially, the fluid that comes to bathe the sperm may be derived from any or all glandular parts of the male system, and its nature may change as secretions are added to or removed from it during insemination and when it reaches the female system.

The following discussion will deal only with our somewhat limited knowledge of the general composition and functions of insect seminal fluid. The nature and functions of specific components will be examined in parts E and F of this Section. The obviously minute quantities of semen together with the high density of sperm make the obtaining of seminal fluid samples for analysis difficult. Further, depending on its source (e.g., ejaculate, spermatophore, or spermatheca), its composition will vary. Perhaps not surprisingly, a variety of potential energy sources for the sperm have been identified. For example, *Apis mellifera* seminal fluid contains trehalose, glucose, and fructose though the latter becomes virtually undetectable 40 min after ejaculation, perhaps because of its use by the sperm.[116] The three sugars occur throughout the reproductive tract with the greatest concentration in the testes and penis bulb. Histochemical studies have demonstrated the existence of glycogen in the seminal fluid of *Periplaneta americana*[117] and *S. gregaria*.[118] The lipid in the semen of *A. mellifera* is almost entirely of spermatozoan origin rather than in the seminal fluid.[119] The amino acids (free and bound) in *A. mellifera* semen apparently resemble those of mammalian semen though their specific origin (sperm or seminal fluid) and functions are undetermined.[120] While Blum et al.[116] report that the semen of *A. mellifera* contains calcium, sodium, manganese, magnesium, copper, and iron (of which only the last three are detectable in sperm), there appear to have been no quantitative studies on the inorganic constituents of seminal fluid despite several reports on the activation of sperm by buffers at various pH values.

E. FECUNDITY-ENHANCING AND RECEPTIVITY-INHIBITING CHEMICALS

In addition to its obvious function of sperm transfer, for many species mating has other important effects, including the stimulation of egg production and the rendering of the female unwilling to remate. In some species (for example, cockroaches[121]) the stimulus given by mating is physical in nature; that is, the insertion of the spermatophore stretches the wall of the bursa copulatrix, which causes the female to become unreceptive. For others, chemicals in the seminal fluid stimulate egg production and/or render females unreceptive. These pheromones have been named fecundity-enhancing (FES) and receptivity-inhibiting substances (RIS), respectively.[122]

Both FES and RIS "inform" the female that she has been inseminated. For the FES, the significance lies in the fact that in most insects unfertilized eggs are inviable; thus, it is critical that oviposition not occur until a supply of sperm is available. For the RIS, the rendering of the female unreceptive "guarantees" that only the first (fittest) male's sperm will be used to fertilize the eggs. It may also provoke males to actively seek virgin females in the population and, for females of some species, to switch their behavior from mate-seeking to food- and/or oviposition-site seeking. Although the FES and RIS play distinct roles, it is appropriate to deal with both concurrently in view of their similar sites of production and chemical nature. Indeed, in some species it seems likely that the same substance may serve as both a FES and a RIS. Gillott and Friedel[122] and Gillott[15] have provided detailed reviews of FES and RIS.

1. Source and Nature

In males of most species where FES or RIS have been identified, the accessory glands are their source. However, in *Musca*, males of which lack such glands (Figure 1C), the FES/RIS is produced in the upper third of the ejaculatory duct.[123] The testes apparently produce these pheromones in *H. cecropia*,[124] crickets,[125,126] and *R. prolixus*,[127] (though in the latter species, an effect of the seminal fluid merely stretching the spermathecal wall cannot be ruled out). It should also be noted that for none of these species is it clear whether glandular cells in the gonads or the sperm are the source of the pheromone. In some *Drosophila* species, there is evidence for the existence of both short- and long-term RIS, the former being produced by the accessory glands, the latter by the gonads.[128,129]

With one exception, all FES and RIS characterized to date are peptides or proteins. For example, the monocoitic substances reported from *M. domestica*, *Cochliomyia hominivorax* and *Phormia regina* are peptides with molecular weights estimated at 750 to 3000 Da.[130] Likewise, the RIS (PS-1) of *D. funebris* has a molecular weight of approximately 2700 Da and comprises 27 amino acid residues.[105] PS-1 exists in two forms, differing in a single amino acid (valine or leucine) at position 2.[131] The sex peptides of *D. melanogaster* and *D. sechellia*, which serve as both FES and RIS, each comprise 36 amino acids and differ from each other at only three positions. Perhaps not surprisingly, the *D. sechellia* sex peptide stimulates oviposition and inhibits receptivity in *D. melanogaster*, as well as in *D. simulans* and *D. mauritania*, all four species belonging to the *melanogaster* species subgroup. However, the *D. melanogaster* peptide has no effect on virgins of *D. funebris* and vice versa.[132]

On the other hand, the FES of *M. sanguinipes*,[133] *L. migratoria*,[134] and *Aedes aegypti*[135] are proteins of molecular weight 30, 13, and 60 kDa, respectively. The FES of the crickets *A. domesticus*[136,137] and *T. commodus*[138] is a prostaglandin synthetase enzyme complex which, after transfer during mating, promotes prostaglandin production in the female reproductive system.

Three *D. melanogaster* accessory gland proteins and the genes which code for them have been studied by Wolfner's group.[139-141] Of these, perhaps the most interesting is *msP 355a*, a basic protein with many features typical of peptide pheromone and hormone precursors; indeed, part of its amino acid sequence is similar to that of the egg-laying hormone (ELH) of the mollusc *Aplysia californica*.[139] In the accessory glands, *msP 355a* exists mainly in a form of molecular weight 37 kDa. During mating, it is transferred to the female genital tract; some of the 37-kDa fraction passes unchanged into the female's hemolymph, but the rest undergoes rapid proteolysis, first to a 29-kDa, then to a 22-kDa fragment. However, the ELH-like segment is retained by both these smaller molecules.[141] The second protein in the trio is *msP 355b*, an acidic protein 90 amino acids in length and having a molecular weight of 11–14 kDa.[141] It, too, is transferred during mating, and some enters the female hemolymph. However, that which remains in the genital tract is not proteolysed and, along with the *msP 355a* fragments, unstored sperm, and other secretions, is expelled from the genital tract some 2–3 h after copulation terminates. DiBenedetto et al.,[140] on the basis of their analysis of the gene that codes for it, have characterized *msP 316*, a small basic protein made up of 52 amino acids.

Prostaglandins have been identified from the male reproductive organs of *B. mori*,[142] and greater levels of these compounds have been noted in mated females compared with virgins.[143] However, it is not clear whether these increases are the result of direct transfer from the male or are an effect of mating.

2. Site and Mode of Action

The most common effect of FES is to stimulate oviposition, though the manner in which this is achieved is not well understood. Extracts of *Locusta* male accessory glands stimulate contractions of the lateral oviduct, an effect that can be partially mimicked by octopamine and forskolin, suggesting the involvement of octopaminergic receptors.[144] And recently, specific peptides with myotropic activity on the oviduct have been isolated from male accessory glands, spermatophores and spermathecae of mated females.[145,146] Obviously, oviposition is a complex process, both mechanically speaking and in terms of when it occurs. Not surprisingly, it is controlled both hormonally and neurally (for review, see Lange[147]). Perhaps the role of these male-derived substances is to prime the oviductal musculature so that, at the appropriate time and place for egg laying, the nervous system can provide the fine control.

In contrast, in *Rhodnius*[127] and *Teleogryllus*[138] the FES acts on the spermatheca, which is stimulated to produce a hormone of unknown nature (the "spermathecal factor") and prostaglandins, respectively. Similarly, in *Hyalophora* the FES is released into the bursa copulatrix where it triggers production of a hormone.[124] For other species, it seems likely that the FES passes through the wall of the female reproductive system to some other site of action. Thus, in *D. funebris*, FES can be detected in all parts of the body 2 h after mating,[148] and in both *M. sanguinipes*[149] and *M. domestica*[150] some male accessory gland proteins pass unchanged into the hemolymph of the female (though in neither species is it clear whether the FES is one of these).

The remaining links in the pathway culminating in egg laying are not well known. For a number of species (references in References 15,147), the cerebral neurosecretory cells are known to produce an ovulation- or oviposition-inducing hormone (a myotropin) whose release may be facilitated by the FES. In *Rhodnius*, for example, ecdysteroids released from the ovary as eggs mature can initiate electrical activity in, and release of myotropin from, specific median neurosecretory cells. However, this occurs only in mated females, that is, in the presence of the spermathecal factor which, it has been proposed,[151a,b] may unmask aminergic receptors within the brain allowing the ecdysteroids to act. In contrast, in *B. mori* the FES may act on the terminal abdominal ganglion, whose spontaneous activity is increased after mating or application of extracts of the male reproductive system.[152a] Further work showed that the sensitivity of the ganglion to Ringer's solutions containing varied amounts of NaCl and KCl is changed after mating, prompting the suggestion that the FES might act by altering the permeability of the neural sheath surrounding the ganglion.[152b]

A second potential role for an FES is to enhance egg development. This has been proposed by several authors, though only in a few species has a direct link been demonstrated. Baumann,[148] using an *in vitro* system, showed that uptake of ^{14}C-labeled amino acids into proteins was increased in the ovaries of virgin *D. funebris* injected with FES. Decapitation immediately after mating prevents this increase, leading Baumann[148] to suggest that the FES acts via the neuroendocrine system.

In mosquitoes, also, the FES promotes vitellogenesis, possibly by inducing the ovaries to produce a hormone whose function is to stimulate release of egg development neurosecretory hormone from the corpora cardiaca.[153] For anautogenous mosquitoes, an additional means of promoting egg development has been suggested,[154] namely, that the FES causes acceleration of blood-meal digestion. Again, the evidence suggests that this effect is exerted via an endocrine pathway. Even in nutritionally stressed female mosquitoes, accessory gland implants trigger egg development, leading Klowden and Chambers[155] to suggest that the FES is acting as a primer pheromone that switches the female's metabolic priorities from self-sustenance to egg production.

In the bean weevil *Acanthoscelides obtectus*, the sole demonstrated role for the FES is stimulation of egg development; that is, the pheromone does not promote egg laying.[156]

The demonstration that mated females carrying an *hsP70-msP 355a* fusion gene lay 20% more eggs than controls has led Monsma et al.[141] to speculate that *msP 355a* may have some role in egg production. The function of *msP 355b* is unknown. Like *msP 355a*, *msP 316* has features common to precursors of peptide hormones though its role remains unknown.[140]

Much less information is available on the site and mode of action of RIS, though, in all instances, the pheromone exerts its influence via the neuroendocrine system. In *H. cecropia*, the RIS, like the FES (with which it may be chemically identical), may stimulate the wall of the bursa copulatrix to release a hormone.[124] Originally, it was proposed that this hormone acted on the corpora cardiaca to stop release of the calling hormone; however, more recent work[157] has ruled out any involvement of the corpora cardiaca in calling behavior, and the remaining steps in the pathway remain unknown. The brain appears to be the site of action of the RIS in *Musca*, as decapitated, decerebrated, or cervically ligated virgins will mate as many as five times in an 8-h period.[158] Further, radiolabeled male material transferred during mating accumulates in the head region,[150,159] leading Leopold et al.[159] to propose that the RIS binds to receptor sites in the head (brain?) to induce refractoriness. Interestingly, in view of the apparently identical nature of the RIS and the FES in *Musca*, refractoriness (which in this insect is marked by withdrawal of the ovipositor) is induced by high concentrations, whereas oviposition (requiring extension of the ovipositor) is triggered by low concentrations of the pheromone.

In contrast, Gwadz's work[160] suggests that in *Aedes aegypti* the RIS acts on the terminal abdominal ganglion, with the brain, the suboesophageal ganglion, and the thoracic ganglionic mass having no direct involvement in the control of sexual behavior.

An aspect worthy of study is the interaction between RIS and JH, the latter enhancing receptivity as females become sexually mature. It is unlikely that RIS directly affects corpus allatum activity as refractoriness is life-long, though a female goes through several cycles of egg production, the latter correlated in many species with changes in corpus allatum activity. As proposed earlier,[15] it may be that the RIS and JH compete for the same receptor sites within the central nervous system, with the RIS being successful in normally monogamous species. In this context, it is noteworthy that when less than normal amounts of RIS are transferred during mating (e.g., by forcibly separating mating pairs or mating females to already multiply-mated males[161]), receptivity returns after a variable period of time.

F. OTHER FUNCTIONS

Several other functions have been ascribed to components of the accessory gland secretion or to secretory products of other parts of the male tract. Most of these relate to maturation, activation (motility), or energy metabolism of the sperm; in addition, for some species, the spermatophore represents a source of nourishment for the female.

How sperm move from the spermatophore to the spermatheca is largely unknown though it presumably involves either active movement on the part of the gametes or peristalsis of the wall of the female tract so as to "squeeze" sperm out of the spermatophore and/or to draw sperm up the spermathecal duct. In *Rhodnius*, there is good evidence for production by the opaque accessory glands of a peristalsis-inducing secretion.[162] Normally, the spermathecae begin to fill with sperm within 5 to 10 min of the end of mating. However, in females mated to males whose opaque glands have been removed, the spermathecae are still empty even 5 h later. This is not due to a malformed spermatophore (which is a product of the transparent accessory glands) or to an action on the sperm (which even if killed can still be moved into the spermathecae). In isolated preparations of the reproductive tract, the opaque gland secretion induces peristalsis, ruling out central nervous control; however, Davey[162] also showed that the secretion acts on the peripheral nervous system rather than directly on the musculature of the reproductive tract. A similar function for the accessory gland secretion has been suggested for several other species; in all cases, however, an effect on the sperm per se rather than the tract musculature cannot be ruled out.

In Saturniidae (Lepidoptera), the noncuticular simplex (see Figure 1D) produces a sperm activator (it is not specified whether apyrene, eupyrene, or both types of sperm are affected).[18] The molecule, which is reported to be a peptide of molecular weight about 3100 Da,[163] apparently diffuses through the spermatophore wall (already formed from secretions emanating further up the simplex) and may work by disrupting the sperm plasma membrane.[164] A detailed study of sperm activation of *Bombyx mori* has been undertaken by Osanai and colleagues, who have demonstrated a remarkable proteolytic cascade within the spermatophore that affects both apyrene and eupyrene sperm.[165-171] During spermatophore and seminal fluid formation, various components are added from different regions of the ejaculatory duct, the final (but key) participant being initiatorin,[165] an endopeptidase released by the proximal segment of the duct.[166] Four distinct roles have been identified for this enzyme. It digests the coating around the apyrene sperm whose resultant activity then serves to stir the seminal fluid;[167] it also digests the intercellular glue that binds together the eupyrene sperm, leading to their release and activation.[167] Its third role is to split proteins (produced elsewhere in the simplex) on the C side of arginine residues,[169] and its fourth role is to activate an arginine carboxypeptidase produced in the ampulla.[168] The arginine then released by the action of this exopeptidase[169] is hydrolyzed to ornithine and urea under the catalytic action of arginase from the seminal vesicles.[170] From the ornithine is derived glutamate, then 2-oxoglutarate, which serves both as a substrate for sperm respiration and as a promoter of pyruvate oxidation.[171] Trehalases have been detected in the accessory glands of male *P. americana*[172] and in the BAGs and spermatophore of *T. molitor*.[173] Characterization of the enzyme[174] shows that it has high specificity towards trehalose. However, the few published analyses do not suggest that this sugar normally occurs in very significant amounts in insect seminal fluid. Thus, a role for this enzyme, for example, in sperm metabolism, remains unclear.

The anterior ejaculatory duct of *D. melanogaster* produces an enzyme, esterase 6, that is transferred to the female in the seminal fluid.[21] Esterase 6 appears to have two purposes: it may be involved in lipid catabolism in the ejaculate, thus affecting sperm motility;[175] it was also proposed[22] that the enzyme catalyzed the conversion of *cis*-vaccenyl acetate (produced in the male's ejaculatory bulb and also transferred during mating) to *cis*-vaccenol, the latter then being released by the female as an antiaphrodisiac. However, attempts to confirm this have been unsuccessful.[176]

In addition to the RIS described earlier, a second sex peptide has recently been identified and characterized in the accessory glands of *D. funebris*.[177] This 63-amino acid-containing molecule has several similar properties and sequence homologies with known protease inhibitors, including the ability to inhibit acrosin, a trypsin-like endopeptidase associated with the acrosome of mammalian sperm. Thus, Schmidt et al.[177] speculate that this peptide temporarily inactivates the acrosomal proteases until the appropriate moment for egg fertilization.

Males of some species produce a very large spermatophore (e.g., in some Gryllidae and Tettigoniidae [Orthoptera] it may be 40% of the male's body weight[178]), part or all of which is eventually eaten by the female; in others, for example *M. sanguinipes*,[179] the male transfers several spermatophores in sequence during a single mating. Though there have been claims that in some species the spermatophore is digested within the female reproductive tract, demonstrations of the existence of hydrolytic enzymes in this region are virtually nonexistent. It has been shown, however, that material from the spermatophore does enter the female's hemolymph and, in some species, the ovaries, leading to the proposal that the male may thereby make a nutritional contribution to the female or, more specifically, to egg development.[149,180-183] The subject of male nutrient investment is discussed by Boggs (Chapter 10).

ACKNOWLEDGMENTS

Original work of the author cited in this review is supported by the Natural Sciences and Engineering Research Council of Canada. Thanks are extended to Dr. J.G. Riemann for provision of a diagram of the male flour moth reproductive system and to Mr. D. Dyck for assistance in the preparation of the figure.

REFERENCES

1. **Cantacuzène, A.-M.,** Recherches morphologiques et physiologiques sur les glandes annexes mâles des Orthoptères. III. Modes d'association des spermatozoides d'Orthoptères, *Z. Zellforsch.*, 90, 113, 1968.
2. **Rojas-Rousse, D.,** Description et fonctionnement de l'appareil génital interne de *Diadromus pulchellus* Wesmael (Hymenoptera: Ichneumonidae). I. Testicules, canaux deferents, glandes annexes, canaux collecteurs, canal ejaculateur du mâle, *Int. J. Insect Morphol. Embryol.*, 1, 225, 1972.
3. **Gerber, G.H., Church, N.S., and Rempel, J.G.,** The anatomy, histology, and physiology of the reproductive systems of *Lytta nuttalli* Say (Coleoptera: Meloidae). I. The internal genitalia, *Can. J. Zool.*, 49, 523, 1971.
4. **Ramamurty, P.S.,** Histological studies of the internal organs of reproduction in *Nezara viridula* Fabr. (Pentatomidae-Heteroptera, Insecta), *Zool. Anz.*, 183, 119, 1968.
5. **Gerber, G.H., Neill, G.B., and Westdal, P.H.,** The anatomy and histology of the internal reproductive organs of the sunflower beetle, *Zygogramma exclamationis* (Coleoptera: Chrysomelidae), *Can. J. Zool.*, 56, 2542, 1978.
6. **Poels, A.,** Histophysiologie des voies génitales mâles de *Tenebrio molitor* L. (Coléoptère: Tenebrionidae), *Ann. Soc. R. Zool. Belg.*, 102, 199, 1972.
7. **Gerber, G.H.,** Reproductive behaviour and physiology of *Tenebrio molitor* (Coleoptera: Tenebrionidae). III. Histogenetic changes in the internal genitalia, mesenteron, and cuticle during sexual maturation, *Can. J. Zool.* 54, 990, 1976.
8. **Riemann, J.G. and Thorson, B.J.,** Ultrastructure of the vasa deferentia of the Mediterranean flour moth, *J. Morphol.*, 149, 483, 1976.
9. **Cantacuzène, A.-M.,** Origine et caractères ultrastructuraux des cellules spermiophages du criquet migrateur *Locusta migratoria*, *J. Microsc. (Paris)*, 10, 179, 1971.
10. **Szöllösi, A. and Landureau, J.-C.,** Imaginal cell differentiation in the spermiduct of *Samia cynthia* (Lepidoptera). Responses *in vitro* to ecdysone and ecdysterone, *Biol. Cell.*, 28, 23, 1977.
11. **Bairati, A.,** Structure and ultrastructure of the male reproductive system in *Drosophila melanogaster* Meig. II. The genital duct and accessory glands, *Monit. Zool. Ital.*, 2, 105, 1968.
12. **Lai-Fook, J.,** The vasa deferentia of the male reproductive system of *Calpodes ethlius* (Hesperiidae, Lepidoptera), *Can. J. Zool.*, 60, 1172, 1982.
13. **Khalifa, A.,** Spermatophore production in Trichoptera and some other insects, *Trans. R. Entomol. Soc. London*, 100, 449, 1949.
14. **Gerber, G.H., Church, N.S., and Rempel, J.G.,** The structure, formation, histochemistry, fate and functions of the spermatophore of *Lytta nuttalli* Say (Coleoptera: Meloidae), *Can. J. Zool.*, 49, 1595, 1971.
15. **Gillott, C.,** Arthropoda-Insecta, in *Reproductive Biology of Invertebrates*, Vol. 3 (*Accessory Sex Glands*), Adiyodi, K.G. and Adiyodi, R.G., Eds., Oxford and IBH, New Delhi, 1988, chap. 12.
16. **Riemann, J.G. and Thorson, B.J.,** Foliate and granule-secreting cells in the ejaculatory duct (simplex) of the Mediterranean flour moth, *J. Ultrastruct. Res.*, 66, 1, 1979.
17. **Omura, S.,** Studies on the reproductive system of the male of *Bombyx mori*. II. Post-testicular organs and post-testicular behaviour of the spermatozoa, *J. Fac. Agric. Hokkaido Imp. Univ.*, 40, 129, 1938.
18. **Shepherd, J.G.,** Activation of saturniid moth sperm by a secretion of the male reproductive tract, *J. Insect Physiol.*, 20, 2107, 1974.
19. **Herman, W.S. and Peng, P.,** Juvenile hormone stimulation of sperm activator production in male monarch butterflies, *J. Insect Physiol.*, 22, 579, 1976.
20. **Wensler, R.J.D. and Rempel, J.G.,** The morphology of the male and female reproductive systems of the midge, *Chironomus plumosus* L., *Can. J. Zool.*, 40, 199, 1962.
21. **Sheehan, K.B., Richmond, R.C., and Cochrane, B.J.,** Studies of esterase 6 in *Drosophila melanogaster*. III. The developmental pattern and tissue distribution, *Insect Biochem.*, 9, 443, 1979.
22. **Mane, S.D., Tomkins, L., and Richmond, R.C.,** Male esterase 6 catalyses the synthesis of a sex pheromone in *Drosophila melanogaster* females, *Science*, 222, 419, 1983.

23. **Riemann, J.G., Moen, D.J., and Thorson, B.J.,** Female monogamy and its control in houseflies, *J. Insect Physiol.*, 13, 407, 1967.
24. **Morrison, P.E., Venkatesh, K., and Thompson, B.,** The role of male accessory-gland substance on female reproduction with some observations of spermatogenesis in the stable fly, *J. Insect Physiol.*, 28, 607, 1982.
25. **Snodgrass, R.E.,** A revised interpretation of the external reproductive organs of male insects, *Smithson. Misc. Collect.*, 132, 1, 1957.
26. **Matsuda, R.,** *Morphology and Evolution of the Insect Abdomen*, Pergamon Press, New York, 1976, 534 pp.
27. **Waage, J.K.,** Sperm competition and the evolution of odonate mating systems, in *Sperm Competition and the Evolution of Animal Mating Systems*, Smith, R.L., Ed., Academic Press, New York, 1984, 251.
28. **Hinton, H.E.,** Sperm transfer in insects and the evolution of haemocoelic insemination, *Symp. R. Entomol. Soc. London*, 2, 95, 1964.
29. **Carayon, J.,** Traumatic insemination and the paragenital system, Entomological Society of America, Lanham, MD, *Thomas Say Foundation Publ.*, 7, 81, 1966.
30. **Carayon, J.,** Parthénogénèse constante prouvée chez deux hétéroptères: le miride *Campyloneura virgula* et l'anthocoride *Calliodis maculipennis*, *Ann. Soc. Entomol. Fr.*, 25, 387, 1989.
31. **Anderson, D.T.,** The development of hemimetabolous insects, in *Developmental Systems: Insects*, Vol. 1, Counce, S.J. and Waddington, C.H., Eds., Academic Press, New York, 1972, chap. 3.
32. **Anderson, D.T.,** The development of holometabolous insects, in *Developmental Systems: Insects*, Vol. 1, Counce, S.J. and Waddington, C.H., Eds., Academic Press, New York, 1972, chap. 4.
33. **Roonwal, M.L.,** Studies on the embryology of the African migratory locust, *Locusta migratoria migratorioides* Reiche and Frm. (Orthoptera, Acrididae). II. Organogeny, *Philos. Trans. R. Soc. London, Ser. B*, 227, 175, 1937.
34. **Cavallin, M.,** Étude descriptive du développement et de la différenciation sexuelle de l'appareil génital chez les embryons des phasmes *Clitumnus extradentatus* Br. et *Carausius morosus* Br., *C. R. Acad. Sci.*, 268, 2189, 1969.
35. **Gehring, W.J. and Nöthiger, R.,** The imaginal discs of *Drosophila*, in *Developmental Systems: Insects*, Vol. 2, Counce, S.J. and Waddington C.H., Eds., Academic Press, New York, 1973, 212.
36. **Leclerq-Smekens, M.,** Organogenèse et différentiation des voies génitales mâles internes des lépidoptères sur la base d'observations effectuées chez *Euproctis chrysorrhea* (Lymantriidae), *Ann. Soc. R. Zool. Belg.*, 104, 131, 1974.
37. **Loeb, M.J.,** Development of isolated spermducts from *Heliothis virescens* (Lepidoptera) *in vitro*, *Invertebr. Reprod. Dev.*, 20, 67, 1991.
38. **Steel, C.G.H. and Davey, K.G.,** Integration in the insect endocrine system, in *Comprehensive Insect Physiology, Biochemistry and Pharmacology*, Vol. 8, Kerkut, G.A. and Gilbert, L.I., Eds., Pergamon Press, Oxford, 1985, 1.
39. **Wigglesworth, V.B.,** Historical perspectives, in *Comprehensive Insect Physiology, Biochemistry and Pharmacology*, Vol. 7, Kerkut, G.A. and Gilbert L.I., Eds., Pergamon Press, Oxford, 1985, 1.
40. **Sehnal, F.,** Growth and life cycles, in *Comprehensive Insect Physiology, Biochemistry and Pharmacology*, Vol. 2, Kerkut, G.A. and Gilbert L.I., Eds., Pergamon Press, Oxford, 1985, 1.
41. **Happ, G.M.,** Maturation of the male reproductive system and its endocrine regulation, *Annu. Rev. Entomol.*, 37, 303, 1992.
42. **Szöllösi, A.,** Imaginal differentiation of the spermiduct in acridids: effects of juvenile hormone, *Acrida*, 4, 205, 1975.
43. **Cantacuzène, A.-M. and Girardie, A.,** Caractères sexuels des adultes prothétéliques mâles obtenus par allatectomie de *Locusta migratoria migratorioides* (Orthoptère), *C. R. Acad. Sci.*, 270, 2465, 1970.
44. **Gallois, D.,** La différenciation fonctionnelle des glandes annexes mâles de *Locusta migratoria migratorioides* R. et F., *Arch. Zool. Exp. Gén*, 122, 109, 1981.
45. **Gallois, D.,** Control of cell differentiation in the male accessory reproductive glands of *Locusta migratoria*: acquisition and reversal of competence to imaginal secretion, *J. Insect Physiol.*, 35, 189, 1989.
46. **Baehr, J.-C., Porcheron, P., Papillon, M., and Dray, F.,** Haemolymph levels of juvenile hormone, ecdysteroids and proteins during the last two larval instars of *Locusta migratoria*, *J. Insect Physiol.*, 25, 415, 1979.
47. **Nowock, J.,** Growth and metamorphosis in the testes of *Ephestia kuhniella in vitro*, *J. Insect Physiol.*, 19, 941, 1973.
48. **Loeb, M.J.,** Growth and development of spermducts of the tobacco budworm moth *Heliothis virescens, in vivo* and *in vitro*, *Invertebr. Reprod. Dev.*, 19, 97, 1991.
49. **Loeb, M.J. and Hakim, R.S.,** Development of genital imaginal discs of *Heliothis virescens* culture *in vitro* with 20-hydroxyecdysone and fat body or testis sheaths, *Invertebr. Reprod. Dev.*, 20, 181, 1991.
50. **Shinbo, H. and Happ, G.M.,** Effects of ecdysteroids on the growth of the post-testicular reproductive organs in the silkworm, *Bombyx mori*, *J. Insect Physiol.*, 35, 855, 1989.
51. **Szopa, T.M., Lenoir Rousseaux, J.-J., Yuncker, C., and Happ, G.M.,** Ecdysteroids accelerate mitoses in accessory glands of beetle pupae, *Dev. Biol.*, 107, 325, 1985.

52. **Yaginuma, T., Kai, H., and Happ, G.M.**, 20-Hydroxyecdysone accelerates the flow of cells into the G_1 phase and the S phase in a male accessory gland of the mealworm pupa (*Tenebrio molitor*), *Dev. Biol.*, 126, 173, 1988.
53. **Grimnes, K.A. and Happ, G.M.**, Ecdysteroids *in vitro* promote differentiation in the accessory glands of male mealworm beetles, *Experientia*, 43, 906, 1987.
54. **Yaginuma, H. and Happ, G.M.**, 20-Hydroxyecdysone acts in the male pupa to commit accessory glands toward trehalase production in the adult mealworm beetle (*Tenebrio molitor*), *Gen. Comp. Endocrinol.*, 73, 173, 1989.
55. **Roosen-Runge, E.C.**, *The Process of Spermatogenesis in Animals*, Cambridge University Press, Cambridge, 1977, 214 pp.
56. **Szöllösi, A.**, Ultrastructural study of the spermatodesm of *Locusta migratoria migratorioides* (R. and F.): acrosome and cap formation, *Acrida*, 3, 175, 1974.
57. **Anderson, J.M.**, A cytological study of the testicular cyst cells in the Japanese beetle, *Physiol. Zool.*, 23, 308, 1950.
58. **Silberglied, R.E., Shepherd, J.G., and Dickinson, J.L.**, Eunuchs: the role of apyrene sperm in Lepidoptera, *Am. Nat.*, 123, 255, 1984.
59. **Lai-Fook, J.**, Testicular development and spermatogenesis in *Calpodes ethlius* Stoll (Hesperiidae, Lepidoptera), *Can. J. Zool.*, 60, 1161, 1982.
60. **Dumser, J.B. and Davey, K.G.**, Endocrinological and other factors influencing testis development in *Rhodnius prolixus*, *Can. J. Zool.*, 52, 1011, 1974.
61. **Dumser, J.B. and Davey, K.G.**, The *Rhodnius* testis: hormones, differentiation of the germ cells, and duration of the molting cycle, *Can. J. Zool.*, 53, 1673, 1975.
62a. **Dumser, J.B. and Davey, K.G.**, The *Rhodnius* testis: hormonal effects on germ cell division, *Can. J. Zool.*, 53, 1682, 1975.
62b. **Dumser, J.B.**, The regulation of spermatogenesis in insects, *Annu. Rev. Entomol.*, 25, 341, 1980.
63. **Koeppe, J.K., Fuchs, M.S., Chen, T.T., Hunt, L.-M., Kovalick, G.E., and Briers, T.**, The role of juvenile hormone in reproduction, in *Comprehensive Insect Physiology, Biochemistry and Pharmacology*, Vol.8, Kerkut, G.A. and Gilbert, L.I., Eds., Pergamon Press, New York, 1985, 165.
64. **Riemann, J.G. and Giebultowicz, J.M.**, Secretion in the upper vas deferens of the gypsy moth correlated with the circadian rhythm of sperm release from the testes, *J. Insect Physiol.*, 37, 53, 1991.
65. **Thorson, B.J. and Riemann, J.G.**, Abdominally entrained periodicities of testis and vas deferens activity in the Mediterranean flour moth, *J. Insect Physiol.*, 23, 1189, 1977.
66. **Giebultowicz, J.M., Riemann, J.G., Raina, A.K., and Ridgway, R.L.**, Circadian system controlling release of sperm in the insect testes, *Science*, 245, 1098, 1989.
67. **Riemann, J.G. and Thorson, B.J.**, Sperm maturation in the male and female genital tracts of *Anagasta kühniella* (Lepidoptera: Pyralididae), *Int. J. Insect Morphol. Embryol.*, 1, 11, 1971.
68. **Hinton, H.E.**, Accessory functions of seminal fluid, *J. Med. Entomol.*, 11, 19, 1974.
69. **Leopold, R.A.**, The role of male accessory glands in insect reproduction, *Annu. Rev. Entomol.*, 21, 199, 1976.
70. **Chen, P.S.**, The functional morphology and biochemistry of insect male accessory glands and their secretions, *Annu. Rev. Entomol.*, 29, 233, 1984.
71. **Happ, G.M.**, Structure and development of male accessory glands in insects, in *Insect Ultrastructure*, Vol. 2, King, R.C. and Akai, H., Eds., Plenum Press, New York, 365, 1984.
72. **Gillott, C. and Gaines, S.B.**, Endocrine regulation of male accessory gland development and activity, *Can. Entomol.*, 124, 871, 1992.
73. **Happ, G.M., Yuncker, C., and Dailey, P.J.**, Cytodifferentiation in the accessory glands of *Tenebrio molitor*. VII. Patterns of leucine incorporation by the bean-shaped glands of males, *J. Exp. Zool.*, 220, 81, 1982.
74. **Fallon, A.M. and Wyatt, G.R.**, Cyclic guanosine 3′, 5′-monophosphate. High levels in the male accessory gland of *Acheta domesticus* and related crickets, *Biochim. Biophys. Acta*, 411, 173, 1975.
75. **Fallon, A.M. and Wyatt, G.R.**, Guanylate cyclase in the accessory gland of the cricket, *Acheta domesticus*, *J. Insect Physiol.*, 23, 1037, 1977.
76. **Shirk, P.D., Dahm, K.H., and Röller, H.**, The accessory sex glands as the repository for juvenile hormone in male cecropia moths, *Z. Naturforsch. Teil C*, 31, 199, 1976.
77. **Shirk, P.D., Bhaskaran, G., and Röller, H.**, The transfer of juvenile hormone from male to female during mating in the cecropia silk moth *Hyalophora cecropia*, *Experientia*, 36, 682, 1980.
78. **Roth, L.M. and Dateo, G.P., Jr.**, Uric acid in the reproductive system of males of the cockroach *Blattella germanica*, *Science*, 146, 782, 1964.
79. **Roth, L.M. and Dateo, G.P., Jr.**, Uric acid storage and excretion by accessory sex glands of male cockroaches, *J. Insect Physiol.*, 11, 1023, 1965.
80a. **Roth, L.M.**, Uricose glands in the accessory sex gland complex of male Blattaria, *Ann. Entomol. Soc. Am.*, 60, 1203, 1967.
80b. **Mullins, D.E. and Keil, C.B.**, Paternal investment of urates in cockroaches, *Nature*, 283, 567, 1980.

81. **Wigglesworth, V.B.,** The function of the corpus allatum in the growth and reproduction of *Rhodnius prolixus*, *Q. J. Microsc. Sci.*, 79, 91, 1936.
82. **Delbecque, J.-P., Weidner, K., and Hoffmann, K.H.,** Alternative sites for ecdysteroid production in insects, *Invert. Reprod. Dev.*, 18, 29, 1990.
83. **Gee, J.D., Whitehead, D.L., and Koolman, J.,** Steroids stimulate secretion by insect Malpighian tubules, *Nature (London)*, 269, 238, 1977.
84. **Hoffmann, K.H. and Behrens, W.,** Free ecdysteroids in adult male crickets, *Gryllus bimaculatus*, *Physiol. Entomol.*, 7, 269, 1982.
85. **Cheeseman, M.T. and Gillott, C.,** Identification and partial characterization of the major secretory protein of the long hyaline gland in the male grasshopper, *Melanoplus sanguinipes*, *Insect Biochem.*, 18, 135, 1988.
86. **Cheeseman, M.T. and Gillott, C.,** Corpus allatum and corpus cardiacum regulation of long hyaline gland protein synthesis in the male grasshopper, *Melanoplus sanguinipes*, *Gen. Comp. Endocrinol.*, 72, 416, 1988.
87. **Odhiambo, T.R.,** Site of action of the corpus allatum hormone at the cellular level in *Schistocerca gregaria*, *Acta Trop.*, 23, 264, 1966.
88. **De Loof, A. and Lagasse, A.,** The ultrastructure of the male accessory reproductive glands of the Colorado beetle, *Z. Zellforsch.*, 130, 545, 1972.
89. **Couche, G.A. and Gillott, C.,** Development of secretory activity in the long hyaline gland of the male migratory grasshopper, *Melanoplus sanguinipes* (Fabr.) (Orthoptera: Acrididae), *Int. J. Insect Morphol. Embryol.*, 16, 355, 1987.
90. **Ogiso, M. and Takahashi, S.Y.,** Trehalases from the male accessory glands of the American cockroach: developmental changes and the hormonal regulation of the enzymes, *Gen. Comp. Endocrinol.*, 55, 387, 1984.
91. **Hartmann, R.,** Der Einfluss endokriner Faktoren auf die männlichen Akzessorischen Drüsen und die Ovarien bei der Keulenheuschrecke *Gomphocerus rufus* L. (Orthoptera, Acrididae), *Z. Vgl. Physiol.*, 74, 190, 1971.
92. **Hartmann, R., Wolf, W., and Loher, W.,** The influence of the endocrine system on reproductive behavior and development in grasshoppers, *Gen. Comp. Endocrinol. Suppl.*, 3, 518, 1972.
93. **Barker, J.F. and Davey, K.G.,** Neuroendocrine regulation of protein accumulation by the transparent accessory reproductive gland of male *Rhodnius prolixus*, *Int. J. Invertebr. Reprod.*, 3, 291, 1981.
94. **Glauser and Chen, unpublished data, in Schmidt, T., Stumm-Zollinger, E., and Chen, P.S.,** Protein metabolism of *Drosophila melanogaster* male accessory glands. III. Stimulation of protein synthesis following copulation, *Insect Biochem.*, 15, 391, 1985.
95. **Yamamoto, K., Chadarevian, A., and Pellegrini, M.,** Juvenile hormone action mediated in male accessory glands of *Drosophila* by calcium and kinase C., *Science*, 239, 916, 1988.
96. **Gold, S.M.W. and Davey, K.G.,** The effect of juvenile hormone on protein synthesis in the transparent accessory gland of male *Rhodnius prolixus*, *Insect Biochem.*, 19, 139, 1989.
97. **Bar-Zev, A. and Kaulenas, M.S.,** Nucleic acid metabolism in adult tissues of *Gromphadorhina* in response to ecdysterone: quantitative effects, *J. Insect Physiol.*, 21, 623, 1975.
98. **Herman, W.S. and Barker, J.F.,** Ecdysterone antagonism, mimicry, and synergism of juvenile hormone action on the monarch butterfly reproductive tract, *J. Insect Physiol.*, 22, 643, 1976.
99. **Sridevi, R., Bajaj, P., and Dutta-Gupta, A.,** Ecdysteroid stimulated protein synthesis in the male accessory reproductive glands of *Spodoptera litura*, *Invertebr. Reprod. Dev.*, 14, 177, 1988.
100. **Sridevi, R., Bajaj, P., and Dutta-Gupta, A.,** Hormonal regulation of macromolecular synthesis in testes and accessory reproductive glands of *Spodoptera litura* during post-embryonic and adult development, *Indian J. Exp. Biol.*, 27, 699, 1989.
101. **Ismail, P. and Dutta-Gupta, A.,** Effect of 20-hydroxyecdysone and inhibitors on the protein synthesis in male accessory reproductive glands of *Chilo partellus*, *Biochem. Arch.*, 6, 321, 1990.
102. **Barker, J.F. and Davey, K.G.,** A polypeptide from the brain and corpus cardiacum of male *Rhodnius prolixus* which stimulates *in vitro* protein synthesis in the transparent accessory reproductive gland, *Insect Biochem.*, 13, 7, 1983.
103. **Friedel, T. and Gillott, C.,** Extraglandular synthesis of accessory reproductive gland components in male *Melanoplus sanguinipes*, *J. Insect Physiol.*, 22, 1309, 1976.

104a. **Ismail, P. and Dutta-Gupta, A.,** *In vitro* uptake of larval haemolymph proteins by male accessory reproductive glands of the stem borer, *Chilo partellus*, *Invertebr. Reprod. Dev.*, 20, 193, 1991.

104b. **Sevala, V.L. and Davey, K.G.,** The transparent accessory reproductive gland secretes a polypeptide into the hemolymph of male *Rhodnius prolixus*, *Insect Biochem.*, 21, 215, 1991.

105. **Baumann, H.,** The isolation, partial characterization, and biosynthesis of the paragonial substances, PS-1 and PS-2, of *Drosophila funebris*, *J. Insect Physiol.*, 20, 2181, 1974.
106. **Schmidt, T., Stumm-Zollinger, E., and Chen, P.S.,** Protein metabolism of *Drosophila melanogaster* male accessory glands. III. Stimulation of protein synthesis following copulation, *Insect Biochem.*, 15, 391, 1985.
107. **Cheeseman, M.T. and Gillott, C.,** Control of copulation-enhanced protein synthesis in the long hyaline gland of the male grasshopper, *Melanoplus sanguinipes*, *Arch. Insect Biochem. Physiol.*, 11, 13, 1989.
108. **Davey, K.G.,** *Reproduction in the Insects*, Oliver and Boyd, Edinburgh, 1965, 96 pp.

109. **Gerber, G.H.**, Evolution of the methods of spermatophore formation in pterygotan insects, *Can. Entomol.*, 102, 358, 1970.
110. **Grimnes, K.A. and Happ, G.M.**, A monoclonal antibody against a structural protein in the spermatophore of *Tenebrio molitor* (Coleoptera), *Insect Biochem.*, 16, 635, 1986.
111. **Grimnes, K.A., Bricker, C.S., and Happ, G.M.**, Ordered flow of secretion from accessory glands to specific layers of the spermatophore of mealworm beetles: demonstration with a monoclonal antibody, *J. Exp. Zool.*, 240, 275, 1986.
112. **Shinbo, H., Yaginuma, T., and Happ, G.M.**, Purification and characterization of a proline-rich secretory protein that is a precursor to a structural protein of an insect spermatophore, *J. Biol. Chem.*, 262, 4794, 1987.
113. **Cheeseman, M.T., Gillott, C., and Ahmed, I.**, Structural spermatophore proteins and a trypsin-like enzyme from the accessory reproductive glands of the male grasshopper, *Melanoplus sanguinipes*, *J. Exp. Zool.*, 255, 193, 1990.
114. **Gregory, G.E.**, On the initiation of spermatophore formation in the African migratory locust, *Locusta migratoria migratorioides* Reiche and Fairmaire, *J. Exp. Biol.*, 42, 423, 1965.
115. **Loher, W.**, Circadian control of spermatophore formation in the cricket *Teleogryllus commodus* Walker, *J. Insect Physiol.*, 20, 1155, 1974.
116. **Blum, M.S., Glowska, Z., and Taber, S., III**, Chemistry of the drone honey bee reproductive system. II. Carbohydrates in the reproductive organs and semen, *Ann. Entomol. Soc. Am.*, 55, 134, 1962.
117. **Vijayalekshmi, V. and Adiyodi, K.G.**, Accessory sex glands of male *Periplaneta americana* (L.). III. Histochemistry of the mushroom-shaped and conglobate glands, Indian *J. Exp. Biol.*, 11, 521, 1973.
118. **Odhiambo, T.R.**, The architecture of the accessory reproductive glands of the male desert locust. I. Types of glands and their secretions, *Tissue Cell*, 1, 155, 1969.
119. **Blum, M.S., Bumgarner, J.E., and Taber, S., III**, Composition and possible significance of fatty acids in the lipid classes in honey bee semen, *J. Insect Physiol.*, 13, 1301, 1967.
120. **Novak, A.F., Blum, M.S., Taber, S., III, and Luizzo, J.A.**, Separation and determination of seminal plasma and sperm amino acids of the honey bee, *Apis mellifera*, *Ann. Entomol. Soc. Am.*, 53, 841, 1960.
121. **Roth, L.M.**, The stimuli regulating reproduction in cockroaches, *Colloq. Int. C.N.R.S.*, 189, 267, 1970.
122. **Gillott, C. and Friedel, T.**, Fecundity-enhancing and receptivity-inhibiting substances produced by male insects: a review, in *Advances in Invertebrate Reproduction*, Vol. 1, Adiyodi, K.G. and Adiyodi, R.G., Eds., Peralam-Kenoth, Karivellur, India, 1977, 199.
123. **Leopold, R.A.**, Cytological and cytochemical studies on the ejaculatory duct and accessory secretion in *Musca domestica*, *J. Insect Physiol.*, 16, 1859, 1970.
124. **Riddiford, L.M. and Ashenhurst, J.B.**, The switchover from virgin to mated behavior in female Cecropia moths: the role of the bursa copulatrix, *Biol. Bull. Woods Hole Mass.*, 144, 162, 1973.
125. **Loher, W. and Edson, K.**, The effect of mating on egg production and release in the cricket, *Teleogryllus commodus*, *Entomol. Exp. Appl.*, 16, 483, 1973.
126. **Destephano, D.B. and Brady, U.E.**, Prostaglandin and prostaglandin synthetase in the cricket, *Acheta domesticus*, *J. Insect Physiol.*, 23, 905, 1977.
127. **Davey, K.G.**, Copulation and egg-production in *Rhodnius prolixus*: the role of the spermathecae, *J. Exp. Biol.*, 42, 373, 1965.
128. **Manning, A.**, The control of sexual receptivity in female *Drosophila*, *Anim. Behav.*, 15, 239, 1967.
129. **Merle, J.**, Fonctionnement ovarien et réceptivité sexuelle de *Drosophila melanogaster* après implantation de fragments de l'appareil génital mâle, *J. Insect Physiol.*, 14, 1159, 1968.
130. **Nelson, D.R., Adams, T.S., and Pomonis, J.G.**, Initial studies on the extraction of the active substance inducing monocoitic behavior in house flies, black blow flies, and screw-worm flies, *J. Econ. Entomol.*, 62, 634, 1969.
131. **Baumann, H., Wilson, K.J., Chen, P.S., and Humbel, R.E.**, The amino acid sequence of a peptide (PS-1) from *Drosophila funebris*: a paragonial peptide from males which reduces the receptivity of the female, *Eur. J. Biochem.*, 52, 521, 1975.
132. **Chen, P.S.**, Biochemistry and molecular regulation of the male accessory gland secretions in *Drosophila* (Diptera), *Ann. Soc. Entomol. Fr. (N.S.)*, 27, 231, 1991.
133. **Friedel, T. and Gillott, C.**, Male accessory gland substance of *Melanoplus sanguinipes*: an oviposition stimulant under the control of the corpus allatum, *J. Insect Physiol.*, 22, 489, 1976.
134. **Lange, A.B. and Loughton, B.G.**, An oviposition-stimulating factor in the male accessory reproductive gland of the locust, *Locusta migratoria*, *Gen. Comp. Endocrinol.*, 57, 208, 1985.
135. **Fuchs, M.S., Craig, G.B., Jr., and Despommier, D.D.**, The protein nature of the substance inducing female monogamy in *Aedes aegypti*, *J. Insect Physiol.*, 15, 701, 1969.
136. **Destephano, D.B., Brady, U.E., and Lovins, R.E.**, Synthesis of prostaglandin by reproductive tissue of the house cricket, *Acheta domesticus*, *Prostaglandins*, 6, 71, 1974.
137. **Destephano, D.B., Brady, U.E., and Woodall, L.B.**, Partial characterization of prostaglandin synthetase in the reproductive tract of the male house cricket, *Acheta domesticus*, *Prostaglandins*, 11, 261, 1976.

138. **Loher, W., Ganjian, I., Kubo, I., Stanley-Samuelson, D., and Tobe, S.S.,** Prostaglandins: their role in egg-laying of the cricket *Teleogryllus commodus*, *Proc. Natl. Acad. Sci. U.S.A.*, 78, 7835, 1981.
139. **Monsma, S.A. and Wolfner, M.F.,** Structure and expression of a *Drosophila* male accessory gland gene whose product resembles a peptide pheromone precursor, *Genes Dev.*, 2, 1063, 1988.
140. **DiBenedetto, A.J., Harada, H.A., and Wolfner, M.F.,** Structure, cell-specific expression, and mating-induced regulation of a *Drosophila melanogaster* male accessory gland gene, *Dev. Biol.*, 139, 134, 1990.
141. **Monsma, S.A., Harada, H.A., and Wolfner, M.F.,** Synthesis of two *Drosophila* male accessory gland proteins and their fate after transfer to the female during mating, *Dev. Biol.*, 142, 465, 1990.
142. **Yamaja Setty, B.N. and Ramaiah, T.R.,** Isolation and identification of prostaglandins from the reproductive organs of male silkmoth, *Bombyx mori* L., *Insect Biochem.*, 9, 613, 1979.
143. **Yamaja Setty, B.N. and Ramaiah, T.R.,** Effect of prostaglandins and inhibitors of prostaglandin biosynthesis on oviposition in the silkmoth *Bombyx mori*, *Indian J. Exp. Biol.*, 18, 539, 1980.
144. **Lafon-Cazal, M., Gallois, D., Lehouelleur, J., and Bockaert, J.,** Stimulatory effects of male accessory-gland extracts on the myogenicity and the adenylate cyclase activity of the oviduct of *Locusta migratoria*, *J. Insect Physiol.*, 33, 909, 1987.
145. **Paemen, L., Schoofs, L., and De Loof, A.,** Presence of myotropic peptides in the male accessory glands of *Locusta migratoria*, *J. Insect Physiol.*, 36, 861, 1990.
146. **Paemen, L., Schoofs, L., Proost, P., Decock, B., and De Loof, A.,** Isolation, identification and synthesis of Lom-AG-myotropin II, a novel peptide in the male accessory reproductive glands of *Locusta migratoria*, *Insect Biochem.*, 21, 243, 1991.
147. **Lange, A.B.,** The neural and hormonal control of locust oviducts and accessory structures, *Adv. Comp. Endocrinol.*, 1, 109, 1992.
148. **Baumann, H.,** Biological effects of paragonial substances, PS-1 and PS-2, in females of *Drosophila funebris*, *J. Insect Physiol.*, 20, 2347, 1974.
149. **Friedel, T. and Gillott, C.,** Contribution of male-produced proteins to vitellogenesis in *Melanoplus sanguinipes*, *J. Insect Physiol.*, 23, 145, 1977.
150. **Terranova, A.C., Leopold, R.A., Degrugillier, M.E., and Johnson, J.R.,** Electrophoresis of the male accessory secretion and its fate in the mated female, *J. Insect Physiol.*, 18, 1573, 1972.
151a. **Ruegg, R.P., Orchard, I., and Davey, K.G.,** 20-Hydroxyecdysone as a modulator of electrical activity in neurosecretory cells of *Rhodnius prolixus*, *J. Insect Physiol.*, 28, 243, 1982.
151b. **Orchard, I., Ruegg, R.P., and Davey, K.G.,** The role of central aminergic neurons in the action of 20-hydroxyecdysone on neurosecretory cells of *Rhodnius prolixus*, *J. Insect Physiol.*, 29, 387, 1983.
152a. **Yamaoka, K. and Hirao, T.,** Stimulation of virginal oviposition by male factor and its effect on spontaneous nervous activity in *Bombyx mori*, *J. Insect Physiol.*, 23, 57, 1977.
152b. **Yamaoka, K.,** The central nervous function in ovipositional behaviour of *Bombyx mori* with special reference to the spontaneous nervous activity, in *Advances in Invertebrate Reproduction*, Vol. 1, Adiyodi, K.G. and Adiyodi, R.G., Eds., Peralam-Kenoth, Karivellur, India, 1977, 414.
153. **Borovsky, D.,** The role of the male accessory gland fluid in stimulating vitellogenesis in *Aedes taeniorhynchus*, *Arch. Insect Biochem. Physiol.*, 2, 405, 1985.
154. **Downe, A.E.R.,** Internal regulation of rate of digestion of blood meals in the mosquito, *Aedes aegypti*, *J. Insect Physiol.*, 21, 1835, 1975.
155. **Klowden, M.J. and Chambers, G.,** Male accessory gland substances activate egg development in nutritionally stressed *Aedes aegypti* mosquitoes, *J. Insect Physiol.*, 37, 721, 1991.
156. **Huignard, J., Quesneau-Thierry, A., and Barbier, M.,** Isolement, action biologique et evolution des substances paragoniales contenues dans le spermatophore d'*Acanthoscelides obtectus* (Coléoptère), *J. Insect Physiol.*, 23, 351, 1977.
157. **Sasaki, M., Riddiford, L.M., Truman, J.W., and Moore, J.K.,** Re-evaluation of the role of corpora cardiaca in calling and oviposition behaviour of giant silk moths, *J. Insect Physiol.*, 29, 695, 1983.
158. **Leopold, R.A., Terranova, A.C., and Swilley, E.M.,** Mating refusal in *Musca domestica*: effects of repeated mating and decerebration upon frequency and duration of copulation, *J. Exp. Zool.*, 176, 353, 1971.
159. **Leopold, R.A., Terranova, A.C., Thorson, B.J., and Degrugillier, M.E.,** The biosynthesis of the male housefly accessory secretion and its fate in the mated female, *J. Insect Physiol.*, 17, 987, 1971.
160. **Gwadz, R.W.,** Neuro-hormonal regulation of sexual receptivity in female *Aedes aegypti*, *J. Insect Physiol.*, 18, 259, 1972.
161. **Smith, P.H., Gillott, C., Barton Browne, L., and van Gerwen, A.C.M.,** The mating-induced refractoriness of *Lucilia cuprina* females: manipulating the male contribution, *Physiol. Entomol.*, 15, 469, 1990.
162. **Davey, K.G.,** The migration of spermatozoa in the female of *Rhodnius prolixus* Stål, *J. Exp. Biol.*, 35, 694, 1958.
163. **Shepherd, J.G.,** A polypeptide sperm activator from male saturniid moths, *J. Insect Physiol.*, 21, 9, 1975.
164. **Shepherd, J.G.,** Sperm activation in saturniid moths: some aspects of the mechanism of activation, *J. Insect Physiol.*, 20, 2321, 1974.

165. **Osanai, M., Aigaki, T., and Kasuga, H.**, Arginine degradation cascade as an energy-yielding system for sperm maturation in the spermatophore of the silkworm, *Bombyx mori*, in *New Horizons in Sperm Cell Research*, Mohri, H., Ed., Japan Scientific Society Press, Tokyo, 1987, 185.
166. **Aigaki, T., Kasuga, H., and Osanai, M.**, A specific endopeptidase, BAEE esterase, in the glandula prostatica of the male reproductive system of the silkworm, *Bombyx mori*, *Insect Biochem.*, 17, 323, 1987.
167. **Osanai, M., Kasuga, H., and Aigaki, T.**, Induction of motility of apyrene spermatozoa and dissociation of eupyrene sperm bundles of the silkworm, *Bombyx mori* by initiatorin and trypsin, *Invertebr. Reprod. Dev.*, 15, 97, 1989.
168. **Aigaki, T., Osanai, M., and Kasuga, H.**, Arginine carboxypeptidase activity in the male reproductive glands of the silkworm, *Bombyx mori*, *Insect Biochem.*, 18, 295, 1988.
169. **Kasuga, H., Aigaki, T., and Osanai, M.**, System for supply of free arginine in the spermatophore of *Bombyx mori*. Arginine-liberating activities of contents of male reproductive glands, *Insect Biochem.*, 17, 317, 1987.
170. **Osanai, M., Aigaki, T., Kasuga, H., and Yonezawa, Y.**, Role of arginase transferred from the vesicula seminalis during mating and changes in amino acid pools of the spermatophore after ejaculation in the silkworm, *Bombyx mori*, *Insect Biochem.*, 16, 879, 1986.
171. **Osanai, M., Aigaki, T., and Kasuga, H.**, Energy metabolism in the spermatophore of the silkmoth, *Bombyx mori*, associated with accumulation of alanine derived from arginine, *Insect Biochem.*, 17, 71, 1987.
172. **Takahashi, S.Y., Higashi, S., Minoshima, S., Ogiso, M., and Hanaoka, K.**, Trehalases from the American cockroach, *Periplaneta americana*: multiple occurrence of the enzymes and partial purification of enzymes from male accessory glands, *Int. J. Invertebr. Reprod.*, 2, 373, 1980.
173. **Yaginuma, T. and Happ, G.M.**, Trehalase from the bean-shaped accessory glands and the spermatophore of the male mealworm beetle, *Tenebrio molitor*, *J. Comp. Physiol. B*, 157, 765, 1988.
174. **Ogiso, M., Shinohara, Y., Hanaoka, K., Kageyama, T., and Takahashi, S.Y.**, Further purification and characterization of trehalases from the American cockroach, *Periplaneta americana*, *J. Comp. Physiol. B*, 155, 553, 1985.
175. **Gilbert, D.G.**, Ejaculate esterase 6 and initial sperm use by female *Drosophila melanogaster*, *J. Insect Physiol.*, 27, 641, 1981.
176. **Vander Meer, R.K., Obin, M.S., Zawistowski, S., Sheehan, K.B., and Richmond, R.C.**, A reevaluation of the role of *cis*-vaccenyl acetate, *cis*-vaccenol and esterase 6 in the regulation of mated female sexual attractiveness in *Drosophila melanogaster*, *J. Insect Physiol.*, 32, 681, 1986.
177. **Schmidt, T., Stumm-Zollinger, E., Chen, P.S., Böhlen, P., and Stone S.R.**, A male accessory gland peptide with protease inhibitory activity in *Drosophila funebris*, *J. Biol. Chem.*, 264, 9745, 1989.
178. **Gwynne, D.T.**, Male nutritional investment and the evolution of sexual differences in Tettigoniidae and other Orthoptera, in *Orthopteran Mating Systems*, Gwynne, D.T. and Morris, G.K., Eds., Westview Press, Boulder, CO, 1983, 337.
179. **Pickford, R. and Gillott, C.**, Insemination in the migratory grasshopper, *Melanoplus sanguinipes* (Fab.), *Can. J. Zool.*, 49, 1583, 1971.
180. **Boggs, C.L. and Gilbert, L.E.**, Male contribution to egg production in butterflies: Evidence for transfer of nutrients at mating, *Science*, 206, 83, 1979.
181. **Huignard, J.**, Transfer and fate of male secretions deposited in the spermatophore of females of *Acanthoscelides obtectus* Say (Coleoptera Bruchidae), *J. Insect Physiol.*, 29, 55, 1983.
182. **Markow, T.A. and Ankney, P.F.**, *Drosophila* males contribute to oogenesis in a multiple mating species, *Science*, 224, 302, 1984.
183. **Boucher, L. and Huignard, J.**, Transfer of male secretions from the spermatophore to the female insect in *Caryedon serratus* (Ol.): analysis of the possible trophic role of these secretions, *J. Insect Physiol.*, 33, 949, 1987.

Chapter 3

SEX DETERMINATION IN INSECTS

Roger L. Blackman

CONTENTS

I. INTRODUCTION

Sex determination is the process by which the gender of a bisexual organism becomes fixed, so that the individual progeny develops either as a son or a daughter. As is the case with other fundamental biological processes, evolution has in the course of time produced a seemingly infinite variety of ways of achieving this one essentially simple objective, and classical genetic and cytogenetic observations have, over the years, combined to display a bewildering diversity of sex-determining mechanisms. Much of this work has been on insects, from the first recognition of sex chromosomes in the heteropteran *Pyrrhocoris apterus*,[1] through the classic experiments of Bridges[2-4] on *Drosophila* and Goldschmidt[5,6] on *Porthetria dispar*, to the recent molecular work elucidating the hierarchy of regulatory genes responsible for the sex of fruit flies.[7,8]

In the general literature on sex determination, two works stand out,[9,10] each with a radically different approach to the subject. Both cover the full range of sex-determining systems, but the cytogeneticist White[9] gives pride of place to evolutionary changes in the sex chromosomes, whereas the evolutionary geneticist Bull[10] pays more attention to the underlying

0-8493-6695-X/95/$0.00+$.50

mechanisms. The problem with reviewing sex determination in insects at the present point in time is that, for all orders except Diptera, the greater part of the evidence is cytological, and even the most basic information about the genetic systems involved is not usually available. It is nevertheless worthwhile to follow the lead of Nöthiger and Steinmann-Zwicky,[11] and look for general principles of sex determination that could be applicable to all insects, and possibly to all biparental organisms.

In this chapter I shall start by reviewing the main types of sex determination found in insects, then outline what is known about sex chromosome systems in each insect order, and end with some discussion of the evolutionary implications of what we now know about sex determination.

II. GENERAL ASPECTS OF SEX DETERMINATION IN INSECTS

The sex of an insect is almost always determined genetically. Hermaphroditism, in the sense of the same genotype producing functional male and female organs in the same individual, seldom occurs in insects; it seems to have evolved only once, in one genus of scale insects (*Icerya*). An oft-quoted second example of insect hermaphroditism, in the termiticolous phorid group Termitoxeriinae, has now been refuted.[12] (True hermaphrodites should not be confused with gynandromorphs and intersexes — genetically abnormal individuals — which are of common occurrence in insects.) Environmental factors may sometimes influence the genetic determination of sex (see Bergerard[13] for review), but the type of environmental sex determination that occurs widely in reptiles, for example, where the sex of an individual is decided by the environment of the egg after it has been fertilized, seems to be rare in insects.

A. XX/XY SYSTEMS

Most bisexual organisms produce a 1:1 sex ratio, and this can be achieved simply by having one sex (the **heterogametic** sex) produce two genetically different types of gamete, and the other sex (the **homogametic** sex) produce gametes of only one of these types. The two types of gamete carry different **sex factors** (here given the notation s_1 and s_2), which segregate from one another in Mendelian fashion in the meiosis of the heterogametic sex:

$$s_1s_1 \times s_1s_2 \text{ (parents)} \rightarrow 1:1\ s_1s_1 \text{ and } s_1s_2 \text{ (progeny)}$$

The heterogametic sex is the male in most insects, but female in Lepidoptera, Trichoptera, and some Diptera.

Although these sex factors are conventionally regarded as alternative alleles at the same locus, it is usually the case that only one sex factor plays an active part in the determination of sex. The other "sex factor" may merely be the corresponding site on a homologous chromosome: e.g., a nonfunctioning (null) allele, or the location at which the sex factor is inserted, in the case of a transposable element.

In some cases, sex factors may be inherited as single genes, recombining freely with other genes on the same chromosome pair, although they are more often than not tightly linked to other genes involved in sex differentiation. Very often, however, chromosomes carrying sex factors are cytologically distinct (heteromorphic), so that the inheritance of sex can be observed cytologically:

$$XX \times XY \text{ (parents)} \rightarrow 1:1\ XX \text{ and } XY \text{ (progeny)}$$

X and Y chromosomes usually pair at meiosis before segregating to opposite poles, but there is normally little or no recombinational exchange between them. This is called an XX/XY, or XY (male) sex determination system. When the female is the heterogametic sex, the sex chromosomes are sometimes termed Z and W, and the system called ZW/ZZ:

$$ZW \text{ (female parent)} \times ZZ \text{ (male parent)} \rightarrow 1{:}1\ ZZ \text{ and } ZW \text{ (progeny)}$$

but this terminology has now been largely abandoned in the literature on insect cytogenetics as an unnecessary complication. The fact that the female is the heterogametic sex can be indicated by putting the heterozygous genotype first; i.e., XY/XX or XY (female) sex determination.

At this stage it is important to address two common misconceptions about sex chromosomes, which can easily hinder understanding of sex-determining mechanisms and their evolution. The first point concerns the common notation of heteromorphic sex chromosomes throughout both plant and animal kingdoms as X and Y, which might be thought to imply some degree of homology, not just at the sex-determining loci but of the chromosome as a whole, across major groups of organisms. On the contrary, there is no doubt that heteromorphic sex chromosomes have evolved many times independently in different taxa.[14] The Y chromosome in particular can sometimes be an extremely labile structure, apparently undergoing cycles of degeneration and regeneration within taxa. These will be discussed further later, and in order to emphasize this lability it is sufficient here to note that what seem to be major changes in sex chromosome constitution, such as the formation of a Y chromosome *de novo* (a "neo-Y"), can be found even within a single species.

The second common misconception is that the Y chromosome always has a dominant, male-determining function. This is certainly the case in mammals and in some insects, but the more general condition in insects with XX/XY systems is for the Y chromosome to have an essentially passive role, influencing sex merely by segregating opposite the X at meiosis. The sex of the zygote is then determined by the balance between the actions of regulatory genes on the X chromosome and on the autosomes. This is roughly equivalent to the "genic balance" model developed by Bridges[3] from his work on *Drosophila*. The zygote must be homozygous for a sex factor in order to be female (XX), so this has also been termed a "recessive-X system".[10]

Genetic studies are obviously needed to establish for certain whether the system operating in any one species is based on a dominant Y or genic balance. For species with heteromorphic sex chromosomes, deductions are possible by observing the sex of individuals with abnormal sex chromosome constitutions. The two most informative abnormalities are XXY and XO. If XXY individuals of a species which normally has an XX/XY system are male, then there is obviously a dominant Y factor operating, but if XXY is female, then the Y chromosome is likely to be sexually inert, and the X has an active, although recessive, role. Frequently, aberrant individuals completely lacking a Y chromosome (i.e., with XO constitution) are at least viable enough to observe their sex; such individuals will be female in dominant-Y systems, but male in genic balance systems, as the latter require XX for female determination. Evidence like this is available for only relatively few insects, mostly Diptera, in which occur both dominant-Y (e.g., *Phormia regina, Lucilia cuprina*[15]) and recessive-X systems (e.g., *Drosophila melanogaster*,[3] *Glossina palpalis*[16]). However, there are good reasons for believing that genic balance is the most general condition in insects, stemming from the widespread occurrence of XX/XO sex determination in many insect orders.[9]

B. XX/XO SYSTEMS

In organisms with heteromorphic sex chromosomes, X-Y recombination is usually suppressed, and the Y chromosome tends to be more degenerate than the X, often having few or no functional alleles. This degeneration of the Y is generally perceived as a progressive evolutionary phenomenon.[10] Various explanations for this have been offered.[14,17] For example, because the Y is permanently heterozygous and nonrecombinant, selection must act at the level of the entire chromosome, so it evolves as "an asexual component of an otherwise sexual genome".[18] Deleterious mutations (often nonfunctional alleles) will tend to accumulate in the absence of recombination by the process known as "Muller's ratchet,"[19] which may be

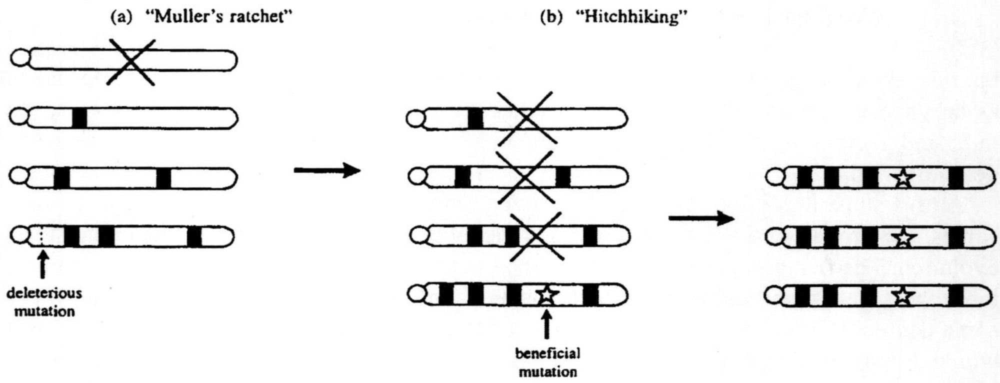

FIGURE 1. Degeneration of a Y chromosome by operation of (a) Muller's "ratchet," and (b) "hitchhiking." In a population of Y chromosomes free from recombination (left), some Ys will have one or more mutations to nonfunctional alleles (black segments), with different Ys mutated at different loci. If the selective disadvantage per locus is small and population size is small enough relative to the mutation rate, the class of Y chromosomes with no mutations may be lost by chance and, because there is no recombination, it cannot be restored. Next the class carrying just one mutation becomes vulnerable to chance loss, and so on, so that as time goes on the mean numbers of mutations per Y chromosome gradually increases. If a favorable Y-linked mutation should occur, however (☆), it could spread rapidly through the population by selection, carrying with it any nonfunctional alleles that happen to be present on the Y chromosome in which it originated ("hitchhiking"), leading to fixation of nonfunctional alleles and accelerating the degenerative process.

accelerated by a "genetic hitchhiking" effect[18] (Figure 1). However, there is still no entirely satisfactory explanation.

The end point of such an evolutionary process may be the complete loss of the Y chromosome, so that males are XO. An XX/XO sex determination system works in exactly the same way as an XX/XY system, with the X moving to one pole in male meiosis, so that sperm are either with or without an X:

$$\text{XX (female parent)} \times \text{XO (male parent)} \rightarrow 1{:}1 \text{ XX and XO progeny}$$

Obviously, since there is no Y, such a system must be based on genic balance. XX/XO sex determination occurs in almost all orders of insects, both primitive and advanced. It is almost certainly the ancestral form of sex determination of orthopteroid insects and probably of other major orders such as Hemiptera and Coleoptera. It is even possible that the ancestors of all insects had XO males, and that all insect Y chromosomes have arisen *de novo*.

Y chromosomes in insects often seem to arise as a result of centric fusion between an X chromosome and an autosome (e.g., Figure 2). A recently formed, autosomally derived Y chromosome (a neo-Y) is often easily recognized because it is likely to be still homologous with the autosomal part of the neo-X, and therefore synapses with it at meiosis. This homology may be gradually lost in the course of evolution as X-Y recombination becomes suppressed, and secondary structural and genetic changes occur independently on both the neo-X and the neo-Y.

Clearly, a neo-Y chromosome cannot carry a dominant factor for male determination, and sex determination in such cases must be based on genic balance, as in the ancestral XX/XO system. Neo-XY systems occur most frequently in groups in which XX/XO systems are common; these include the orthopteroid orders, Odonata, the hemipteroid orders, and Coleoptera. Each of these groups will be discussed later. In all of them, there are also species with multiple sex chromosomes.

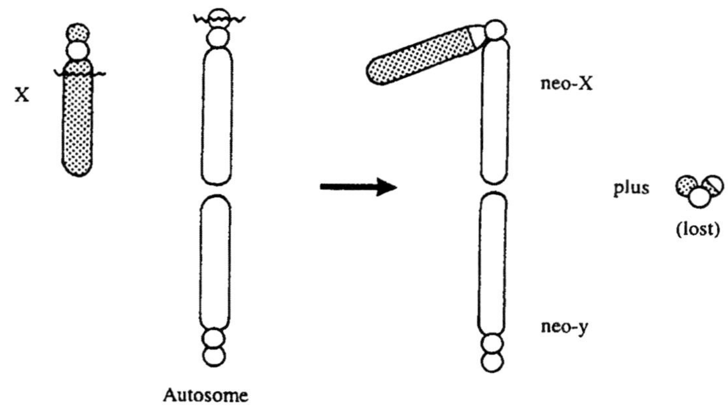

FIGURE 2. White's[9] model of the origin of neo-XY sex determination from an XO condition in the heterogametic sex. After breakage near the centromeres of an X and an autosome, centric fusion occurs, creating a neo-X chromosome. When this fusion has reached fixation in the population (i.e., when the original unfused Xs no longer occur), the original homologue of the autosome involved in the fusion will be confined to the male line and will act as a "neo-Y", segregating opposite the neo-X at meiosis. (White[9] believed that centromeres were never situated at the extreme ends (telomeres) of chromosomes, and therefore his models assume arm breaks and exchanges, followed by loss of a minute chromosome. Alternative models of centric fusion involving breakage *within* centromeres are discussed by John and Hewitt.[19a])

C. MULTIPLE SEX CHROMOSOME SYSTEMS

Chromosomal differences between male and female insects normally involve only one chromosome pair (X and Y, or just the X in XX/XO systems), but there are numerous cases in most insect orders where the difference between the sexes involves a larger number of chromosomes. The most common "multiple" sex chromosome systems have just two Xs (the notation used for the male condition X_1X_2Y, or X_1X_2O, and the notation for the sex determination system is $X_1X_1X_2X_2/X_1X_2Y$, or $X_1X_1X_2X_2/X_1X_2O$), although species are known with almost any number of Xs from 1 to 6 (and in one extreme case, 12), and any number of Ys from 1 to 6.

The origins of $X_1X_1X_2X_2/X_1X_2Y$ sex determination are usually fairly simply explained, with good evidence provided by the way in which the X_1, X_2, and Y associate at metaphase I of spermatogenesis. Figure 2 showed the origin of a neo-XY condition by centric fusion, resulting in a metacentric X (i.e., with the centromere near the center of the chromosome) and an acrocentric Y (with the centromere at one end). If a further centric fusion should occur, this time between the Y chromosome and another acrocentric autosome, then an X_1X_2Y condition will arise, with a metacentric Y (Figure 3).

Alternatively, an X_1X_2Y condition may be derived directly from an XO, by a reciprocal translocation between the X and one member of an autosome pair, the other member of the autosome pair becoming a neo-Y (Figure 4).

The other common derivation of a multiple sex chromosome system is that found in organisms with holocentric chromosomes (i.e., chromosomes with diffuse centromeric activity), such as the Hemiptera and Dermaptera, where X_1, X_2, etc., have almost certainly arisen simply by dissociation (fission) of the single original X chromosome into two or more parts, which still segregate to the same pole at meiosis (Figure 5). Such X dissociations have occurred in both XX/XY and XX/XO systems, a single X dissociation giving $X_1X_1X_2X_2/X_1X_2Y$ and $X_1X_1X_2X_2/X_1X_2O$, respectively. Multiple Ys have also arisen by dissociation in some species. It is characteristic of multiple sex chromosome systems formed by dissociation that the X_1, X_2, etc. are much smaller than the original X, and that there is no accompanying change in the number of autosomes.

Multiple sex chromosomes are cytogenetically interesting, because they show very clearly the fixation of different types of chromosomal rearrangement, but they have little or no

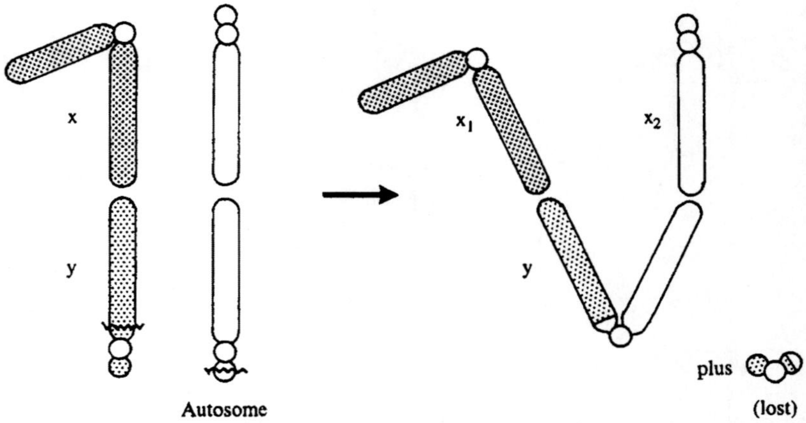

FIGURE 3. Origin of an X_1X_2Y condition in the heterogametic sex (leading at fixation to an $X_1X_1X_2X_2/X_1X_2Y$ sex determination mechanism), by centric fusion between an acrocentric Y and an acrocentric autosome, to give a metacentric "neo-Y" ("neo" because part of it is recently derived from an autosome), and a neo-X_2. The arrangement of sex chromosomes on the spindle of the first meiotic division, with homologous sections associated, is shown diagrammatically. (Adapted from White, M. *Animal Cytology and Evolution,* 3rd ed., Cambridge University Press, Cambridge, U.K., 1973.)

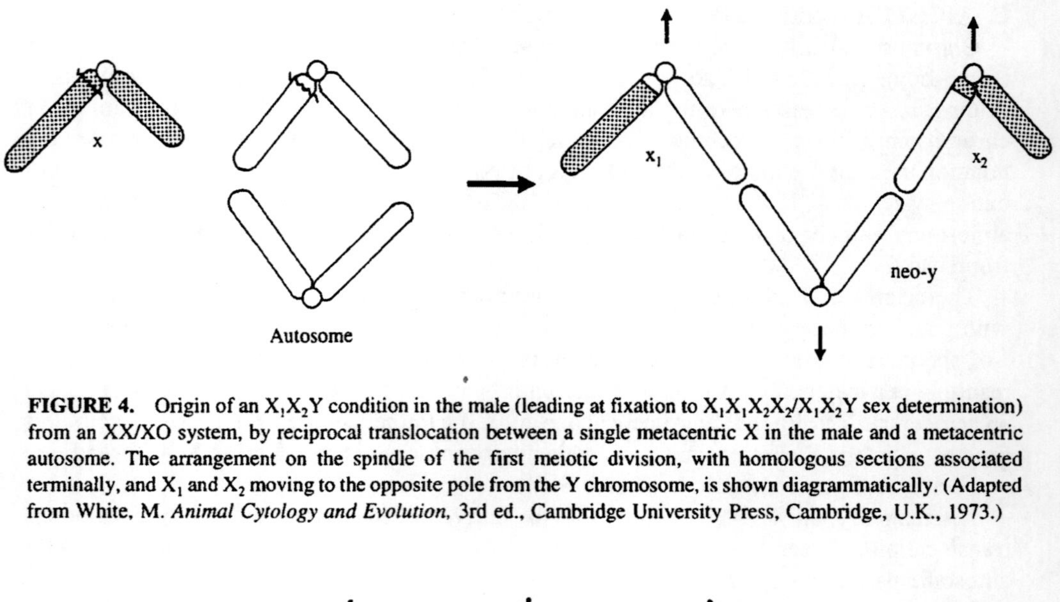

FIGURE 4. Origin of an X_1X_2Y condition in the male (leading at fixation to $X_1X_1X_2X_2/X_1X_2Y$ sex determination) from an XX/XO system, by reciprocal translocation between a single metacentric X in the male and a metacentric autosome. The arrangement on the spindle of the first meiotic division, with homologous sections associated terminally, and X_1 and X_2 moving to the opposite pole from the Y chromosome, is shown diagrammatically. (Adapted from White, M. *Animal Cytology and Evolution,* 3rd ed., Cambridge University Press, Cambridge, U.K., 1973.)

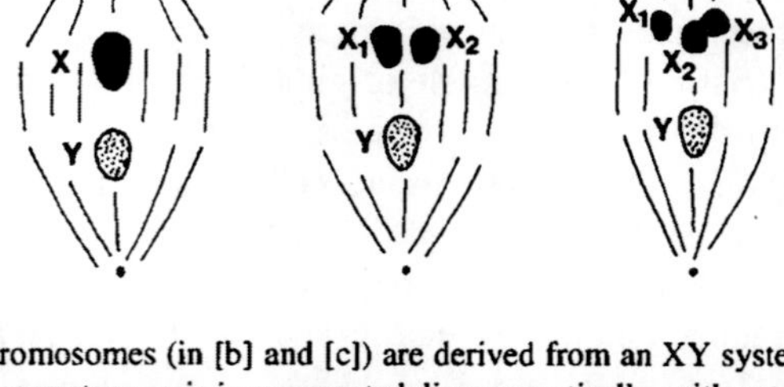

FIGURE 5. Multiple X chromosomes (in [b] and [c]) are derived from an XY system (a) by simple dissociation. Segregation of Xs and Y at spermatogenesis is represented diagrammatically, with autosomes not shown. Usually in such systems the Xs and Y do not associate, but show "distance" (or "touch-and-go") pairing.

significance in relation to the molecular genetic basis of sex determination. Of more interest in this respect are the so-called "multiple factor" systems of certain Diptera.

D. "MULTIPLE FACTOR" SYSTEMS

Species with multiple sex chromosomes are nevertheless still likely to have only a single sex-determining locus — for example, on just one of the Xs in a multiple X system. But more complex sex-determination systems are known where one species may have sex-determining factors at several different loci, and sometimes even different individuals within the same population have their sex factor on different chromosomes. Such complex systems cannot be detected cytogenetically, and they have only been recognized in genetically well-studied organisms with an abundance of genetic markers. In insects, this means certain species of Diptera.

The type of multiple factor system most commonly observed in Diptera and found in members of both the Nematocera and the Cyclorrhapha, can be symbolized in a general way as follows

Females	**Males**
$s_1s_1\ s_2s_2$	$S_1s_1\ s_2s_2$
	$s_1s_1\ S_2s_2$

S_1 and S_2 behave as dominant male-determining factors, segregating, respectively, at the two different loci 1 and 2, and always restricted to the male line. This 2-locus system, with S_1 and S_2, can be generalized to any number of loci:

Females	**Males**
$s_1s_1\ s_2s_2\ s_3s_3....$	$S_1s_1\ s_2s_2\ s_3s_3....$
	$s_1s_1\ S_2s_2\ s_3s_3....$
	$s_1s_1\ s_2s_2\ S_3s_3....$
	⋮

So for any species with a multiple factor system, there is always just one female genotype, homozygous for all sex factor loci, and *n* distinct male genotypes, each heterozygous at just one of the *n* loci. Sometimes a sex factor locus may be on a cytologically distinct sex chromosome, so that S_1, for example, is manifestly a Y chromosome; but sex factors may equally occur on chromosomes that are in all other respects autosomal, in which case there is no obvious cytological difference between the sexes. Dipteran geneticists have termed the autosomal male-determining factors "*M* factors". Green[20] suggested that *M* factors at different loci in any one species were perhaps all actually the same gene, transposed to several different sites in the genome. The evidence now strongly favors this interpretation in several cases (discussed later under Diptera). This explains the exclusive nature of their occurrence in individual males, but makes the term "multiple sex factors" something of a misnomer.

This kind of sex determination may occur in other insects, especially where heteromorphic sex chromosomes have not been detected or do not occur regularly, but the necessary genetic evidence is lacking for other orders apart from Diptera.

E. HAPLODIPLOID SEX DETERMINATION

No general survey of sex determination in insects would be complete without mention of haplodiploidy, of which Hymenoptera are, of course, the leading exponents. Haplodiploidy may also be general to Thysanoptera, and is found in some species of Homoptera and Coleoptera (as well as occurring widely in mites and ticks). The genetic mechanisms involved have been the subject of much speculation (reviewed by Crozier[21,22]). Indeed, at first sight, it is difficult to see how any genetic mechanism at all can be operating, as haploid males simply have half the female dose of all genes. However, several hypotheses based on multiple sex factors have been suggested, and one has some experimental foundation.

It has long been known that, in certain Hymenoptera, inbreeding results in diploid males,[23-25] indicating that not only haploids but diploid homozygotes are male. This can be explained if there are multiple sex factors ($S_1S_2S_3$...) — possibly alternative alleles at a single locus — segregating in opposition:

Females	Males
S_1S_2, S_1S_3, S_2S_3, ...	S_1, S_2, S_3, ... (if eggs unfertilized)
	or S_1S_1, S_2S_2, S_3S_3, ... (in inbred populations)

The sex factors appear to complement one another, and this has therefore been termed a complementary sex-determining mechanism.

While this explanation fits members of several groups of Hymenoptera very well (e.g., *Habrobracon, Apis, Neodiprion,* and *Solenopsis*), it cannot apply generally, because some other Hymenoptera inbreed considerably, yet fail to produce diploid males. To accommodate this problem, the hypothesis can be modified to involve multiple loci.[21] The theory is that diploids would then have to be homozygous at all loci in order to be male, and this would only be likely after long-term intensive inbreeding.

F. THE MOLECULAR BASIS OF SEX DETERMINATION

Thus there is a variety of ways in which sex can be determined in insects. Sex factors can apparently determine either maleness or femaleness, can be dominant or recessive in their action, can be single or multiple, and can occur on sex chromosomes or autosomes. How can all this be explained in molecular terms?

Our knowledge of the molecular biology of sex determination is almost entirely restricted to *D. melanogaster*, but at least in this one species some of the details of the mechanism are now worked out. The key gene is *Sexlethal* (*Sxl*), which is located on the X chromosome. This gene is essential for determination of females, but completely functionless in males, so that it can be eliminated by mutation without affecting the male phenotype. There is a maternal gene, *daughterless* (*da*), that has to be active for *Sxl* to function, because a mutation at the *da* locus causes mothers to produce only sons; however, *da* is not normally involved in sex determination. Activation of the female-determining function of *Sxl* in *Drosophila* is in fact dependent in some way on the ratio between the number of X chromosomes and the number of sets of autosomes (henceforth X:A). Thus, if X:A is 1.0 (as in diploid eggs with two X chromosomes), *Sxl* produces an active product that causes the embryo to develop as female, but if X:A is 0.5 (e.g., diploid eggs with XY or XO) then *Sxl* is silent and the embryo develops as male.

The genes repressing *Sxl* in male eggs have not been identified, but in *Drosophila* they must be located on the autosomes, because a genotype with one X chromosome and one set of autosomes (X + A) is female, whereas the addition of another autosome set (X + AA) results in a male. The molecular nature of the X:A signal is a source of continuing speculation.[26] Chandra[27] proposed that the normal, diploid forms of both sexes produce the same limited number of repressor (R) molecules; two X chromosomes can bind all the R molecules so that *Sxl* can be transcribed, but one X chromosome leaves sufficient R molecules to repress *Sxl*. One currently favored form of this hypothesis[28] has the X:A signal produced by two or more X-linked genes (e.g., *sis-a* and *sis-b* in Figure 6) whose products act as "numerator elements," and are titrated by certain autosomal products ("denominator elements"), so that a sufficient concentration of *sis* products to promote transcription of an active product by *Sxl* will only be achieved in females (Figure 6).[29,30]

Sxl regulates the differentiation of the female tissues, in *Drosophila* acting entirely through its control of a locus on chromosome 3, *transformer* (*tra*), which is also only functional in

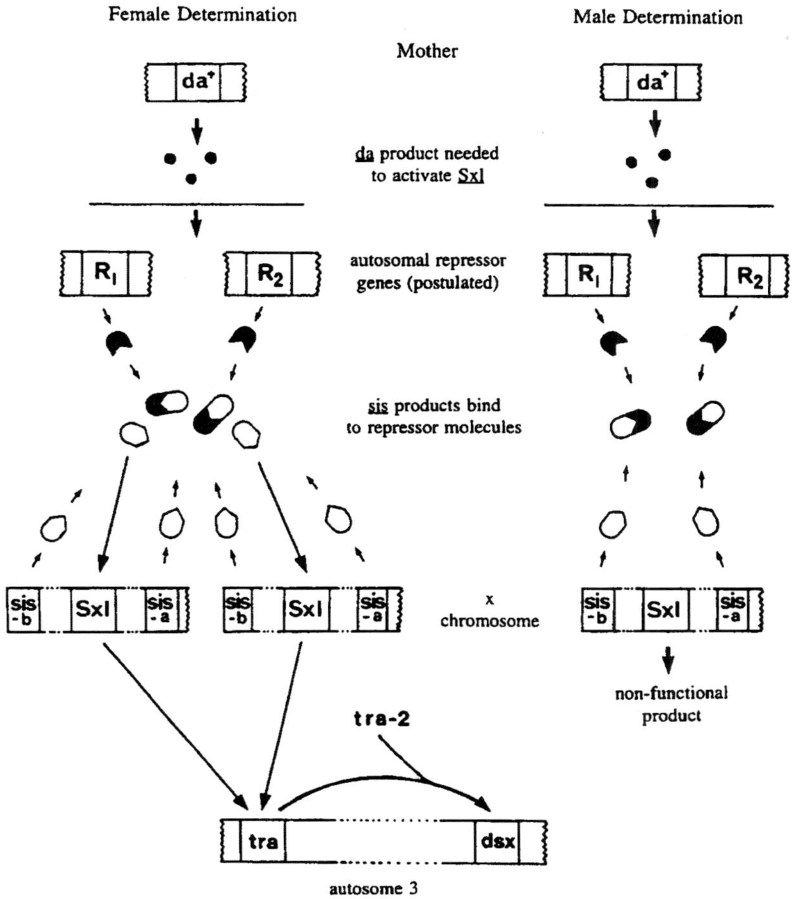

FIGURE 6. Simplified model of a possible mechanism for genetic control of sex determination in *Drosophila melanogaster*. Female determination and differentiation (left) depends on the product of the key gene *Sex lethal* (*Sxl*). The product of the maternal gene *daughterless* (*da*) needs to be present for *Sxl* to be active, but this is normally supplied to both male and female eggs. It is thought that unidentified autosomal genes produce a similar concentration of repressor molecules ("denominator elements": R) in both sexes. Female eggs, with two X chromosomes, produce twice as many "numerator elements" (i.e., products of X-linked loci such as *sis-a* and *sis-b*) as male eggs, so that there is an excess of unbound molecules to promote the female-determining activity of *Sxl*. The active product of *Sxl* influences the transcription of the product of the autosome 3 gene *transformer* (*tra*), which in turn acts on the *doublesex* locus (*dsx*: see text and Figure 7).

females. The *tra* product collaborates with the product of another gene (*tra-2*) to control the expression of another locus on chromosome 3, *doublesex* (*dsx*) (Figure 6). The *dsx* locus is active in both sexes and provides the double switch mechanism necessary to ensure that development proceeds only as either one sex or the other; it consists of two cistrons, dsx^m and dsx^f, only one of which functions in each sex. In female eggs (i.e., when both *tra* and *tra-2* are active), dsx^f is active and its products repress the male sex differentiation genes, whereas in male eggs the products of dsx^m repress female sex differentiation genes.

It is now known that regulation at all the three main stages occurs at the level of RNA splicing;[8] that is to say, the primary gene products are the same in both sexes, but they are "edited" by the splicing out of different sections (introns) of RNA to produce the male- and female-specific messenger RNAs (Figure 7). The male-specific messenger RNAs of both *Sxl* and *tra* include a stop codon which truncates the open reading frame so that the transcript is nonfunctional. This explains why mutational loss of these genes has no effect in males.

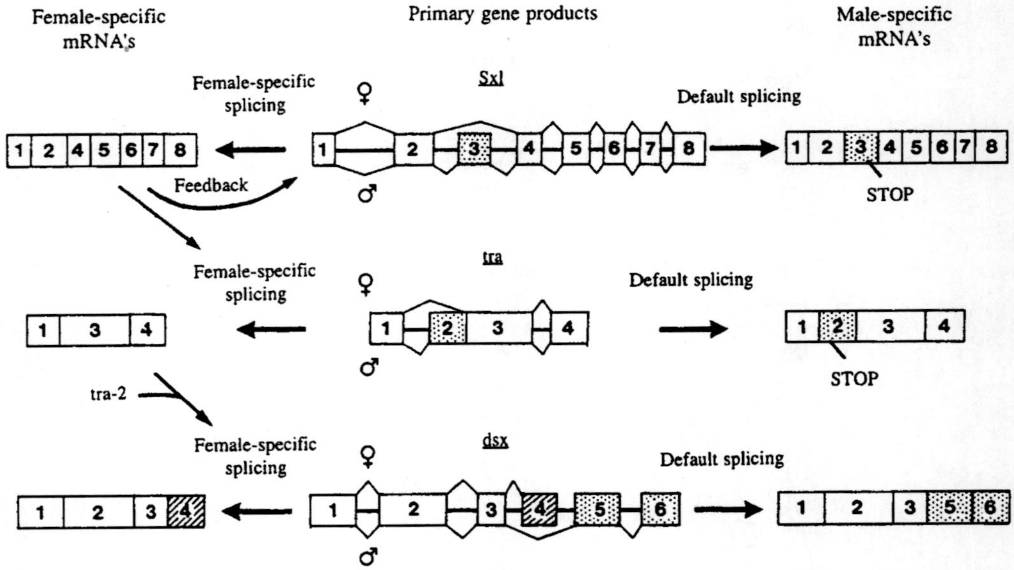

FIGURE 7. Production of sex-specific messenger RNAs (mRNAs) from the primary gene products of the *Sexlethal* (*Sxl*), *transformer* (*tra*), and *doublesex* (*dsx*) genes of *Drosophila melanogaster* by differential splicing. Primary transcripts of these genes are shown in the center; the boxes represent coding regions (exons), the horizontal lines joining them represent introns (which do not form part of an active product), and the female-specific and male-specific patterns of splicing are depicted, respectively, above and below the structures of the primary transcripts. The mRNAs generated by this process are depicted to left (female) and right (male) of the primary transcripts. In males the mRNAs result from the default pattern of splicing, which in the cases of *Sxl* and *tra* includes a stop codon rendering the mRNA nonfunctional. The female-specific product of *Sxl* regulates its own activity by positive feedback, and regulates *tra* activity by promoting the female-specific product of that gene, which in turn plays its part in directing the female-specific pattern of splicing of the *dsx* gene. (Adapted from Baker, B. *Annu. Rev. Genet.*, 17, 345, 1983.)

Thus, sex determination in *Drosophila* depends on a hierarchical system of regulatory genes. Can such a mechanism be generally applicable, given the apparently diverse systems of sex determination found in other organisms? Nöthiger and Steinmann-Zwicky[11] speculated on various mutations in the regulatory system that could, together with changes in the sex chromosomes, explain most of the variations observed in insects. They regarded the action of the double switch gene, *dsx*, as likely to be basic to the sex determination of all insects, and therefore not capable of functional mutation. They postulated that the primitive system in insects was probably represented by species in which the male is heterozygous at a single sex-determining locus. Using the notation introduced earlier in this chapter, in dominant-Y systems the sex factor s_2 acts as the male determiner by repression of the key gene *Sxl*, whereas the sex factor s_1 does not code for functional product, thus allowing *Sxl* to be active. If heteromorphic sex chromosomes are involved, then the male sex factor s_2 (or strictly S_2, since its effect is dominant) would be located on the Y chromosome, but it could be transposed to different locations in the genome, or copied to different locations to form a "multiple factor" (M) system. Recessive-X systems, which occur commonly and widely in insects, are explicable in terms of a genic balance as already discussed for *Drosophila*; the repressor gene is on an autosome, and *Sxl* is only activated if two X chromosomes are present, i.e., in individuals homozygous for s_1.

It is possible to interpret other, less common, types of sex determination in terms of mutations of the key gene *Sxl*, of its repressor, or of the maternal gene *da*. Some of these special cases will be referred to later in this chapter.

G. DOSAGE COMPENSATION

Organisms in which the Y has little or no homology with the X or does not exist at all (XO), have a gene dosage problem. An XX female has two copies of every X-linked gene, while an XY or XO male has only one copy. Mammals compensate for this by inactivation of one of the two X chromosomes in female somatic tissues. However, in *Drosophila*, where polytene chromosomes make it easy to study the level of transcriptional activity, dosage compensation is achieved in a different way. Both X chromosomes are active in female tissues, but the transcription rate is only half that of the single X chromosome in males, which produces just as much RNA as the two Xs in females put together.[7,31] The hyperactivity of X-linked genes in male *Drosophila* appears to be due to a set of genes (*msl*) which are inhibited when the key gene *Sxl* is active and are therefore only functional in males.[8]

The only other information on dosage compensation in insects is in Orthoptera. Rao and Ali[32] showed that both X chromosomes in hepatic cecal cells of female *Acheta domesticus* were euchromatic (i.e., transcriptionally active), and provided some evidence — using an indirect measure of transcriptional activity of unproven reliability — that the single X chromosome in the male may be hyperactive, as in *Drosophila*. On the other hand, females of the mole cricket *Gryllotalpa fossor* (=*africana*?) seem to have only one arm of one X chromosome transcriptionally active in hepatic cecal cells,[33,34] which resembles the system in mammals. However, there is evidence that activity or inactivity of the X chromosomes in Orthoptera may differ among tissues.[35]

In Lepidoptera, the limited evidence available from the differential activity of sex-linked loci suggests that members of this order may manage without a dosage compensation mechanism. Indeed, Johnson and Turner[36] suggested that in mimetic butterflies the dosage differential may be used to advantage, in order to limit expression of a polymorphism to the female sex.

III. SEX DETERMINATION IN DIFFERENT GROUPS OF INSECTS

A. APTERYGOTA

Of the four most primitive extant orders of insects, only the Collembola have been studied sufficiently to warrant generalization, and in these the male is the heterogametic sex and is normally XO,[37,38] but in Neanuridae, species with XO and others with XY are known.[39] Presumably the XY species are neo-XY, but there is no cytological evidence to confirm this. In *Neanura monticola*, with XO males, the X chromosome shows considerable polymorphism with large amounts of heterochromatin (probably repetitive, noncoding DNA) in high altitude populations.[39] In the Thysanura, *Thermobia domestica* possibly has X_1X_2O males.[40] In Protura, on the other hand, no instances of an XX/XO system have been reported; very few of the species examined had morphologically differentiated sex chromosomes.[41,42] No representatives of the Diplura seem to have been examined cytologically.

B. PRIMITIVE EXOPTERYGOTA

XX/XO sex determination predominates in the Odonata, possibly in the Ephemeroptera[43] (although these are poorly studied), and certainly in the main Orthopteroid orders (Dictyoptera-Phasmida-Orthoptera). Where an XX/XY system occurs in these groups, it is usually clear that it is a neo-XY system, formed by fusion of an X with an autosome (Figure 2), so that the neo-Y is homologous with a large part of the neo-X. In the anisopteran families of Odonata, for example (reviewed by Kiauta,[44,45]) most species are XX/XO, but there are apparent neo-XY systems in 15 species scattered through the families Gomphidae, Aeschnidae, Cordaliidae, and Libellulidae, representing about 4% of the dragonflies then studied. Species with neo-XY generally have, as might be expected, one less autosome pair than related species with an XO system; e.g., *Aeshna crenata* has 2n = 28 and XO males, whereas *A. grandis* has 2n = 26 and neo-XY males.[46] Kiauta,[44] followed by Tyagi,[47] explained an evolutionary decrease in the number of autosomes in the family Gomphidae as a succession of fusions and translocations between the neo-Y, the neo-X, and autosomes, the outcome of each step being a secondarily derived XO system with one fewer autosome pairs. However, there is no clear cytogenetic evidence that the sex chromosomes are involved in these changes of karyotype.

C. THE ORTHOPTEROID ORDERS

In the Plecoptera, which are generally thought to be an orthopteroid order that retains primitive features, several species in different genera have XO males, always with a very large metacentric X chromosome that moves in a highly characteristic way in the first meiotic division.[48] XY males (presumably neo-XY) are only recorded for one species (*Perla (=Paragnetina) immarginata*), but some species of *Perla* have a multiple X chromosome system apparently derived from XX/XO, males being X_1X_2O, and in *Perlodes* there are three species known with $X_1X_2X_3O$ males. The two or three Xs in these species are much smaller than the single X of XO and XY males in related species, and associate together in the first meiotic division. Their mode of origin is a mystery, as there is no simple way in which a large metacentric chromosome can give rise to several smaller elements.

The cytogenetics of the other orthopteroid orders has been comprehensively reviewed by Hewitt[49] (Orthoptera) and White[50] (Grylloblattodea, Dictyoptera, Isoptera, Phasmida, Dermaptera, and Embioptera), so information about the sex chromosome systems of these groups will only be summarized here and up-dated. Dictyoptera (Blattodea + Mantodea), Phasmida, Orthoptera, and Embyoptera all seem to be primitively XX/XO, whereas in Isoptera and Dermaptera XX/XY predominates and is possibly the primitive condition.

In Blattodea (cockroaches), males seem to be invariably XO where both sexes have been karyotyped.[51] White[50] suggested that this stability of the sex determination system could be due to the fact that the X is almost always metacentric, and therefore not so readily available for centric fusion with an autosome to generate a neo-XY system. Yet it is difficult to see why this argument does not equally apply to Phasmida, which also have a metacentric X yet frequently develop a neo-XY system.

Isoptera (termites) are generally thought to have arisen from primitive Blattodea but, whereas the most primitive cockroach examined cytologically has XX/XO,[52] the most primitive extant termites seem to be mainly XX/XY.[53] Nevertheless, some species do have XX/XO,[54] and fusions and translocations between sex chromosomes and autosomes are so common in termites[53,55] that XX/XO could still be the primitive condition.[56]

Several species of Kalotermitidae in southern U.S. and the Caribbean form remarkable chains or rings of up to 19 chromosomes in male meiotic nuclei,[57-60] often involving more than half the total chromosome complement. The chains are thought to be due to a series of reciprocal translocations involving both the sex chromosomes and the autosomes of one chromosome set, these changes being restricted entirely to the male line, so that all the chromosomes involved function together as a multiple Y chromosome complex (Figure 8). Females are structurally homozygous and form normal bivalents at meiosis. The genetic consequences of such an arrangement are quite profound; for example, they restrict many alleles to males and increase the genetic similarity of offspring to the same-sex parent and to same-sex siblings. The idea that this unusual system has played a significant part in the development of eusociality in termites[61] is, however, somewhat undermined by the fact that the most extreme rearrangements are found in only a few of the more primitive termites.

Mantodea also seem to have an XX/XO mechanism with a metacentric X chromosome as the primitive condition, but members of the largest subfamily Mantinae consistently have an $X_1X_1X_2X_2{:}X_1X_2Y$ mechanism that has aroused considerable interest among cytogeneticists. White[62] proposed that this was derived from XX/XO by translocation between the X and a metacentric autosome (Figure 4), and all subsequent evidence has been consistent with this

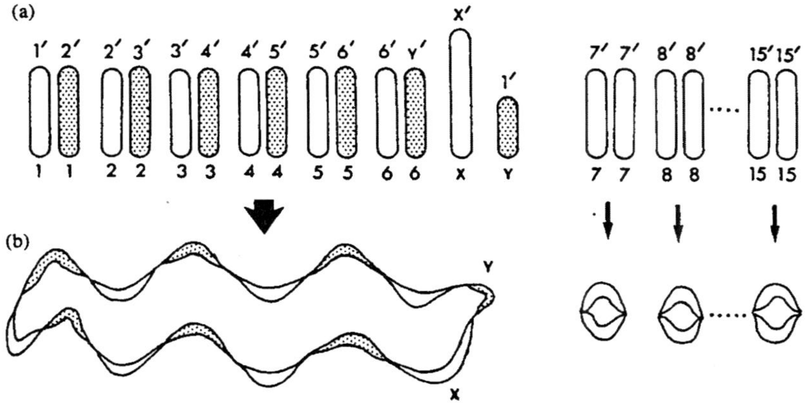

FIGURE 8. Diagram illustrating how a series of reciprocal translocations, involving one member of each of six autosome pairs and the Y chromosome (a), could lead to a ring of linked chromosomes in male meiosis of the termite *Incisitermes schwartzi*. The reciprocal interchange set is stippled, and only occurs in males. Chromosome pairs not involved in translocations (right) form normal bivalents. Sizes of chromosomes are arbitrary; the X and Y chromosomes have not actually been distinguished from the autosomes or from each other in this species. (Adapted from Syren, R. and Luykx, P., *Nature (London)*, 266, 167, 1987.)

hypothesis.[50] All Mantinae have a remarkably consistent chromosome complement, with 2n (male) = 27 (only one exception is known — see below). Presumably the X_1X_2Y system had a single origin in the evolution of this subfamily, and some of the species with X_1X_2Y males currently placed in other subfamilies are perhaps wrongly classified. However, the African mantid genus *Compsothespis* has an X_1X_2Y system with much smaller sex chromosomes, and 2n (male) = 23; at least in this case, an independent origin seems likely (see White[50] for further details). It is not at all apparent why this system has proved so successful for mantids. The mechanism itself does not seem very efficient; complications often seem to arise in correctly forming an X_1X_2Y trivalent in the first division of spermatogenesis, and consequently in correctly segregating the X_1 and X_2 into one daughter spermatocyte and the Y into the other. Callan and Jacobs[63] showed for *Mantis religiosa* that the first meiotic division of those spermatocytes that fail to form the X_1X_2Y trivalent is inhibited, thus preventing the formation of aneuploid sperm. Liebenberg et al.[64] reported a single male of *Polyspilota aeruginosa* (Mantinae) with 2n = 28 instead of the usual 2n = 27, and an $X_1X_2Y_1Y_2$ mechanism. The origin of the extra Y (neo-Y) in this one aberrant case is unclear.

White[50] listed 57 species of Phasmida (stick insects) with identified sex chromosomes, of which 49 are reported to have XO males — undoubtedly the primitive condition — and 7 species (in 6 separate genera) have a neo-XY system arising through fusion of an X with an autosome (see Figure 2). The one other species studied, *Didymuria violescens*, occurs in Southeast Australia, where it has at least 10 chromosomal races, occupying contiguous distribution areas, and including both XO and neo-XY forms.[65] Several independent origins of a neo-XY system can be traced from XO ancestors.[50]

In the well-known parthenogenetic stick insect *Carausius morosus*, males and masculinized females (intersexes or sex mosaics) appear occasionally in laboratory cultures, and their numbers can be enhanced by various treatments, e.g., subjecting the eggs to high (30°C) temperature,[66] centrifuging the eggs,[67] X-irradiating egg or oocytes,[68] or injecting the mother with pterine derivatives.[69] Females and masculinized females have three metacentric chromosomes that are regarded as sex chromosomes because of their behavior in meiosis. Males lack one of these sex chromosomes or a segment of one of them.[70] The method of sex determination is difficult to work out because the female karyotype is highly aberrant due to its long history of parthenogenesis. It has been suggested that *C. morosus* originated as a triploid or tetraploid.[71] However, whatever their origins, both sex chromosome and autosome complements are now aneuploid, and cannot be regarded as comprising any particular number of chromosome sets. Male determination presumably occurs because of a change in the genic balance between factors on the sex chromosomes and on the autosomes (so that, assuming that the molecular model established for *Drosophila* applies, the key female-determining gene *Sxl* is repressed). It is not clear why intersexes, which retain the female karyotype, arise under certain conditions; one possibility is that high temperatures, etc. prevent splicing of female-specific messenger RNAs. General inactivation of the sex chromosomes by heterochromatinization has been suggested[70] to function in sex determination, but such heterochromatinization has only been observed in germ-line (spermatogonial) interphase nuclei, and it is not known whether it occurs in embryonic somatic cells.

Only eight species of Embioptera have been studied cytologically (four in each of the families Oligotomidae and Embiidae), and all have an odd number of chromosomes in male somatic cells, indicating that sex determination is probably XX/XO, with the X chromosomes large and metacentric.[50] Nothing is known about sex determination in Zoraptera.

Hewitt[49] comprehensively reviewed the extensive cytogenetic studies that have been carried out on the Orthoptera proper (Saltatoria). Since Hewitt's review, there have been significant contributions on the sex chromosome systems of neotropical Acridoidea (about 200 species[72]), the acridoid subfamilies Catantopinae,[73] and Pamphaginea,[74] Indian Orthoptera (30 species[75]), and certain Tettigonoidea.[76-78] XX/XO sex determination is found in the great majority of species in all subdivisions of the order, both primitive and advanced, and is undoubtedly the primitive condition for the Orthoptera as a whole. The only exception is the relic group Grylloblattodea, with XY males in the only two species studied,[50] but in the face of all the other evidence, this must be regarded as a derived state. About 8% of species have XX/XY or $X_1X_1X_2X_2/X_1X_2Y$ systems, which occur in every major subdivision of the group and are usually clearly evolved secondarily from an XX/XO condition by centric fusion (Figures 1, 2). Two cases are known, one in Eumastacoidea ("Morabinae species P45b")[79] and the other in Tettigonoidea (*Callicrania seoanei*)[76] of the neo-X being formed by "tandem fusion" of an autosome to the centromeric end of the original X. In both these cases, the neo-Y forms a terminal connection with the neo-X at meiosis, and this neo-XY bivalent divides equationally at first meiotic division, so that the X and Y do not segregate until the second division ("postreductional meiosis"). The mantid type of origin of an X_1X_2Y system, directly from XX/XO by translocation between an X and an autosome, is not known to occur in Saltatoria, perhaps because autosomes in this order are predominantly acrocentric,[9] making centric fusions a more likely occurrence.

The neo-X produced by centric fusion between an acrocentric X and an acrocentric autosome is likely to be large and metacentric, and the neo-Y (the original autosome) is acrocentric (see Figure 1); the majority of cases of neo-XY systems in Saltatoria have sex chromosomes of this form (see Table 8 in Hewitt[49]). Likewise, neo-X_1X_2Y males produced as a result of a Y-autosome fusion have a metacentric X_1 and Y and an acrocentric X_2 (Figure 2); again, the majority of X_1X_2Y systems in Saltatoria conform to this pattern. This may reflect the recent origin of many of these systems because, once a neo-Y is formed, it is subject to very different evolutionary pressures from the original autosome. Several species have been studied that have both XO and neo-XY populations;[73,77,79-82] presumably the neo-XY system is only very recently established in such populations, and in some cases the early stages of differentiation of the neo-Y from its homologue, now part of the neo-X, can be observed. The neo-Y may acquire heterochromatic segments, and pairing between the neo-X and neo-Y may become restricted to terminal regions, so that crossing-over is limited, paving the way for further differentiation of the genetic role of the neo-Y from that of its former homologue.

In time, as discussed earlier, the neo-Y is likely to degenerate; an example of this may be the Gryllacridoid genus *Dolichopoda*, where the "neo"-XY system is possibly as old as the genus itself, and all species studied have a large metacentric X and a small dot-like Y.[83] However, no instance has yet been identified in Orthoptera of the complete loss of a neo-Y, to revert to an XO system, which suggests that the neo-Y may acquire and retain some functional male-linked loci.[49]

The earwigs (Dermaptera) seem to stand somewhat apart from the other orthopteroid orders, and this is reflected in their chromosomes, which have diffuse centromeric activity like those of Hemiptera, and in their sex determination system, as the primitive condition for the group seems to be XX/XY rather than XX/XO. Only two species with XO males are recorded, belonging to different families.[84] Multiple sex chromosomes are very common, occurring in about half the species that have been karyotyped, with similar frequency of incidence in all families. Multiple Xs have probably arisen by simple dissociation of the existing X chromosomes, as in other insects with holocentric chromosomes (Figure 5). They form a close cluster on the spindle at first meiotic division, and all move together to one pole, while the X moves to the opposite pole. The ubiquitous earwig *Forficula auricularia* is unusual in having two alternative Y chromosomes, one of which ("Y_2") is mitotically unstable so that it tends to accumulate in number, and individual males may have up to four copies ($XY_2Y_2Y_2Y_2$). Mosaic males have been recorded with different numbers of Y chromosomes in the cells of each testis.[84]

D. THE HEMIPTEROID ORDERS

The Psocoptera are generally regarded as close to the basal hemipteroid stock, and all the 32 species so far examined cytologically[85] seem to have XX/XO sex determination. Nothing is known about the sex-determining mechanisms of biting and sucking lice (Mallophaga and Siphunculata), as no sex chromosomes have been identified in any of them. When a female human louse (*Pediculus humanus*) is mated with a single male, the sex ratio of the progeny is strongly biased toward one or other sex, and unisexual broods are common.[86] Contrary to White,[9] no information is available about the progeny of females mated more than once, and it seems likely on the available evidence that maternal factors are involved in the determination of sex in lice, as in certain Diptera (e.g., *Chrysomya*).

Most species of Heteroptera have XX/XY sex determination,[87] but there are some groups — e.g., 124 species in the related families Coreidae and Alydidae — that are almost exclusively XX/XO.[88] XO males also predominate in the supposedly more primitive Heteroptera (Gerromorpha; but see Calabrese and Tallerico[89]), and Ueshima[87] concluded that the XY system in Heteroptera, despite its widespread occurrence, is derived from a primitive XO condition. Nokkala and Nokkala,[90] on the other hand, argued that XY was ancestral. Clearly, XY systems are ancient and well-established in terrestrial Heteroptera; the X and Y chromosomes generally show little or no evidence of the homology expected of a neo-XY system and undergo a characteristic pattern of meiotic behavior in which they usually segregate at the second division (for details, see White[9] (pp. 620–621) or Ueshima[87]). The scattered occurrence of XO species within genera must surely be due to secondary loss of the Y chromosome, and such loss may have occurred early in the evolution of many families of terrestrial Heteroptera. However, this does not rule out the possibility that the common ancestor of all Heteroptera was XO, as in Psocoptera, and that XX/XO sex determination may be the primitive condition in some families of Gerromorpha. The problem can only be resolved when the cytology of members of the most primitive groups, Enicocephalomorpha and Dipsocoromorpha, now thought to have a sister-group relationship with all other Heteroptera,[91] as well as of the relic family Peloridiidae (suborder Coleorrhyncha), have been examined. The only information for these groups so far is for one species of Dipsocoromorpha, males of which were tentatively recorded as XO.[92] Multiple sex chromosomes are common in Heteroptera, and may be derived from either XY or XO systems. Apparently they are in most cases due to dissociation of the X chromosome into two or more smaller parts, which group together on the spindle of the second meiotic division and move *en bloc* to one pole (see Figure 5). In some species, the number of X chromosomes varies; the best-known example is the bedbug *Cimex lectularius*, where the number of separate X elements varies from 2 to 15.

Messthaler and Traut[93] showed that the Y chromosome was heterochromatic and therefore transcriptionally inactive in all stages of spermatogenesis of the milkweed bug, *Oncopeltus fasciatus*. Despite the reservations of Thomas,[94] there can be little doubt that the Y chromosome in Heteroptera is genetically inert, and that sex determination is based, as in most insects, on a "recessive-X" (i.e., genic balance) system. Otherwise it would be impossible to explain how the secondary loss of the Y chromosome could occur in so many groups without concomitant loss of genetic viability.

Sex determination in Homoptera-Auchenorrhyncha is predominantly XX/XO.[95,96] A few species with XY males occur within genera and subfamilies that are otherwise exclusively XO, and in such cases it is often clear that the Y is a neo-Y; i.e., the homologue of an autosome that has recently fused with the X. Such a neo-Y pairs with the autosomal part of the neo-X in meiosis and segregates from it at the first division.[97] In several species of *Oncopsis*, both XO and neo-XY males occur in the same or different populations; the XY state results from fusion of the X chromosome with a different autosome in each species.[98]

The Homoptera-Sternorrhyncha include some of the most specialized hemipteroid families, and the basic system of sex determination is often obscured, especially in groups with well-developed parthenogenesis. The Psylloidea are the least reproductively specialized, and here again sex determination is predominantly XX/XO. Only three species with XY males have been found in a total of 39 species examined, all apparently recently derived from XO by X chromosome-autosome fusion.[99-101] The Aleyrodoidea have received very little attention from cytogeneticists. On the basis of early cytological work on three species,[102,103] and the observation that males are only produced in laboratory populations by unmated females, the general presumption is that all male aleyrodids are haploid.[9] It would be preferable to have this confirmed for more species before assuming that haplodiploidy is of general occurrence in this group. The factors invoking male determination are unclear, but the cytological mechanism in those species studied seems to be the same as in Hymenoptera, with meiosis replaced by a single mitotic division, each primary spermatocyte giving rise to only two spermatids. Populations of *Trialeurodes vaporariorum* seem to have an approximately 1:1 sex ratio in field populations in both Europe and North America,[104] which is unusual for a haplodiploid system. However, thelytoky is a complicating factor in interpreting sex ratios in this species. The populations originally introduced from North America to England consisted almost exclusively of thelytokous females,[103] and it is not known what proportion (if any) of females reproduce thelytokously in present-day populations. In *Bemisia tabaci*, which has not been studied cytologically, thelytoky is unknown and the number of males produced seems to be temperature dependent.[105]

It is impossible to do justice here to the remarkable sex determination systems of scale insects (Coccoidea), and for details the reader is referred to the authoritative reviews by Nur.[106,107] All the different systems are believed to have evolved from an ancestral XX/XO system which is still found in some members of the more primitive families (Ortheziidae, Margarodidae, Phenaeolaechiidae). Some of the margarodids (*Icerya* and four closely related genera) have evolved male haploidy, and in some species of *Icerya* there is the further development of hermaphroditism, with morphologically female individuals maturing haploid sperm and diploid ova in an ovotestis. In hermaphrodite *Icerya*, fertilization is usually between eggs and sperm of the same individual; nevertheless, some eggs apparently remain unfertilized and give rise to functional haploid males. In all the more advanced families of Coccoidea, the paternal set of chromosomes is rendered inactive in most tissues by heterochromatinization during the development of male embryos (the "lecanoid" and "Comstockiella" systems; see Nur[106]). It seems that sex in these families is determined maternally rather than by the genotype of the zygote, because the sex ratio is greatly affected by the age of the female at mating and by environmental conditions such as temperature.[107] It is not clear, however, whether the inactivation occurs after, and as a consequence of, the embryo already having been determined as male, as in Sciaridae (see Diptera, below), or whether the inactivation process itself provides the mechanism for male determination.[107]

Bull[10] pointed out that the evolution of these advanced coccoid systems from XX/XO is something of a mystery, because the heterochromatinized paternal chromosomes are eliminated in spermatogenesis, so that all sperm carry only the maternal genome. There is thus no genetic polymorphism among sperm to serve as a basis for sex determination, which effectively means that the advanced coccoid systems can never have coexisted with a system such as XX/XO, and must therefore have evolved through a form of sex determination without male heterogamety. There are a few coccid species (e.g., *Lachnodius eucalypti*) without identifiable sex chromosomes, and which do not undergo heterochromatinization of one chromosome set in the male (*2N-2N* of Nur[106]). Nur thought that these were probably derivatives from forms with heterochromatinization, but Bull's argument makes it more likely that they are representative of this intermediate stage, evolved from XX/XO prior to the origin of heterochromatinization, which is in line with the original views of Brown.[108] Haig has developed a model for the evolution of the advanced coccoid systems based on sex ratio theory.[109]

Aphids (Aphididae) all have XX/XO sex determination. An XX/XY system would be an impossibility for these cyclically parthenogenetic insects, because most species exist through the summer as all-female, thelytokous populations, and during this period the Y chromosome would have "nowhere to go". Aphids produce males parthenogenetically. To develop as XO males, oocytes have to lose half the sex chromatin of the parent female. This is achieved in a single egg maturation division, as in the thelytokous production of females, but the X chromosomes pair during prophase[110] and then undergo a sort of "mini-meiosis" on their own, first separating the products of pairing and then dividing equationally with the autosomes, all on the spindle of the single maturation division.[111,112] This peculiar cytological mechanism for male determination is of special interest because it is normally triggered by environmental conditions and mediated by a low level of juvenile hormone in the haemolymph; males can be induced by treatment with precocene, which destroys the corpus allatum, and inhibited by the juvenile hormone analogue kinoprene.[113] The environmental factors are normally photoperiod (actually the length of the dark phase) and temperature in Aphidinae, but may be nutritional in other groups, and in some species, males appear spontaneously or after a genetically programmed number of thelytokous generations.[114]

The spermatogenesis of aphids is also relevant to their sex determination, because the fertilized eggs must all develop as thelytokous females, so all the sperm from XO males must carry an X chromosome. This is achieved by a peculiar first meiotic division in which the X is stretched on the spindle before passing into one of the daughter spermatocyte nuclei, after which the daughter nucleus without an X degenerates.[115]

Multiple X chromosome systems occur in some aphids, apparently as a result of dissociation of the original X, and the separate elements all behave in the same way in meiosis.[116,117] The greenideine species *Schoutedenia ralumensis* (=*lutea*) has what was presumably originally an $X_1X_1X_2X_2/X_1X_2O$ system, but it has become modified in a remarkable way by consistent association or temporary fusion of one member of an autosome pair with X_1 and the other member of the same pair with X_2.[118] Male determination necessarily retains both these autosomal homologues (AA), so males receive the two elements $X_1 + A$ and $X_2 + A$. One of the X chromosomes (it is not clear whether it is X_1 or X_2) then has to lose its connection with the autosome at anaphase I of spermatogenesis, so that males can transmit one X + A and one X to the next generation (Figure 9). How this peculiar system evolved as a stable mechanism for sex determination is something of a mystery.

Multiple X chromosome systems with X_1X_2 males also occur in the primitive aphidoid families Phylloxeridae and Adelgidae (see Blackman[119] for review), and show some unusual features in the few species studied. In particular, there seem to be species in each group which have evolved a potential for male-linked inheritance "by proxy," which overcomes the total

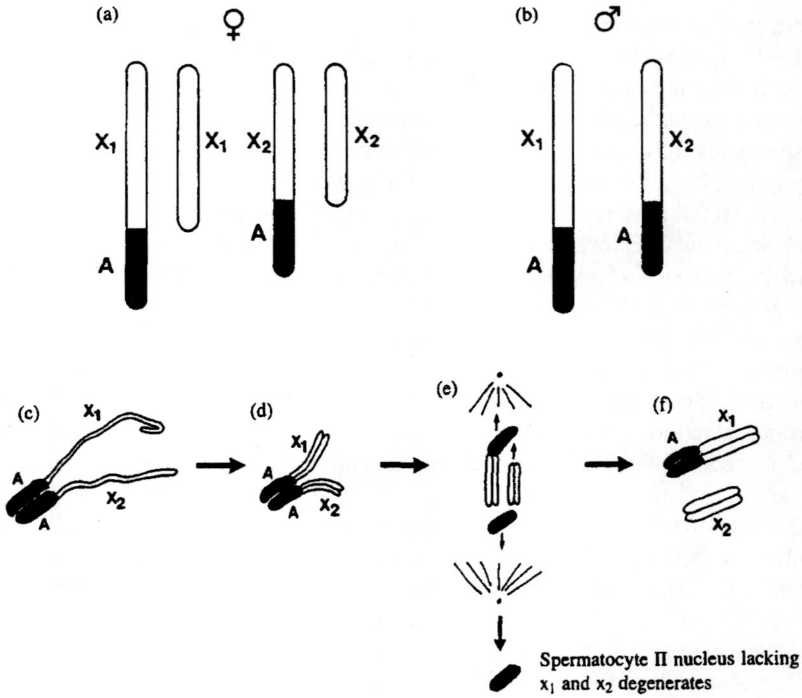

FIGURE 9. Sex chromosome-autosome associations in the aphid *Schoutedenia ralumensis* (=*S. lutea*). For simplicity only, the X chromosomes and the pair of autosomes (AA) associated with them are shown. Female somatic cells (a) have four long chromosomes of unequal length, representing X_1, (X_1 + A), X_2 and (X_2 + A). Male somatic cells and spermatogonia (b) have the longest two chromosomes, which are (X_1 + A) and (X_2 + A). In spermatogenesis, at prophase of the first meiotic division, the autosomes attached to X_1 and X_2, being homologous, pair in parallel (c,d). When the cell divides (anaphase), either X_1 or X_2 loses its connection with the autosome (e; shown here as X_2, but it is uncertain which). The lost autosome passes into one daughter cell which lacks both X chromosomes and degenerates. The other daughter cell has both X chromosomes, with one autosome still attached to one of them (f); it divides equationally to give spermatids with the same chromosome constitution. Presumably, for the system to be stabilized, oogenesis must somehow result in oocytes with the complementary arrangement; i.e., if sperm have (X_1+ A) and X_2, oocytes will have X_1 and (X_2 + A). It is not known how this is achieved. (Based on Hales, D. *Chromosoma* 98, 295, 1989.)

absence of males during the parthenogenetic (thelytokous) part of the life cycle, by having two cytologically distinct types of all-female line; one leading eventually to male production and the other to sexual females. In *Phylloxera caryaecaulis*, studied by the pioneer cytogeneticist T. H. Morgan,[120] one member of the smaller "pair" of X chromosomes seems to be limited to the male-producing line, and behaves differently from the other in its pairing relationships during sex determination and spermatogenesis (Figure 10). In the adelgid *Gilletteella* (=*Adelges*) *cooleyi*, Steffan[121] found one member of the longer pair of X chromosomes dissociated into two parts in about 50% of thelytokous females, and in the somatic cells of males, but not in sexual females. Further work is needed on these groups to confirm and extend these findings.

The last hemipteroid order to be considered is the Thysanoptera (thrips), both suborders of which (Terebrantia, Tubulifera), on the basis of the few species that have been studied cytologically, have haploid males.[122,123] The cytological mechanism involved is not very clear, but seems to differ from that of Aleyrodoidea and Hymenoptera, and must be independently derived. Instead of meiosis being replaced by a single mitotic division, as in other insects with haploid males, two meiotic divisions are retained; the first is apparently equational, giving rise to two similar-sized spermatocytes, but the second produces one large functional spermatid

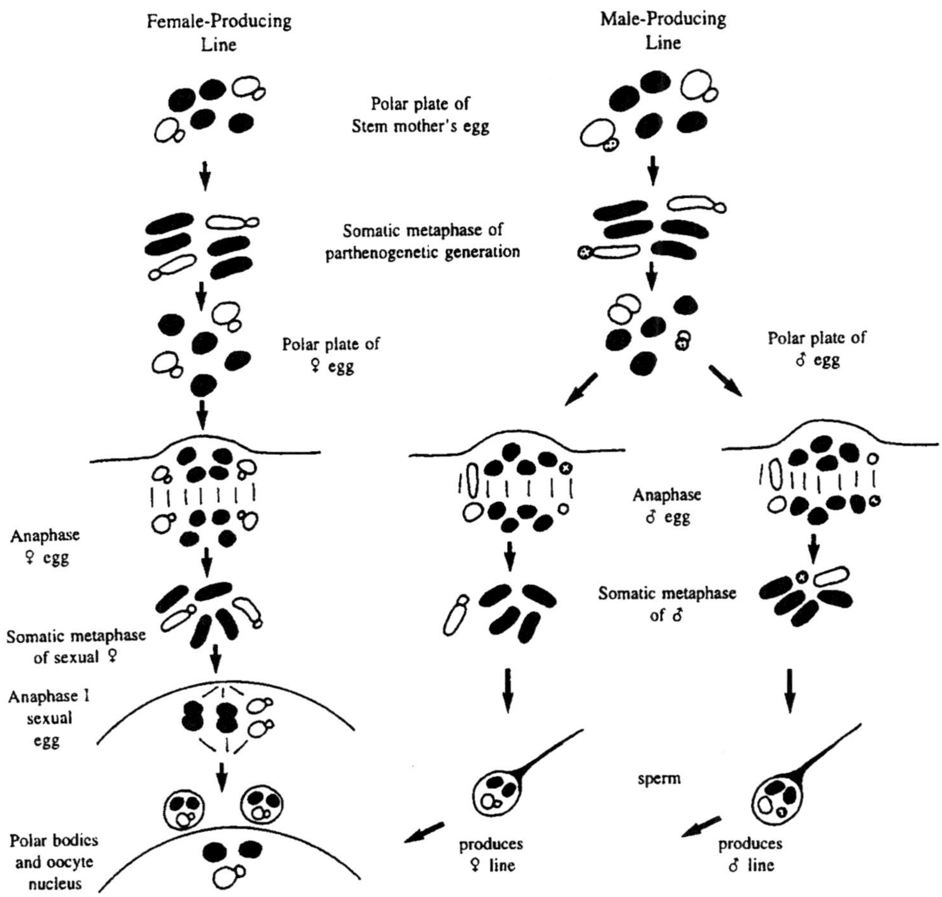

FIGURE 10. Chromosome cycle of *Phylloxera caryaecaulis*, redrawn from Morgan.[120] Autosomes are shown black, X chromosomes white, except for one member of the smaller pair (X_2) in the male line, which is stippled to show its differential behavior and possible role in sex determination. In the line leading to production of sexual females (left), small and large X chromosomes seem to be consistently associated, in the somatic cells of both parthenogenetic and sexual females and throughout oogenesis. In the line leading to male production (right), the small and large X chromosomes are likewise associated throughout the parthenogenetic phase, but during maturation of eggs destined to become male, the Xs exchange partners, so that the two large X chromosomes form one pair, and the small Xs another. Consequently, at maturation division of male eggs, the small and large X chromosomes segregate from each other independently. Males have X_1X_2O; half of them apparently have X_1 and X_2 associated together as in females, and half have them separate. Sperm with separate X_1 and X_2 are believed to give rise to the parthenogenetic line that will produce the males of the next bisexual generation.

and one much smaller one that rapidly degenerates.[122] As in the Aleyrodoidea, the factors invoking male determination are unclear; sex ratios show considerable variation within and between species,[124] but the interpretation of these in genetic terms is complicated by the occurrence of thelytokous parthenogenesis in many of the best-studied species.[125] In *Elaphrothrips tuberculatus*, females have unisexual broods, the males being produced viviparously and the females oviparously; more males seem to be produced when the offspring are larger and fitter in the favorable nutritional conditions of spring.[126]

E. NEUROPTEROIDEA AND COLEOPTERA

The Neuropteroidea (Megaloptera, Raphidioptera, and Plannipennia) are generally regarded as an early branch in the phylogeny of the endopterygote insects, but no species have

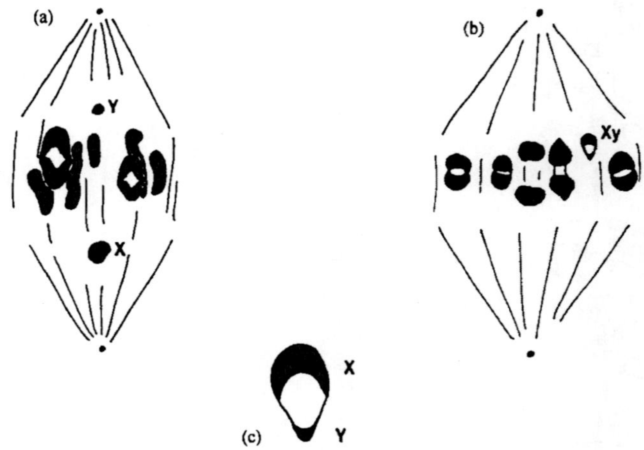

FIGURE 11. Diagrammatic drawings of first meiotic metaphase of male of (a) the neuropteran *Macroneurus appendiculatus*, showing "distance pairing" of X and Y chromosomes, and (b) the megalopteran *Neohermes filicornis*, showing X and Y forming a bivalent like the "parachute bivalent" (Xy_p) of Coleoptera. Structure of the parachute bivalent is shown in (c). (Based on Hughes-Schrader, S. *Chromosoma*, 81, 307, 1980.)

been found with XX/XO. Almost all species studied seem to have XX/XY sex determination, with a few showing multiple X systems. The X and Y chromosomes of Raphidioptera and Plannipennia (=Neuroptera *sensu stricto*) behave in a very consistent fashion during spermatogenesis (Figure 11a). They are both small chromosomes that apparently lack any homology, because they never pair to form a bivalent in the first meiotic division, and regularly take up positions in opposite halves of the spindle before segregating into the daughter spermatocytes.[9,127] In the two species of Megaloptera that have been studied, however, the X and Y chromosomes form a bivalent that positions itself with the autosomes on the equator of the spindle and segregates synchronously with them at the first meiotic division.[128] The Y chromosome is much smaller than the X, and in one species the bivalent looks very like the "parachute" bivalent (Xy_p) found in Coleoptera (Figure 11b,c, and see below).

Thus, the sex chromosome systems of the Neuropteroidea seem to provide useful phylogenetic evidence pointing to a sister-group relationship between Raphidioidea and Plannipennia, and also supporting the often-held view (e.g., Henning[129]) that the Megaloptera are the sister group to the Coleoptera.

The Coleoptera show a great diversity of sex chromosome systems, although the underlying genetics of sex determination may well be far less variable, and is likely to be based on a recessive-X mechanism, except where male haploidy has evolved. Coleopteran cytogeneticists have accumulated information about the sex chromosome systems of over 2500 species. Fortunately, the comprehensive reviews by Smith and Virkki[130] and Virkki[131] mean that only a brief overview and some updating are necessary here.

The peculiar symbols used in the literature on beetle sex chromosomes are somewhat daunting to the nonspecialist, but can be simply explained. They symbolize the appearance and behavior of the sex chromosomes in the first meiotic division of the male beetle. Sex chromosome symbols are written together if there is any sort of pairing between them to form a bivalent (e.g., XY), but separated by a plus sign (e.g, X+Y) in the much rarer cases where they do not pair. The Y chromosome is usually very small in Coleoptera, and this is indicated by writing Xy instead of XY. In most Polyphaga with Xy, the minute Y is attached by both its arms to the larger X, so that it resembles a parachutist suspended below the "canopy" formed by the X (Figure 11c). The formation and structure of the parachute has recently been studied by silver staining;[132] it is believed to have a role in assisting the regular segregation of the X and Y at first meiotic division. When the Xy bivalent takes this form, then a subscript "p" (for parachute) is added: Xy_p. XX/XO systems in Coleoptera are represented by a single X, rather than as XO. Systems with multiple small Y (=y) chromosomes involved in a single parachute are written Xyy_p, $Xyyy_p$, etc.

Two main types of neo-XY system occur in beetles; those with a large Y, probably derived from an XO system by X-autosome fusion (e.g., Figure 2), and those where an autosome has apparently undergone a reciprocal translocation with either the X_p or the y_p of an Xy_p system to give an "$X_p$neoX-neoY_p", or some other complex system in which the original parachute has elements (neo-X, neo-Y) associated with it in the first division of meiosis.[131,133]

The more primitive beetles (Adphaga) differ from the Polyphaga in that Xy_p systems are virtually absent except in a few Dytiscidae (records of Carabidae with Xy_p are apparently questionable[134]). XX/XO is most frequent in Adephaga, occurring in about 53% of species, with 29% having XX/XY (or XX/Xy).[134] XY systems predominate, however, in the carabid genus *Bembidion* (176 out of the 205 species examined[135]). Tiger beetles (Cicindelidae) mostly seem to have multiple X systems, with 2, 3, or 4 X chromosomes.[136]

More than half of over 2000 species of Polyphaga examined cytologically have Xy_p sex chromosome systems, which are well represented in every major family, and are generally thought to be the ancestral condition for the whole suborder. Whether Xy_p is the primitive condition for all beetles is not quite so clear, because of its rarity in Adephaga, although the recent finding of a sex parachute in a megalopteran makes it more likely. Only single species have been examined in each of the two primitive beetle suborders Archostemmata and Myxophaga, and they may both be untypical.[130,137]

Since Virkki's 1984 review,[131] there have been significant studies on the sex chromosome systems of Chrysomelidae,[138,139] Histeridae,[140] Indian Staphylinidae,[141] Indian Curculionidae,[142] 32 other Indian beetle species,[143] Tenebrionidae,[144] Bruchidae,[145] and 50 Russian beetle species.[146]

Apparently the related order Strepsiptera is still cytologically unknown.

F. HYMENOPTERA

All the Hymenoptera except the few species that are thelytokous have haploid males, produced from unfertilized eggs. The origin of haplodiploidy in this group presumably therefore dates back to its inception in the early Mesozoic or late Palaeozoic. There seems little doubt that this form of sex determination has been the key factor enabling the development of eusociality in the higher groups of the order.[147]

Possible genetic mechanisms underlying haplodiploid sex determination have already been outlined. At least two different models are necessary to fit the observed facts, one involving multiple alleles at a single locus, and the other involving multiple loci.[21,148,149] The single-locus mechanism (see p. 64) results in up to 50% of the fertilized eggs in inbred populations developing as diploid males, which generally have low viability and fertility.[21,150] Diploid males have been reported from several species of Tenthredinoidea, Ichneumonoidea, Apoidea, and Formicoidea.[150] Not all the reported instances can be attributed to inbreeding, but the single-locus model seems to be established for one or more species in each of the above-cited subfamilies, suggesting that it is ancestral to the Hymenoptera as a whole; e.g., the sawfly (*Neodiprion nigroscutum*, the braconid *Habrobracon hebetor*, the ichneumonid *Diadromus pulchellus*,[151] the honeybee *Apis mellifera*, and the fire ant *Solenopsis invicta*.[152]

If single-locus sex determination occurs generally in the ichneumonid and braconid parasitoids reared and released as biological control agents, they may suffer from reduced viability if inbred populations are used, because of diploid male production.[150] In several species of Chalcidoidea, however, in which sibmating is common in nature, inbreeding has not led to the male-biased sex ratios that would be expected if diploid males were being produced, and a multiple-locus model seems to be necessary. A single-locus scheme also does not seem to explain sex determination in six species of meliponine bees, which did not produce diploid males when sibmated,[153] although diploid males were later obtained in another meliponine species.[154] Neither single-locus nor multiple-locus models seem applicable to the bethyloid *Goniozus nephantidis*, which typically has within-brood mating and hence marked inbreeding.[155] Generalizations would be unwise in the present state of knowledge, but it seems that single-locus sex determination is likely to occur in Hymenoptera that generally practice outbreeding, or in the higher social groups where the production of diploid males can be controlled; for example, diploid male honeybee larvae are eaten by workers about 72 h after eclosion.[156]

There have been estimates of the number of sex-determining alleles for several species, either by crossing different lines (9 alleles, in *H. hebetor*), or by a statistical calculation based on the incidence of diploid males in natural populations (99–19 in *A. mellifera*, depending on population size;[157] 20 in *Melipona compressipes fasciculata*;[154] 15 in *S. invicta*[152]).

The pteromalid (chalcidoid) wasp *Nasonia vitripennis* has on occasions produced fully fertile diploid males in laboratory cultures, but will not do so in response to intensive inbreeding.[158] It is difficult to explain sex determination in this species, even as a multiple-locus mechanism.[10] *Nasonia* has been studied particularly with regard to the ability of the female wasp to manipulate sex ratios by controlling sperm access to eggs, and in the course of those studies several apparently extrachromosomal factors were discovered that influence sex. One of these is of particular interest because it is transmitted paternally, but then inactivates the paternal chromosome set by heterochromatinization in the fertilized egg, so that genomically haploid, all-male broods are obtained.[159] It thus mimics the normal mode of sex determination of some scale insects and of sciarid flies. The transmitting agent has now been identified as an accessory (B) chromosome, termed the paternal sex ratio or PSR chromosome.[160,161] In effect, the PSR chromosome "jumps" from one haploid set to another at the expense of the chromosomes with which it is associated and is thus an extreme example of "selfish" DNA.

Most gall-forming Cynipoidea have two generations per year, one thelytokous and the other bisexual. For several common species it has been shown that females of the unisexual, thelytokous generations differ in the eggs that they lay, producing either only haploid (male) eggs or only diploid (female) eggs.[162] The females of the bisexual generation are also of two types, one giving rise only to the male producers of the next unisexual generation, the other only to the female producers. Thus each female of the bisexual generation has grandchildren of only one sex. Possible underlying genetic mechanisms were discussed by Crozier.[21]

G. THE PANORPOID ORDERS

It is generally agreed that the remaining insect orders — including Lepidoptera and Diptera — form a monophyletic group, with the Mecoptera close to its main stem. It is therefore of interest that, whereas only a minority of species in the higher panorpoid orders have an XX/XO system, XO males are found in all species of Mecoptera so far examined except one (which has a clearly derived multiple sex chromosome system, with X_1X_2Y males).[9] It seems likely, therefore, that the variety of sex chromosome systems found in the remaining orders were all derived from an XX/XO system.

Very little is known about sex determination in the highly specialized order Siphonaptera (fleas), which is placed somewhat uncertainly in the panorpoid complex. Of the four species examined, two were probably XX/XY, and two apparently had multiple sex chromosomes;[9,163] males of one of these latter were X_1X_2Y, and of the other possibly $X_1Y_1X_2Y_2$.

At some stage in one of the two main branches of panorpoid evolution — that leading to the Trichoptera and Lepidoptera — the XX/XO system was replaced by a system involving female heterogamety (XY/XX or ZW/ZZ). Suomalainen[164] discussed the similarity between trichopteran and lepidopteran sex chromosome systems. Both Trichoptera and Lepidoptera have numerous small holocentric chromosomes, with sex chromosomes that are hardly distinguishable in either mitotic or meiotic cell divisions, so much of the early work on sex determination in *Lymantria dispar*[6] and *Bombyx mori*[165] was done without any cytological information. Robinson's[166] list of chromosome numbers of over 1000 species of Lepidoptera has no information on sex chromosomes. However, Smith's[167] discovery that all or part of the unpaired Y (or W) chromosome in females is heterochromatic in interphase nuclei — and hence that nuclei of female Lepidoptera with XY sex chromosome constitution contain a dense "sex chromatin body" that is not found in male cells — provided a simple method for determining the sex chromosome system. Traut and Mosbacher[168] and Ennis[169] between them found a sex chromatin body in cell nuclei of females of 151 out of a total of 185 species studied. Where females of a species had no sex chromatin body, female somatic or oogonial cells of that species generally had an odd number of chromosomes, one less than in the male, indicating an XO/XX (or ZO/ZZ) sex chromosome system. XO females are more common in some families (e.g., Lasiocampidae, Yponomeutidae, Noctuidae) than others. Genera in several families contain both XY and XO species.[169] A very few species have one or two heterochromatic bodies in male as well as female nuclei, presumably due to presence of constitutive heterochromatin on other chromosomes apart from the Y.

Multiple sex chromosomes (females with XY_1Y_2) have been found in two species of Pyraloidea and two of Tortricoidea.[170] The two Y chromosomes, probably arising by simple dissociation, both associate with the X in the first meiotic division of oocytes to form a trivalent. In female somatic cell nuclei of three of the four species, it was possible to observe two sex chromatin bodies, corresponding to Y_1 and Y_2.

Some species of *Solenobia* (Psychidae) have thelytokous races, and a special mechanism is necessary to ensure that progeny are all heterogametic like their mother. In *S. triquetrella*, which has both diploid and tetraploid thelytokous races, the nucleus that develops as an embryo is derived by fusion of two of the four nuclei resulting from meiosis; as this fusion is always between nonsister nuclei, the sex chromosome heterozygosity is preserved.[171]

Even though very few species have been studied in any detail, it is clear that the genetic mechanisms involved in sex determination in Lepidoptera must be almost as variable as they are in Diptera (see below). In the silkworm *B. mori*, it has long been known that the Y chromosome of the female carries a dominant female-determining gene.[165] Diploid, triploid and tetraploid individuals are female as long as there is at least one Y chromosome, even when the X:Y ratio is 3:1. Intersexes do not occur. Thus any male-determining factors on the X or on autosomes must be very weak in their effect. On the other hand, there is abundant evidence from Goldschmidt's work that sex determination in *L. dispar* depends on a delicate and evolutionarily labile balance between the strengths of a male-determining factor or factors on the X chromosomes and one or more female-determining factors, probably on the Y (reviewed by White;[9] but see also Clarke and Ford[172]).

In several lepidopteran families, including some Lasiocampidae[169] and Saturnidae,[169,173] in the same superfamily as *Bombyx*, the occurrence of XO/XX sex chromosome systems seems to rule out any mechanism based on a dominant Y-borne sex factor and, where closely-related XO and XY species occur, it seems likely that the Y has little or no role in sex determination. In *Ephestia*, at least half the Y chromosome can be lost without any effect on sex determination[174] (although the female-determining factor could still be located on the remaining half). It seems possible, therefore, that the vital and dominant role of the silkworm Y chromosome is somewhat unusual.

The other main branch of the panorpoid complex — that leading to the higher Diptera — also seems to have undergone major changes in the mechanism of sex determination early in its evolution. The most primitive group of Diptera-Nematocera, the Tipuloidea, includes species with XY males, such as *Pales (=Nephrotoma) ferruginea*, in which the X and Y behave in an almost identical fashion to those of Neuroptera-Plannipennia; i.e., they do not form a bivalent in the first division of meiosis, and take up positions in opposite halves of the spindle ("distance pairing"), before segregating to the poles.[175] Even in Tipuloidea, however, there is evidence of a change in the sex determination mechanism which forms the basis for the variety of systems in the Diptera as a whole. In *P. ferruginea*, XXY individuals are male,[176] indicating that sex determination is based on a dominant-Y mechanism, which is unlike all the other insect groups covered so far.

The X and Y in most tipulids are very small, and in *Tipula caesia* and *T. pruinosa* they have "disappeared"; it seems probable that the sex chromosomes in these two species, or at least the Y chromosome, have fused with members of a pair of autosomes, so that one of the three pairs of autosomes now bears the sex-determining locus.[9] A similar change seems to have occurred in the two species of the tipulid subfamily Limnobidae (=Limoniinae) studied by Wolf:[177] *Dicranomiya (=Limonia) trinotata* and *Thaumastoptera calceata*. A species of *Liriope* in the family Ptychopteridae, which is placed phylogenetically somewhere between Tipulidae and Psychodidae, provides support for this idea;[178] it has heteromorphic sex chromosomes (X and Y), but they have "acquired" homologous regions so that they pair to form a bivalent in meiosis, suggesting that they are a neo-X and neo-Y formed by translocation or fusion with a pair of autosomes. In this case, the size of the original Y, or the size differential between the original X and Y, was presumably large enough to ensure that the neo-X and neo-Y are recognizably heteromorphic.

In the Psychodoidea, which appear to be a branch of the dipteran phylogeny arising between the Tipuloidea and Culicoidea, only a few species of sand flies (Psychodidae) have been studied cytologically,[179-181] but these provide a similar picture to the Tipuloidea. Most species have $2n = 6$ or $2n = 8$ without recognizable sex chromosomes, but one (*Phlebotomus perniciosus*) had $2n = 10$ including a small heteromorphic pair of sex chromosomes[179] — presumably the more primitive condition.

In Culicoidea, where many more species have been studied cytogenetically, only the Chaoboridae and the culicid subfamily Anophelinae have heteromorphic sex chromosomes.[182,183] The Chaoboridae and one anopheline species (*Chagasia bathana*) have $2n = 8$, with acrocentric X and Y, whereas other Anophelinae and all other Culicidae studied have $2n = 6$, perhaps as a result of sex chromosome-autosome fusion. X and Y chromosomes in most mosquito species can, however, be distinguished by their different patterns of staining with Giemsa or quinacrine ("G, C, or Q bands"; e.g., Newton et al.[184]), and slight intraspecific or interspecific differences size can often be attributed to different-sized blocks of constitutive heterochromatin (repetitive, noncoding DNA) (e.g., Mezzanotte et al.[185]). *Anopheles* X and Y chromosomes have extensive heterochromatic regions.[183]

The dominant, male-determining locus (M) of *Aedes aegypti* is on one member of the shortest chromosome pair near the centromere,[186] and has been similarly located in several species of *Culex*,[183] but in one strain of *C. tritaeniorhynchus* in Japan, "M" is on one of the longest chromosome pair.[187] Thus, the sex determinant can alter its position in the genome, a phenomenon which comes into its own in the next superfamily, Chironomoidea.

Members of the other family usually included in Culicoidea, the Dixidae, lack heteromorphic sex chromosomes,[188] as in the Chironomoidea.

Sex determination in Chironomoidea — or at least, in Simuliidae and Chironomidae, since the Ceratopogonidae are little studied — is characterized by two features: (1) there are almost always three chromosome pairs, none of which are heteromorphic; and (2) there is usually a dominant male-determining factor that can apparently occur almost anywhere in the genome and often differs in location between closely related species (Figure 12). In the *Eusimulium vernum* complex alone, for example, five out of the six chromosome arms are involved in sex determination in different species and sibling species.[189] In the *E. aureum* species group, which is unusual in having $2n = 4$, either of the two chromosome pairs may serve as the sex chromosomes, and sex factors may occur in any of the four chromosome arms.[190]

Sometimes the location of the sex factor varies within species, in which case sex determination operates as a multiple factor system. For example, the Australian *Chironomus oppositus* species complex includes one form, *whitei*, which is apparently polymorphic for four different sex factor locations, with up to three locations occurring in any one population.[191] Many other examples are now available which support the idea that the sex-determining locus in

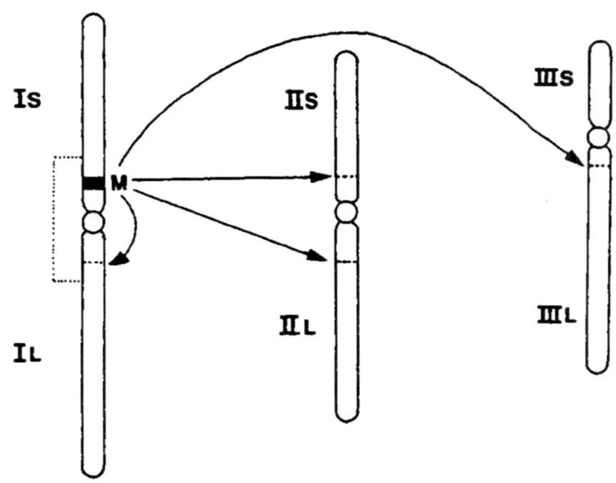

FIGURE 12. Diagrammatic illustration of the mobility of the male-determining factor in blackflies (Simuliidae). The diagram also shows the standard notation used by blackfly cytogeneticists for the six arms of the three chromosome pairs (I, II, and III) of the normal blackfly chromosome complement. Any of these six arms can function as the sex chromosomes, due to transposition of the male-determining factor (M) between chromosomes. The location of M could also be switched from one arm to the other of the same chromosome by a pericentric inversion; for example, an inversion of the section bracketed by a dotted line on chromosome I (although paracentric inversions — those not involving the centromere — are much more common in blackflies).

Chironomoidea — and in many of the higher Diptera discussed below — acts as, or is associated with, a transposable element, and can thus be excised and moved to multiple locations in the genome.[20]

The location of the male sex factor can sometimes be detected cytologically in polytene chromosomes by minor differences in the banding pattern, or by sex-linked inversions. Inversions arise when sections of the chromosome of varying length are excised and then reinserted in the chromosome the "wrong way around". Individuals heterozygous for an inversion can be detected by examining the sequence of bands in the polytene chromosomes, which is inverted in a section of one chromosome in comparison with its homologue. If an inversion is close to or encompasses the sex factor locus, as seems to happen very frequently in blackflies, then it will be partially or completely sex-linked.[192] The frequency of sex-linked inversions also fits the idea of a transposable element being involved in sex determination, because the breakpoints for inversions can also be the sites of excision or insertion of transposable elements.[193]

Rothfels[192] considered that the ancestral condition for Simuliidae and related groups was a complete absence of differentiation of the chromosome carrying the male-determining factor (the notation used for undifferentiated, homomorphic sex chromosomes by simuliid cytogeneticists is X_0Y_0). While this may well be true, it is also possible for the X_0Y_0 condition to be secondary; if, for example, a sex locus associated with an inversion is transposed out of the inversion to a new genomic site, so that the inversion is no longer sex-linked.[194,195]

In *Chironomus tentans,* males are normally heterozygous for a dominant male sex factor, but one population was found that seemed to have female heterogamety; this was interpreted, on the basis of crosses between populations, as due to a dominant female sex factor.[10,196] Nöthiger and Steinmann-Zwicky[11] postulated that this situation might arise by a null mutation of the key gene *Sxl*, accompanied by loss of the dominant male sex factor (M). It has also been suggested, however, that a model involving a weakened male determiner could provide a better explanation of the published results.[197,198]

The remaining groups of Nematocera all have an achiasmate male meiosis, a feature that links them cytogenetically with the higher Diptera. The sex chromosomes of Thaumaleidae and Bibionoidea are usually small, do not form a bivalent, and show "distance pairing" in the first meiotic division, as in tipulids.[176] The Mycetophilidae also have XY males, but the related families Sciaridae and Cecidomyidae have developed remarkably aberrant chromosome systems, with more chromosomes in the germ line than in the soma (reviewed in detail by White[9]). Neither of these families have Y chromosomes, so that the genic balance between X chromosomes and autosomal factors must be the basis for sex determination.

In *Sciara*, male somatic cells are XO, but the germ line is XX and, after passing through a highly peculiar spermatogenesis, the sperm are homogametic and all carry *two* X chromosomes. Oocytes are normal, with a single X, so all zygotes have *three* X chromosomes, with the potential to develop as either sex. Either one or two of the three X chromosomes are eliminated from presumptive somatic cells at the seventh or eighth cleavage division, to determine the soma of the embryo as either female (XX) or male (XO), respectively. (Germ-line cells later lose one X chromosome, irrespective of the sex of the embryo, so that they are XX in both sexes.) The sex of the offspring — i.e., whether one or two X chromosomes are eliminated from somatic cells — depends entirely on the genetic constitution of the mother. Certain species of *Sciara* are monogenous, i.e., they invariably have unisexual progenies, so that there are two kinds of mother, male producing and female producing. The latter are thought to be heterozygous for a dominant factor (F), presumably acting through the cytoplasm of the egg to cause the soma of the embryo to develop as female. The ratio of male- to female-producing mothers in such species, and hence the resulting sex ratio, is approximately 1:1. Thus, sex is inherited genetically, but the inheritance is displaced back to the maternal generation. The genetic mechanism could be a null mutation of the *daughterless* gene or its equivalent, as discussed for *Chrysomya* below. Other species of *Sciara* have females that normally produce progeny of both sexes, however, and sex determination in other genera of Sciaridae has still hardly been studied, so it would be unwise to generalize. Haig[199] reviewed the chromosome system of *Sciara coprophila* and developed a model for its evolution based on sex ratio theory.

Cecidomyidae have even more aberrant chromosome systems, with numerous extra ("E") chromosomes in the germ line that are eliminated from somatic cells in early cleavage divisions.[9] In most species studied, there are six chromosomes in male somatic cells and eight in female somatic cells, so that the sex chromosome system is $X_1X_1X_2X_2/X_1X_2O$. This is the case in the hessian fly, *Mayetiola destructor*,[200,201] despite early reports of eight chromosomes in the somatic cells of both sexes. As in *Sciara*, sex is determined by a maternal factor rather than by male heterogamety. Males although X_1X_2O are homogametic, producing only X_1X_2 sperm, so that zygotes are all $X_1X_1X_2X_2$. In male embryos, two X chromosomes (one X_1 and one X_2) are eliminated from presumptive somatic cells at a separate, later cleavage division than that at which the E chromosomes are eliminated; e.g., in *Wachtliella persicariae*, E chromosomes are eliminated at the fourth cleavage division and X chromosomes at the seventh.[202]

As in Sciaridae, many cecidomyids have unisexual progenies,[200,203] but others have the same mothers producing both male and female progeny, and the system by which sex is controlled is unclear. *Heteropeza pygmaea* is best studied in this respect, but Heteropezinae differ from other cecidomyids in that the male somatic cells appear to be haploid, with five chromosomes, whereas female somatic cells have ten chromosomes. However, this only applies when the progeny are produced pedogenetically; in *H. pygmaea*, females reproducing as adults lay only female-determined eggs, but these have five chromosomes as in pedogenetically produced male eggs.[204] Sex determination therefore cannot be based on haplodiploidy, and does not seem to have a genetic basis at all. Went and Camenzind[205] cultured larval ovaries of *H. pygmaea in vitro*, using as culture medium the hemolymph of larvae that had been previously kept in different nutritional environments, and were able to show that the sex of the progeny was dependent on the nutritional conditions experienced by the mother during development.

The more primitive groups of Brachycera have received very little attention from cytogeneticists. In the Tabanoidea, Rhagionidae and Stratiomyidae have XY males where these have been studied,[206] whereas in Asiloidea, the asilid *Dasyllis* (=*Laphria*) *grossa* is reported to have an XX/XO system.[207] In the more advanced groups of Brachycera (=Cyclorrhapha), there has been detailed work on sex determination mechanisms of representatives of five families: Phoridae (*Megaselia scalaris*), Muscidae (*Musca domestica*), Calliphoridae (*Chrysomya rufifacies, Lucilia cuprina*), Tephritidae (*Ceratitis capitata*), and, of course, Drosophilidae (*D. melanogaster, D. miranda*). These five families span four superfamilies of Cyclorrhapha, so may be fairly representative of the range of mechanisms in the higher Diptera as a whole.

In the phorid fly *M. scalaris*, X and Y chromosomes are not morphologically differentiated, and the male-determining factor (M) is capable of being located on any of the three chromosome pairs,[208] much as in chironomids. In laboratory strains the chromosome (Y) carrying the M factor could be distinguished from its homologue (X) using a combination of cytogenetic and molecular techniques.[209] The segment of the Y chromosome carrying M was found to be conserved in comparison with the homologous region of X, when two unrelated strains were compared. Nevertheless, when the two strains were crossed, four cases were found where the M factor had moved to a different chromosome. The frequency of this change was about 0.06%, which is comparable with known rates of movement of transposable elements in other organisms.[210] The conservation of the M-containing chromosomal region observed in pure strains perhaps indicates that a specific location is favored under certain circumstances, and this could be the first step in the differentiation of new heteromorphic sex chromosomes.

The housefly *M. domestica* provides some particularly interesting examples of evolution of sex-determining systems in progress. It was fully reviewed by Bull,[10] but since then there have been further interesting developments. Earlier European work established that sex determination in houseflies was XX/XY, with heteromorphic sex chromosomes and a presumably dominant male sex factor on the Y. Apart from the Y-borne sex factor, X and Y chromosomes seem to have few or no functional coding regions and are heterochromatic. In strains of non-European origin, however, various sex factors have been found on the autosomes, especially a male determiner (M) near the centromere on autosome 3, and a female determiner (F) on autosome 4 which is epistatic to (i.e., overrides) any number of male-determining factors.[211] In continental Europe, samples from Denmark to Sicily taken in 1975–1981 showed a latitudinal cline: north European populations were all XX/XY, whereas in south and central Italy all populations were XX/XX with sex determined autosomally, the X being totally neutral with regard to sex determination. In southern France, Yugoslavia, and northern Italy, intermediate, mixed populations occurred with all combinations of X and Y in either sex.[212] A very similar north-south cline was found in Japan.[213] The changes in sex determination mechanisms in both southern Europe and Japan are believed to be recent. Various models (e.g., Jayakar[214]) have been advanced to explain this phenomenon; possibly climatic influences are involved, or perhaps the driving force is selective insecticide pressure, as there is now good evidence that pyrethroid/DDT resistance (the "knockdown factor," *Kdr*) is genetically linked with the male-determining locus on autosome 3.[215] However, recent changes have also occurred in the sex-determination system of housefly populations in southeast England, involving an apparent increase in frequency of a male factor on the X chromosome,[216,217] and there is no evidence that loci associated with insecticide resistance (or, in fact, any other functional genes) occur on the housefly X chromosome. Or has the resistance-conferring gene also been transposed to the X from autosome 3, along with the male sex factor? This seems quite possible, since a laboratory housefly strain in Australia was shown to have DDT resistance linked to a male sex factor, but in this case on autosome 2.[218]

Few other Muscidae have been studied cytogenetically. Recent work on *Hydrotaea meriodionalis* indicates a similar story to *M. domestica*, with a dominant autosomal male determiner in some populations and others with XY males.[219] In the closely related Anthomyiidae, the cabbage root fly, *Delia radica* (=*Hylemyia brassicae*), is now known to have a male-determining factor on an autosome (chromosome 6), whereas *D. antiqua* has a small heteromorphic X and Y.[220]

Sex determination in houseflies, with male determiners at various locations in the genome, and the presence of a dominant female determiner in some populations, seems a very different mechanism to the *Drosophila* system, based on an X:A ratio, but there are ways of deriving one from the other fairly simply. Nöthiger and Steinmann-Zwicky[11] suggested, for example, that the dominant female determiner (F) could be a mutation of the key gene *Sxl* to an irrepressible condition, so that it cannot be turned off by M. A similar conclusion was reached by Inoue and Hiroyishi;[221] their model for housefly sex determination incorporates the discovery of a mutation *tra*, closely linked with F on autosome 4; when this is present in the mother, it causes progeny to develop as males even in the absence of any M factors.

In fact, a genic balance/recessive-X system does operate in another muscid, the tsetse fly *G. palpalis*, which also resembles *Drosophila* in that the Y chromosome carries some loci that are necessary for sperm viability, but is not involved in sex determination.[16]

Blowflies (Calliphoridae) generally have small, heteromorphic, and mainly heterochromatic, sex chromosomes. The Y chromosome in *L. cuprina* carries a dominant male sex factor, located near its centromere.[222,223] In *Calliphora erythrocephala*, however, the small heterochromatic pair are no longer sex chromosomes, and the male-determining locus is on one of the other chromosomes, where it is recognizable as a small heterozygosity of the chromomere pattern of the polytene chromosome.[224] And in the monogenic blowfly *C. rufifacies*, sex is controlled by a dominant female determining factor (F') in the mother, $F'f$ mothers producing only daughters and *ff* mothers producing only sons, which are therefore also of *ff* genotype. Ullerich,[225] in some elegant experimental work, transplanted pole cells (primordial germ cells) between female embryos of $F'f$ and *ff* genotypes. The resultant mothers were germ-line mosaics for $F'f$ and *ff*, and both the donor and recipient genotypes were expressed, resulting in a mixture of male and female progeny. Thus, the F' gene product is synthesized by the germ-line cells themselves, rather than by maternal somatic cells. Ullerich also did pole cell transplantations between male and female embryos. These resulted in germ-line mosaics that were completely fertile and heterosexual; the donor cells underwent sex reversal and developed as male or female according to their mother's genotype and irrespective of their own genotype. Thus, a genotypically male germ cell can develop as a functional oocyte in a female host, a genotypically female germ cell can develop as a functional sperm in a male host, and sex is determined solely by regulatory factors provided by maternal somatic cells.

Nöthiger and Steinmann-Zwicky[11] postulated that F' in *Chrysomya* is similar or identical to the *daughterless* gene (*da*) of *Drosophila*, which is necessary in the mother in order for the key gene *Sxl* to be active (Figure 6). If *f* is the null (mutant) allele da^-, then in homozygous condition it will render *Sxl* of embryonic germ cells inactive, so that all progeny will be sons. DNA sequence homology has now been demonstrated between the *da* gene of *Drosophila* and a polytene band on the *Chrysomya* chromosome that carries the F' locus,[226] strongly supporting this hypothesis.

Tephritidae mostly have heteromorphic sex chromosomes (XX/XY), and in several cases a dominant-Y system has been demonstrated, e.g., in the medfly *C. capitata*.[227] X_1X_2Y males occur in some species,[9,228] Indian species in four genera of Trypetinae apparently have homomorphic sex chromosomes,[229] and female heterogamety (XY/XX) has been demonstrated in some Australian species.[230] In *C. capitata*, the sex chromosomes are almost entirely heterochromatic, and the Y chromosome can suffer large deletions without any obvious ill effect; the male-determining factor is located on its long arm close to the centromere.[227] Some repetitive DNA sequences that are specific to or concentrated in the Y chromosome of *C. capitata* were recently isolated.[231]

The sex determination mechanism of *D. melanogaster* was discussed earlier in this chapter, and there are numerous recent reviews.[8,26,232-236] As regards *Drosophila* other than *D. melanogaster*, the most interesting developments have been with *D. miranda*, a species in the *obscura* group which has an X_1X_2Y system, the X_2 and the Y being recently derived (i.e., a neo-X and neo-Y) by translocation to the original Y chromosome of one member of the third autosome pair found in the closest relatives (*D. pseudobscura, D. persimilis*), leaving its homologue as a neo-X. Chromosome 3 of *D. pseudobscura/persimilis* is also homologous to the right arm of chromosome 2 of *D. melanogaster*. Thus, a very comprehensively mapped chromosome segment has quite recently become a neo-Y, providing considerable scope for study of the degenerative changes that follow from its permanent heterozygosity, and the consequent accumulation of nonfunctional alleles. Comparisons of the neo-Y and its recently homologous neo-X have particularly shown that the neo-Y has acquired inserted DNA sequences that are not present in the neo-X, and appear to represent a novel transposable element that may be involved in the degenerative process.[237-239]

IV. EVOLUTION OF SEX CHROMOSOMES AND SEX DETERMINATION IN INSECTS

It has taken quite a lot of words to provide an outline review of the many and various methods by which sex determination is achieved in the different orders of insects. It is clearly important to distinguish between the sex chromosome systems, which display to the cytogeneticist a remarkable diversity in their form, behavior, and extent of evolutionary change within and between groups, and the underlying molecular mechanisms, which may perhaps show less variation. However, for most insect groups, the molecular genetics of the sex-determining process is still merely a matter of speculation or extrapolation from the paradigm of *Drosophila*.

Much of the discussion on sex chromosome evolution has centered on the Y chromosome. The ideas about the progressive evolutionary degeneration of the Y chromosome discussed earlier in this chapter were developed primarily with regard to vertebrate systems, particularly mammals where the Y chromosome bears a dominant male-determining locus.[19,240,241] The various models that have been proposed[241] all assume a primitive condition where the sex chromosomes are undifferentiated, homologous, and in fact essentially autosomal except at the sex-determining locus, but in time become progressively differentiated as the Y acquires noncoding DNA and the X acquires a system of dosage compensation.[14] Bull[10] reviewed the evidence for progressive sex chromosome differentiation in other groups including insects, and concluded that it could be applied more generally. Nöthiger and Steinmann-Zwicky[11] postulated that the various genetic mechanisms for sex determination in higher Diptera arose by mutations from a primitive state with undifferentiated sex chromosomes and a dominant male determiner, as in some mosquitoes.

However, when insects are looked at as a group, certain qualifications to the model of progressive sex chromosome differentiation and Y degeneration are necessary. First, a model that is applicable to the genic balance systems that seem to predominate in insects has not yet been developed.[14] Second, the XX/XO system predominates in the lower insect orders, and must be the ancestral condition for several major groups, if not for the class Insecta as a whole. Thus, for many insects with XX/XY, if not the majority, the XY condition has arisen by a major chromosomal rearrangement, rather than by a progressive, gradual change from a primitively undifferentiated state. The subsequent evolution of the neo-Y and the homologous region of the neo-X may be comparable in many ways to the process of mammalian sex chromosome differentiation, but the presence from the start of a fully differentiated X chromosome, coupled with the lack of sex determiners on the Y, must have consequences that need to be fully addressed. It would be instructive to compare the molecular changes in the recently acquired neo-XY system of *Drosophila miranda* with the changes following a recent acquisition of "autosomal" sex determination due to the transposition of a male determiner (M) to a new location, e.g., in certain mosquito species or, very recently, in certain populations of *M. domestica*.

Third, although there are examples of loss of homology between a neo-Y and a neo-X, and evidence of accumulation of nonfunctional alleles and repetitive DNA on the Y chromosome, there is a lack of information about the circumstances which determine whether a neo-XY becomes an evolutionarily stable XY system, or proceeds inevitably towards complete degeneration and eventual loss of the Y chromosome. In the cytologically well-studied Orthoptera, which show much evolutionary change in the sex chromosomes including numerous examples of *de novo* acquisition of XY systems, there are no clear cases where the Y chromosome has been secondarily lost. It seems that a stable condition may sometimes be reached, where it is advantageous to have sex-linked genes retained on the Y chromosome. In Coleoptera, Xy_p systems with a "degenerate Y" are very ancient and show great evolutionary stability. In Heteroptera, secondary loss of Y chromosomes seems to have occurred many times in the course of evolution at the family level, but not between closely related species, indicating that it does not happen fast or frequently.

This leads to the fifth and final point of qualification, which was discussed by Feraday et al.[195] specifically with respect to the evolution of the sex chromosomes of Simuliidae. There has been a tendency to regard sex chromosome differentiation as an inevitable sequence of events, under the influence of mutation and random drift, rather than as an adaptive process. In Simuliidae, any of the three chromosome pairs can be heterozygous for the male-determining sex factor. Usually the only cytological differentiation between the "X" and the "Y" is in the form of inversions, which may be sex-linked but do not form part of any progressive evolutionary sequence of sex chromosome differentiation.[195] White[9] pointed out that if certain autosomal alleles are polymorphic and exert different selective pressures in the two sexes, then it is advantageous to have them linked to the sex chromosomes. In Orthoptera this may be accomplished by centric fusions between sex chromosomes and autosomes to give neo-XY systems.[242] In those Diptera which have single locus, dominant male sex factors, the linkage may be more easily obtained by transposing the sex locus to another position in the genome. Selective advantage is thus important in establishing a new sex chromosome system, and presumably continues to influence the nature and extent of any subsequent differentiation of the X and Y chromosomes.

REFERENCES

1. **Henking, H.,** Untersuchungen über die ersten Entwichlungsvorgange in die Eiern der Insekten. II. Über spermatogénèse und deren Beziehung zur Entwicklung bei *Pyrrhocoris apterus, Z. Wiss. Zool. Abt. A.*, 51, 685, 1891.
2. **Bridges, C. B.,** The origin of variations in sexual and sex-limited characters, *Am. Nat.*, 56, 51, 1922.
3. **Bridges, C. B.,** Sex in relation to genes and chromosomes, *Am. Nat.*, 59, 127, 1925.
4. **Bridges, C. B.,** Cytological and genetic basis of sex, in *Sex and Internal Secretions*, 2nd ed., Allen, C., Ed., Williams & Wilkins, Baltimore, 1939, 15.
5. **Goldschmidt, R. B.,** *The Mechanism and Physiology of Sex Determination*, Methuen, London, (translation by W. J. Dakin), 1923.
6. **Goldschmidt, R. B.,** Die Sexuellen Zwischenstufen, *Monogr. Gesamtgeb. Pflanzen Tiere*, Vol. 23, Springer, Berlin, 1931.
7. **Baker, B. S. and Belote, J. M.,** Sex determination and dosage compensation in *Drosophila melanogaster, Annu. Rev. Genet.*, 17, 345, 1983.
8. **Baker, B. S.,** Sex in flies: the splice of life, *Nature (London)*, 340, 521, 1989.
9. **White, M. J. D.,** *Animal Cytology and Evolution*, 3rd ed., Cambridge University Press, Cambridge, U.K., 1973.
10. **Bull, J. J.,** *Evolution of Sex-Determining Mechanisms*, Benjamin/Cummings Publishing, Menlo Park, CA, 1983.
11. **Nöthiger, R. and Steinmann-Zwicky, M.,** A single principle for sex determination in insects, *Cold Spring Harbor Symp. Quant. Biol.*, 50, 615, 1985.
12. **Disney, R. H. L. and Cumming, M. S.,** Abolition of Alamirinae and ultimate rejection of Wasmann's theory of hermaphroditism in Termitoxeniinae (Diptera, Phoridae), *Bonn. Zool. Beitr.*, 43, 145, 1992.

13. **Bergerard, J.,** Environmental and physiological control of sex determination and differentiation, *Annu. Rev. Entomol.*, 17, 57, 1972.
14. **Charlesworth, B.,** The evolution of sex chromosomes, *Science*, 251, 1030, 1991.
15. **Ullerich, F.-H.,** Geschlechtschromosomen und Geschlechtsbestimmung bei einigen Calliphoren (Calliphoridae, Diptera), *Chromosoma*, 14, 45, 1963.
16. **Southern, D. I.,** Chromosome diversity in tsetse flies, in *Insect Cytogenetics*, Blackman, R. L., Hewitt, G. M., and Ashburner, M., Eds., Blackwell, Oxford, 1980, 225.
17. **Steinemann, M. and Steinemenn, S.,** Degenerating Y chromosome of *Drosophila miranda*: a trap for retrotransposons, *Proc. Natl. Acad. Sci. U.S.A.*, 89, 7591, 1992.
18. **Rice, W. R.,** Genetic hitchhiking and the evolution of reduced genetic activity of the Y chromosome, *Genetics*, 116, 161, 1987.
19. **Charlesworth, B.,** Model for evolution of Y chromosomes and dosage compensation, *Proc. Natl. Acad. Sci. U.S.A.*, 75, 5618, 1978.
19a. **John, B. and Hewitt, G. M.,** Patterns and pathways of chromosome evolution within the Orthoptera, *Chromosoma*, 25, 40, 1968.
20. **Green, M. M.,** Transposable elements in *Drosophila* and other Diptera, *Annu. Rev. Genet.*, 14, 109, 1980.
21. **Crozier, R. H.,** Hymenoptera, *Animal Cytogenetics* 3, *Insecta* 7, Gebrüder Borntraeger, Berlin/Stuttgart, 1975.
22. **Crozier, R. H.,** Evolutionary genetics of Hymenoptera, *Annu. Rev. Entomol.*, 22, 263, 1977.
23. **Whiting, P. W.,** Multiple alleles in complimentary sex determination of *Habrobracon*, *Genetics*, 28, 365, 1943.
24. **Mackensen, O.,** Viability and sex determination in the honey bee *Apis mellifica*, *Genetics*, 36, 500, 1951.
25. **Smith, S. G. and Wallace, D. R.,** Allelic sex determination in a lower hymenopteran, *Neodiprion nigroscutum* Midd., *Can. J. Genet. Cytol.*, 13, 617, 1971.
26. **Steinmann-Zwicky, M., Amrein, H., and Nöthiger, R.,** Genetic control of sex determination in *Drosophila*, *Adv. Genet.*, 27, 189, 1990.
27. **Chandra, H. S.,** Sex determination: a hypothesis based on noncoding DNA, *Proc. Natl. Acad. Sci. U.S.A.*, 82, 1165, 1985.
28. **Parkhurst, S. M., Bopp, D., and Ish-Horowicz, D.,** X:A ratio, the primary sex-determining signal in *Drosophila*, is transduced by helix-loop-helix proteins, *Cell*, 63, 1179, 1990.
29. **Torres, M. and Sanchez, L.,** The sisterless-b function of the *Drosophila* gene *scute* is restricted to the stage when the X:A ratio determines the activity of *Sex-lethal*, *Development*, 113, 715, 1991.
30. **Keyes, L. N., Cline, T. W., and Schedl, P.,** The primary sex determination signal of *Drosophila* acts at the level of transcription, *Cell*, 68, 933, 1992.
31. **Stewart, B. and Merriam, J.,** Dosage compensation, in *The Genetics and Biology of Drosophila*, Ashburner, M. and Wright, T. R. F., Eds., 2d, Academic Press, London/New York, 1980, 107.
32. **Rao, S. R. V. and Ali, S.,** Insect sex chromosomes 6: a presumptive hyperactivation of the male X chromosome in *Acheta domesticus*, *Chromosoma*, 86, 325, 1982.
33. **Rao, S. R. V. and Arora, P.,** Insect sex chromosomes. III: differential susceptibility of homologous X chromosomes of *Gryllotalpa fossor* to ^{3}H-Urd-induced aberrations, *Chromosoma*, 74, 241, 1979.
34. **Sarkar, S. and Rao, S. R. V.,** Insect sex chromosomes XI. ^{3}H-TdR induces random aberrations in the X chromosomes of *Gryllotalpa fossor* (Orthoptera), *Mutat. Res.*, 282, 113, 1992.
35. **Nagl, W.,** The sex chromatin of *Tachycines asynamorus* (Orthoptera) and its implications, *Cytologia*, 38, 107, 1973.
36. **Johnson, M. S. and Turner, J. R. G.,** Absence of dosage compensation for a sex-linked enzyme in butterflies (*Heliconius*), *Heredity*, 43, 71, 1979.
37. **Nuñez, O.,** Cytology of Collembola, *Nature (London)*, 194, 946, 1962.
38. **Kiauta, B.,** Review of the germ cell chromosome cytology of Collembola, with a list of chromosome numbers and data on two species new to cytology, *Genen Phaenen*, 13, 89, 1970.
39. **Cassagnau, P.,** Les chromosomes polytènes de *Neanura monticola* Cassagnau (Collembola). I. Polymorphisme écologique du chromosome X, *Chromosoma*, 46, 343, 1974.
40. **Charlton, H. H.,** The spermatogenesis of *Lepisma domestica*, *J. Morphol.*, 35, 381, 1921.
41. **Fratello, B.,** Citotassonomia dei Proturi (Insecta, Apterygota), *Atti Congr. Naz. Ital. Entomol. Siena*, 9, 267, 1972.
42. **Fratello, B. and Sabatini, M. A.,** Chromosome studies in Protura Eosentomoidea, in *Third Int. Sem. Apterygota*, Dallai, R., Ed., University of Siena, Italy, 1989, 167.
43. **Mol, A. W. M.,** Notes on the chromosomes of some western European Ephemeroptera, *Chromosome Information Service*, 24, 10, 1978.
44. **Kiauta, B.,** Sex chromosomes and sex-determining mechanisms in Odonata, with a review of the cytological conditions in the family Gomphidae, and references to the karyotype evolution in the order, *Genetica*, 40, 127, 1969.
45. **Kiauta, B.,** Synopsis of the main cytotaxonomic data in the order Odonata, *Odonatologica (Utrecht)*, 1, 73, 1972.

46. **Oksala, T.**, Zytologische Studien an Odonata. I. Chromosomenverhältnisse bei der Gattung *Aeschna* mit besonderer Berücksichtigung der postreduktionellen Teilung der Bivalente, *Ann. Acad. Sci. Fenn. A IV*, 4, 1, 1943.
47. **Tyagi, B. K.**, Cytotaxonomy of the genus *Onychogomphus* Selys (Odonata: Anisoptera, Gomphidae), with special reference to the evolution of the sex-determining mechanism and the reduced chromosome number in the family Gomphidae, in *Proc. 1st Indian Symp. Odonatology*, Mathavan, S., Ed., Madurai Kamaraj University, Madurai, India, 1985, 217.
48. **Matthey, R. and Aubert, J.**, Les chromosomes des Plécoptères, *Bull. Biol. Fr. Belg.*, 81, 202, 1947.
49. **Hewitt, G. M.**, Orthoptera, *Animal Cytogenetics* 3, *Insecta* 1, Gebrüder Borntraeger, Berlin/Stuttgart, 1979.
50. **White, M. J. D.**, Blattodea, Mantodea, Isoptera, Grylloblattodea, Phasmatodea, Dermaptera, Embioptera, *Animal Cytogenetics* 3, *Insecta* 2, Gebrüder Borntraeger, Berlin/Stuttgart, 1976.
51. **Cohen, S. and Roth, L. M.**, Chromosome numbers of the Blattaria, *Ann. Entomol. Soc. Am.*, 63, 1520, 1970.
52. **Luykx, P.**, XO:XX sex chromosomes and Robertsonian variation in the autosomes of the wood-roach *Cryptocercus punctulatus* (Dictyoptera: Blattaria: Cryptocercidae), *Ann. Entomol. Soc. Am.*, 76, 518, 1983.
53. **Luykx, P.**, A cytogenetic survey of 25 species of lower termintes from Australia, *Genome*, 33, 80, 1990.
54. **Tyagi, B. K.**, Review of the cytotaxonomy of Isoptera (Insecta) with a description of the male germ cell chromosomes of *Microcerotermes beesoni* Snyder from the Dehradun Vallet, India, *Indian Rev. Life Sci.*, 7, 263, 1987.
55. **Vincke, P. P. and Tilquin, J. B.**, A sex-linked ring quadrivalent in Termitidae (Isoptera), *Chromosoma*, 67, 151, 1978.
56. **Fontana, F.**, Cytological analysis of the chromosome complement of *Kalotermes flavicollis* Fabr. (Isoptera, Kalotermitidae). The Sex determining mechanism, *Cytologia*, 47, 147, 1982.
57. **Syren, R. M. and Luykx, P.**, Permanent segmental interchange complex in the termite *Incisitermes schwarzi*, *Nature (London)*, 266, 167, 1987.
58. **Syren, R. M. and Lyukx, P.**, Geographic variation of sex-linked translocation heterozygosity in the termite *Kalotermes approximatus* Snyd. (Insecta: Isoptera), *Chromosoma*, 82, 65, 1981.
59. **Syren, R. M. and Luykx, P.**, Experimental hybridization between chromosomal races in *Kalotermes approximatus*, a termite with extensive sex-linked translocation heterozygosity, *Chromosoma*, 83, 563, 1981.
60. **Luykx, P. and Syren, R. M.**, Multiple sex-linked reciprocal translocations in a termite from Jamaica, *Experientia*, 37, 819, 1981.
61. **Luykx, P. and Syren, R. M.**, The cytogenetics of *Incisitermes schwarzi* and other Florida termites, *Sociobiology*, 4, 191, 1979.
62. **White, M. J. D.**, The evolution of sex chromosomes I. The XO and X_1X_2Y mechanisms in praying mantids, *J. Genet.*, 42, 143, 1941.
63. **Callan, H. G. and Jacobs, P. A.**, The meiotic process in *Mantis religiosa* males, *J. Genet.*, 55, 200, 1957.
64. **Liebenberg, H., Fossey, A., and Jacobs, D. H.**, An unexpected sex chromosome mechanism in a South African mantid *Polyspilota aeriginosa* Goez, *Caryologia*, 44, 195, 1991.
65. **Craddock, E.**, Chromosomal evolution and speciation in *Didymuria*, in *Genetic Mechanisms of Speciation in Insects*, White, M. J. D., Ed., Australia and New Zealand Book Co., Sydney, 1974, 24.
66. **Bergérard, J.**, Intersexualité experimentale chez *Carausius morosus* Br. (Phasmidae), *Bull. Biol. Fr. Belg.*, 95, 273, 1961.
67. **Pijnacker, L. P.**, Effect of centrifugation of the eggs on the sex of *Carausius morosus* Br., *Nature (London)*, 210, 1184, 1966.
68. **Pijnacker, L. P.**, Effects of X-rays on different meiotic stages of oocytes in the parthenogenetic stick insect *Carausius morosus* Br., *Mutat. Res.*, 13, 251, 1971.
69. **l'Helias, C. and Boulanger-Sandrin, J.**, Influence des ptérines réduites et de l'hormone juvenile sur l'intersexualité du Phasme *Carausius morosus*, *Ann. Endocrinol.*, 37, 189, 1976.
70. **Pijnacker, L. P. and Ferwerda, M. A.**, Sex chromosomes and origin of males and sex mosaics of the parthenogenetic stick insect *Carausius morosus*, *Chromosoma*, 79, 105, 1980.
71. **Pijnacker, L. P. and Harbott, J.**, Structural heterozygosity and aneuploidy in the parthenogenetic stick insect *Carausius morosus* Br. (Phasmatodea: Phasmatidae), *Chromosoma*, 76, 165, 1980.
72. **Mesa, A., Ferreira, A., and Carbonell, C. S.**, Cariologie de los acridoides neotropicales, estado actual de su conocimiento y nuevas contribucones, *Ann. Soc. Entomol. Fr.*, 18, 507, 1982.
73. **Yadav, J. S. and Yadav, A. S.**, X-autosome fusion in catantopine grasshoppers (Acridoidea: Orthoptera), *Cytobios*, 61, 21, 1990.
74. **Bugrov, A. G.**, [Neo-XY sex chromosome determination in grasshoppers *Asiotmethis heptaopamicus* (Zub.) and *Atrichotmethis semenovi* (Zub.) (Orthoptera: Pamphagidae)], *Tsitologiya*, 28, 117, 1986.
75. **Yadav, J. S. and Yadav, A. S.**, Chromosome number and sex-determining mechanisms in 30 species of Indian Orthoptera, *Folia. Biol. (Krakow)*, 34, 277, 1986.
76. **Fernandez-Piqueras, J., Rojo Garcia, E., and Sentis Castano, C.**, A tandem fusion origin of a neo XY sex determining mechanism in the long-horned *Callicrania seoanei* (Bol.), *Heredity*, 47, 397, 1981.
77. **Fernandez-Piqueras, J., Rodriguez Campos, A., Sentis Castano, C., and Rojo Garcia, E.**, Sex chromosome evolution in the polytypic species *Pycnogaster cucullata* (Charp.), *Heredity*, 50, 217, 1983.

78. **Sentis Castano, C. and Fernandez-Piqueras, J.**, Nature and distribution of heterochromatinized regions in the chromosomal races of *Pycnogaster cucullata* (Insecta, Orthoptera), *Genetica*, 72, 127, 1987.
79. **White, M. J. D., Blackith, R. E., Blackith, R. M., and Cheney, J.**, Cytogenetics of the *viatica* group of morabine grasshoppers: I. The "coastal" species, *Austr. J. Zool.*, 15, 263, 1967.
80. **Hewitt, G. M. and John, B.**, Inter-population sex chromosome polymorphism in the grasshopper *Podisma pedestris*. II. Population parameters, *Chromosoma*, 37, 23, 1972.
81. **Bella, J. L., Westerman, M., Lopez Fernandez, C., Delatorre, J., Rubio, J. M., and Gosalvez, J.**, Sex chromosome and autosome divergence in podisma (Orthoptera) in Western Europe, *Genet. Sel.*, 23, 5, 1991.
82. **Sands, V. E.**, The neo-XY system of *Catantops humilis* (Acrididae: Catantopinae) in Malaysia, *Biol. J. Linn. Soc.*, 39, 269, 1990.
83. **Saltet, P.**, Les Dolichopodes de Corse (Orthoptera — Raphidophoridae) 1° étude cytologique preliminaire, *Bull. Soc. Hist. Nat. Toulouse*, 104, 165, 1967.
84. **Webb, G. C. and White, M. J. D.**, A new interpretation of the sex-determining mechanism of the European earwig, *Forficula auricularia*, *Experientia*, 26, 1387, 1970.
85. **Meinander, M., Halkka, O., and Söderland, V.**, Chromosomal evolution in the Psocoptera, *Not. Entomol.*, 54, 81, 1974.
86. **Buxton, P. A.**, *The Louse*, Edward Arnold, London, 1939.
87. **Ueshima, N.**, Hemiptera. II: Heteroptera, *Animal Cytogenetics* 3, *Insecta* 6, Gebrüder Borntraeger, Berlin/Stuttgart, 1979.
88. **Muramoto, N.**, A chromosomal study of thirty Japanese heteropterans (Heteroptera), *Genetica*, 49, 37, 1978.
89. **Calabrese, D. M. and Tallerico, P.**, Cytogenetic study in males of Nearctic genera of Gerridae (Hemiptera: Heteroptera), *Proc. Entomol. Soc. Wash.*, 86, 354, 1984.
90. **Nokkala, S. and Nokkala, C.**, The occurrence of the XO sex chromosome system in *Dictyonota tricornis* (Schr.) Tingidae Hemiptera and its significance for concepts of sex chromosome system evolution in Heteroptera, *Hereditas*, 100, 299, 1984.
91. **Schuh, R. T.**, The influence of cladistics on heteropteran classification, *Annu. Rev. Entomol.*, 31, 67, 1986.
92. **Cobben, R. H.**, *Evolutionary Trends in Heteroptera. I. Eggs, Architecture of the Shell, Gross Embroylogy and Eclosion*, Center for Agricultural Publication and Documentation, Wageningen, 1968, 374.
93. **Messthaler, H. and Traut, W.**, Phases of sex chromosome inactivation in *Oncopeltus fasciatus* and *Pyrrhocoris apterus* (Insecta, Heteroptera), *Caryologia*, 28, 501, 1975.
94. **Thomas, D. B., Jr.**, Chromosome evolution in the Heteroptera (Hemiptera): Agmatoploidy versus aneuploidy, *Ann. Entomol. Soc. Am.*, 80, 720, 1987.
95. **Emeljanov, A. F. and Kirillova, V. I.**, Trends and modes of karyotype evolution in the Cicadina (Homoptera). I. Karyotypic peculiarities and evolutionary changes in the karyotypes of cicadas of suprafamily Cicadelloidea, *Entomol. Rev. (USSR)*, 69, 62, 1990.
96. **Emeljanov, A. F. and Kirillova, V. I.**, [Trends and types of karyotype evolution in Cicadina (Homoptera). II. Peculiarities and evolutionary changes of the karyotypes in the superfamilies Cercopoidea, Cicadoidea, Fulgoroidea and in the Cicadina as a whole, *Entomol. Rev. (USSR)*, 71, 59, 1992.
97. **den Hollander, J.**, The chromosomes of *Niloparvata lugens* Stal. amd some other Auchenorrhyncha, *Cytologia*, 47, 227, 1982.
98. **John, B. and Claridge, M. F.**, Chromosome variation in British populations of *Oncopsis* (Hemiptera: Cicadellidae), *Chromosoma*, 46, 77, 1974.
99. **Maryańska-Nadachowska, A., Kuznetsova, V. G., and Warchalowska-Sliwa, E.**, Karyotypes of Psyllina (Homoptera). I. New data and check list, *Folia Biol. (Kraków)*, 40, 15, 1992.
100. **Maryańska-Nadachowska, A. and Hodkinson, I. D.**, Karyotypes of Psylloidea. II. Chromosome numbers of nine Mediterranean species from Mallorca (Spain), *Folia Biol. (Kraków)*, 41, 1, 1993.
101. **Maryańska-Nadachowska, A., Glowacka, E., and Warchalowska-Sliwa, E.**, Karyotypes of Psylloidea (Homoptera). III. Chromosome numbers of eight species belonging to the families Aphalaridae, Psyllidae, Homotomidae and Triozidae, *Folia Biol. (Kraków)*, 41, 7, 1993.
102. **Schrader, F.**, Sex determination in the white-fly (*Trialeurodes vaporariorum*), *J. Morphol.*, 34, 267, 1920.
103. **Thomsen, M.**, Studien über die Parthenogenese bei einigen Cocciden und Aleurodiden, *Z. Zellforsch. Mikrosk. Anat.*, 5, 1, 1927.
104. **van Lenteren, J. C. and Noldus, P. J. J.**, Whitefly-plant relationships: behavioural and ecological aspects, in *Whiteflies: Their Bionomics, Pest Status and Management*, Gerling, D., Ed., Intercept, Andover, U.K., 1990, 47.
105. **Sharaf, N. and Batta, Y.**, Effect of some factors on the relationship between the whitefly *Bemisia tabaci* Genn. (Homopt., Aleyrodidae) and the parasitoid *Eretmocerus mundus* Mercet (Hymenopt., Aphelinidae), *Z. Angew. Entomol.*, 99, 267, 1985.
106. **Nur, U.**, Unusual chromosome systems in scale insects, in *Insect Cytogenetics*, Blackman, R. L., Hewitt, G. M., and Ashburner, M., Eds., Blackwell, Oxford, 1980, 97.
107. **Nur, U.**, Chromosomes, sex-ratios and sex determination, in *Armored Scale Insects, Their Biology, Natural Enemies and Control*, Rosen, D., Ed., Elsevier, Amsterdam, 1990, 179.

108. **Brown, S. W.,** Adaptive status and genetic regulation in major evolutionary changes of coccid chromosome systems, *Nucleus,* 20, 145, 1977.
109. **Haig, D.,** The evolution of unusual chromosomal systems in coccoids: extraordinary sex ratios revisited, *J. Evol. Biol.,* 6, 69, 1993.
110. **Blackman, R. L. and Hales, D. F.,** Behaviour of the X chromosomes during growth and maturation of parthenogenetic eggs of *Amphorophora tuberculata* (Homoptera, Aphididae), in relation to sex determination, *Chromosoma,* 94, 59, 1986.
111. **Orlando, E.,** Sex determination in *Megoura viciae* Buckton (Homoptera, Aphididae), *Monit. Zool. Ital. (N.S.),* 8, 61, 1974.
112. **Blackman, R. L.,** Chromosomes and parthenogenesis in aphids, in *Insect Cytogenetics,* Blackman, R. L., Hewitt, G. M., and Ashburner, M., Eds., Blackwell, Oxford, 1980, 133.
113. **Hales, D. F. and Mittler, T. E.,** Chromosomal sex determination in aphids controlled by juvenile hormone, *Genome,* 29, 107, 1987.
114. **Kawada, K.,** Polymorphism and morph determination, in *Aphids, Their Biology, Natural Enemies and Control,* 2A, Minks, A. K. and Harrewijn, P., Eds., Elsevier, Amsterdam, 1987, 255.
115. **Blackman, R. L.,** Spermatogenesis in the aphid *Amphorophora tuberculata* (Homoptera, Aphididae), *Chromosoma,* 92, 357, 1985.
116. **Blackman, R. L.,** Stability of a multiple X chromosome system and associated B chromosomes in birch aphids (*Euceraphis* spp., Homoptera: Aphididae), *Chromosoma,* 96, 318, 1988.
117. **Blackman, R. L.,** The chromosomes of Lachnidae, *Acta Phytopathol. Entomol. Hung.,* 25, 273, 1990.
118. **Hales, D. F.,** The chromosomes of *Schoutedenia lutea* (Homoptera, Aphidoidea, Greenideinae), with an account of meiosis in the male, *Chromosoma,* 98, 295, 1989.
119. **Blackman, R. L.,** Aphid cytology and genetics, in *Evolution and Biosystematics of Aphids,* Szelegiewicz, H., Ed., Ossolineum, Warsaw, 1985, 171.
120. **Morgan, T. H.,** The predetermination of sex in phylloxerans and aphids, *J. Exp. Zool.,* 19, 285, 1915.
121. **Steffan, A. W.,** Zum Generation- und Chromosomenzyklus der Adelgidae (Homoptera: Aphidina), *Verh. Dtsch. Zool. Ges.,* 1967, 762.
122. **Bournier, A.,** Contribution à l'étude de la parthénogenèse des Thysanoptères et de sa cytologie, *Arch. Zool. Exp. Gen.,* 93, 219, 1956.
123. **Risler, H. and Kempter, E.,** Die Haploidie der Männchen und die Endopolyploidie in einigen Geweben von *Haplothrips* (Thysanoptera), *Chromosoma,* 12, 351, 1961.
124. **Mound, L. A.,** Patterns of sexuality in Thysanoptera, in 1991 Conference on Thrips (Thysanoptera): Insect and Disease Considerations in Sugar Maple Management, U.S.D.A. Forest Service, North Eastern Forest Experimental Station General Tech. Rep. NE-161, University Park, PA, 1991, 2.
125. **Ananthakrishnan, T. N.,** *Bioecology of Thrips,* Indira Publishing House, Oak Park, MI, 1984.
126. **Crespi, B. J.,** Sex-ratio selection in a bivoltine thrips. I. Conditional sex-ratio manipulation and fitness variation, *Evolution,* 42, 1199, 1988.
127. **Hughes-Schrader, S.,** Distance segregation and compound sex chromosomes in mantispids (Neuroptera: Mantispidae), *Chromosoma,* 27, 109, 1969.
128. **Hughes-Schrader, S.,** Segregational mechanisms of sex chromosomes in Megaloptera (Neuropteroidea), *Chromosoma,* 81, 307, 1980.
129. **Hennig, W.,** *Insect Phylogeny* (English translation by Pont, A. C.), Wiley, Chichester, U.K., 1981.
130. **Smith, S. G. and Virkki, N.,** Coleoptera, *Animal Cytogentics* 3, *Insecta* 5, Gebrüder Borntraeger, Berlin/Stuttgart, 1978.
131. **Virkki, N.,** Chromosomes in evolution of Coleoptera, in *Chromosomes in Evolution of Eukaryotic Groups,* Vol. 2, Sharma, A. K. and Sharma, A., Eds., CRC Press, Boca Raton, FL, 1984, 41.
132. **Virkki, N., Mazzella, C., and Denton, A.,** Silver staining of the coleopteran Xy_p sex bivalent, *Cytobios,* 67, 45, 1991.
133. **Postiglioni, A., da Silva, A., de Leon, R., and de Vaio, E. S.,** Three species of *Helipodus* (Coleoptera, Curculionidae) with different karyotypes and sex chromosome systems, *Genetica,* 75, 213, 1987.
134. **Serrano, J. and Yadav, J. S.,** Chromosome numbers and sex-determining mechanisms in adephagan Coleoptra, *Coleopt. Bull.,* 38, 335, 1984.
135. **Maddison, D. R.,** Chromosomal diversity and evolution in the ground beetle genus *Bembidion* and related taxa (Coleoptera: Carabidae: Trechitae), *Genetica,* 66, 93, 1985.
136. **Yadav, J. S., Kondal, K., and Yadav, A. S.,** Cytology of *Cicindela (Myriochile) undulata* and *C. (M.) fastidiosa* with a summary of chromosomal data on the Cicindelidae, *Cicindela,* 17, 1, 1985.
137. **Mesa, A. and Fontanetti, C. S.,** The chromosomes of a primitive species of beetle, *Ytu zeus* (Coleoptera, Myxophaga, Torridincolidae), *Proc. Acad. Nat. Sci. Philadelphia,* 137, 102, 1985.
138. **Petitpierre, E. and Segarra, C.,** Chromosomal variability and evolution of Chrysomelidae (Coleoptera), particularly that of Chrysomelinae and palaearctic Alticinae, *Entomography,* 3, 403, 1985.
139. **Petitpierre, E.,** Recent advances in the evolutionary cytogenetics of the leaf beetles (Coleoptera, Chrysomelidae), *Entomography,* 6, 433, 1989.

140. **Yadav, J. S. and Dange, M. P.**, Chromosomal investigations on eight species of histerids (Coleoptera: Histeridae), *Elytron (Barcelona)*, 3, 103, 1989.
141. **Yadav, J. S. and Dange, M. P.**, Chromosome number and sex determining mechanisms in twenty species of Indian rove beetles (Staphylinidae: Polyphaga), *Cell Chromosome Res.*, 10, 23, 1987.
142. **Gill, T. K., Gulati, M., and Pajni, H. R.**, Chromosome numbers in Indian weevils (Coleoptera: Curculionidae), *Coleopt. Bull.*, 44, 437, 1990.
143. **Yadav, J. S., Burra, M. R., and Dange, M. P.**, Chromosome number and sex determining mechanism in 32 species of Indian Coleoptera (Insecta), *Nat. Acad. Sci. Lett. (India)*, 12, 93, 1989.
144. **Juan, C. and Petitpierre, E.**, Chromosome numbers and sex-determining systems in Tenebrionidae (Coleoptera), in *Advances in Coleopterology*, Zunino, M., Belles, X., and Blas, M., Eds., European Association of Coleopterology, Barcelona, 1992.
145. **Handa, S. M. and Kochhar, N.**, Cytology of bruchids: I. Chromosome number and sex mechanism of seven cytologically unknown species of bruchids, *Res. Bull. Panjab Univ. Sci.*, 37, 145, 1987.
146. **Yadav, J. S., Lyapunova, E. A., and Vorontsov, N. N.**, Chromosome numbers and sex-chromosome mechanisms in fifty species of Coleoptera from USSR, *Folia Biol. (Krakow)*, 34, 269, 1986.
147. **Trivers, R. L. and Hare, H.**, Haplodiploidy and the evolution of social insects, *Science*, 191, 249, 1976.
148. **Luck, R. F., Stouthamer, R., and Nunney, L.**, Sex determination and sex ratio patterns in parasitic Hymenoptera, in *Evolution and Diversity of Sex Ratio in Haplodiploid Insects and Mites*, Wrensch, D. L. and Ebbert, M. A., Eds., Chapman and Hall, Engelwood Cliffs, NJ, 1993.
149. **Cook, J. M.**, Experimental tests of sex determination in *Goniozus nephantidis* (Hymenoptera: Bethylidae), *Heredity*, 71, 130, 1993.
150. **Stouthamer, R., Luck, R. F., and Werren, J. H.**, Genetics of sex determination and the improvement of biological control using parasitoids, *Environ. Entomol.*, 21, 427, 1992.
151. **Periquet, T. G., Hedderwick, M. P., El Agoze, M., and Poirie, M.**, Sex determination in the hymenopteran *Diadromus pulchellus* (Ichneumonidae): validation of the one-locus multi-allele model, *Heredity*, 70, 420, 1993.
152. **Ross, K. G. and Fletcher, D. J. C.**, Genetic origin of male diploidy in the fire ant *Solenopsis invicta* (Hymenoptera: Formicidae), and its evolutionary significance, *Evolution*, 39, 888, 1985.
153. **Kerr, W. E. and Nielsen, R. A.**, Sex determination in bees (Apinae), *J. Apic. Res.*, 6, 3, 1967.
154. **Kerr, W. E.**, Sex determination in bees. XXI. Number of XO-heteroalleles in a natural population of *Melipona compressipes fasciculata* (Apidae), *Insectes Soc.*, 34, 274, 1987.
155. **Cook, J. M.**, Sex determination in the Hymenoptera: a review of models and evidence, *Heredity*, 71, 421, 1993.
156. **Woyke, J.**, Sex determination, in *Bee Genetics and Breeding*, 1st edition, Rinderer, T. E. (Ed.), Academic Press, Orlando, FL, 1986, 91.
157. **Adams, J., Rothman, E. D., Kerr, W. E., and Paulino-Simões, Z. L.**, Estimation of the number of sex alleles and queen matings from diploid male frequencies in a population of *Apis mellifera*, *Genetics*, 86, 583, 1977.
158. **Skinner, S. W. and Werren, J. H.**, The genetics of sex determination in *Nasonia vitripennis*, *Genetics*, 94, 598, 1980.
159. **Werren, J. H., Nur, U., and Eickbush, D.**, An extrachromosomal factor causing loss of paternal chromosomes, *Nature (London)*, 327, 75, 1987.
160. **Werren, J. H.**, The paternal sex-ratio chromosome of *Nasonia*, *Am. Nat.*, 137, 392, 1991.
161. **Reed, K. M.**, Cytogenetic analysis of the paternal sex ratio chromosome of *Nasonia vitripennis*, *Genome*, 36, 157, 1993.
162. **Folliot, R.**, Contribution à l'étude de la biologie des cynipides gallicoles (Hyménoptères, Cynipoidea), *Ann. Sci. Nat. Zool. Biol. Anim. Zool. Ser.*, 12(6), 407, 1964.
163. **Bayreuther, K.**, Die Cytogenetik zweier norddeutscher Populationen von *Nosopsyllus fasciatus* Bosc. (Aphaniptera), *Chromosoma*, 27, 20, 1969.
164. **Suomalainen, E.**, Achiasmatische Oogenese bei Trichopteren, *Chromosoma*, 18, 201, 1966.
165. **Tazima, Y.**, *Genetics of the Silkworm*, Logos Press, Bristol, 1964.
166. **Robinson, R.**, *Lepidoptera Genetics*, Pergamon Press, Oxford, 1971.
167. **Smith, S. G.**, Heteropycnosis as a means of diagnosing sex, *J. Hered.*, 36, 194, 1945.
168. **Traut, W. and Mosbacher, G. C.**, Geschlechtschromatin bei Lepidopteren, *Chromosoma*, 25, 343, 1968.
169. **Ennis, T. J.**, Sex chromatin and chromosome numbers in Lepidoptera, *Can. J. Genet. Cytol.*, 18, 119, 1976.
170. **Suomalainen, E.**, On the sex chromosome trivalent in some lepidopteran females, *Chromosoma*, 28, 298, 1969.
171. **Seiler, J. and Schäffer, K.**, Untersuchungen über die Entstehung der Parthenogenese bei *Solenobia triquetrella* F.R. (Lepidoptera, Psychidae). II. Analyse der diploidparthenogenetischen *S. triquetrella*, Verhalten, Aufzuchtresultate und Zytologie, *Chromosoma*, 11, 29, 1960.
172. **Clarke, C. and Ford, E. B.**, Intersexuality in *Lymantria dispar* (L.). A reassessment, *Proc. R. Soc. London, Ser. B*, 206, 381, 1980.

173. **Gupta, M. L. and Narang, R. C.,** Karyotype and meiotic mechanism in muga silkmoths, *Antheraea compta* Roth. and *A. assamensis* (Helf.) (Lepidoptera: Saturnidae), *Genetica*, 57, 21, 1981.
174. **Traut, W., Weith, A., and Traut, G.,** Structural mutants of the W chromosome in *Ephestia* (Insecta, Lepidoptera), *Genetica*, 70, 69, 1986.
175. **Fuge, H.,** The 3-dimensional architecture of chromosome fibres in the cranefly. II. Amphitelic sex univalents in meiotic anaphase I, *Chromosoma*, 91, 322, 1985.
176. **Ullerich, F. H., Bauer, H., and Dietz, R.,** Geschlechtsbestimmung bei Tipuliden (Nematocera; Diptera), *Chromosoma*, 15, 591, 1964.
177. **Wolf, E.,** Die Chromosomen in der Spermatogenese einiger Nematocera, *Chromosoma*, 2, 192, 1941.
178. **White, M. J. D.,** Cytological evidence on the phylogeny and classification of the Diptera, *Evolution*, 3, 252, 1949.
179. **White, G. E. and Killick-Kendrick, R.,** Polytene chromosomes of the sandfly *Lutzomyia longipalpis* and the cytogenetics of Psychodidae in relation to other Diptera, *J. Entomol. Ser. A*, 50, 187, 1975.
180. **Troiano, G.,** Heterozygous heterochromatin in giemsa C-banded chromosomes of *Clogmia albipunctata* (*Telmatoscopus albipunctatus*) (Diptera: Psychodidae), *Caryologia*, 41, 201, 1988.
181. **Kreutzer, R. D., Modi, G. B., Tesh, R. B., and Young, D. G.,** Brain cell karyotypes of six species of New and Old World sand flies (Diptera: Psychodidae), *J. Med. Entomol.*, 24, 609, 1987.
182. **Kitzmiller, J. B.,** Genetics, cytogenetics and evolution of mosquitoes, *Adv. Genet.*, 18, 315, 1976.
183. **White, G. B.,** Academic and applied aspects of mosquito cytogenetics, in *Insect Cytogenetics*, Blackman, R. L., Hewitt, G. M., and Ashburner, M., Eds., Blackwell, Oxford, 1980, 245.
184. **Newton, M. E., Southern, D. I., and Wood, R. J.,** X and Y chromosomes of *Aedes aegypti* (L.) distinguished by Giemsa C-banding, *Chromosoma*, 49, 41, 1974.
185. **Mezzanote, R., Ferrucci, L., and Contini, C.,** Identification of sex chromosomes and characterization of the heterochromatin in *Culiseta longiaerolata* (Macquart 1838), *Genetica*, 50, 135, 1979.
186. **Newton, M. E., Wood, R. J., and Southern, D. I.,** Cytological mapping of the M and D loci in the mosquito, *Aedes aegypti* (L.), *Genetica*, 48, 137, 1978.
187. **Baker, R. H. and Sakai, R. K.,** Male-determining factor on chromosome 3 in the mosquito, *Culex tritaeniorhynchus*, *J. Hered.*, 67, 289, 1976.
188. **Frizzi, G., Contini, C., and Mameli, M.,** Ulteriori recherche citogenetiche sui Dixidae della Sardegna, *Atti Assoc. Genet. Ital.*, 11, 286, 1966.
189. **Brockhouse, C.,** Sibling species and sex chromosomes in *Eusimulium vernum* (Diptera: Simulidae), *Can. J. Zool.*, 63, 2145, 1985.
190. **Leonhardt, K. G. and Feraday, R. M.,** Sex chromosome evolution and population differentiation in the *Eusimulium aureum* group of black flies, *Genome*, 32, 543, 1989.
191. **Martin, J. and Lee, B. T. O.,** A phylogenetic study of sex determiner location in a group of Australasian *Chironomus* species (Diptera, Chironomidae), *Chromosoma*, 90, 190, 1984.
192. **Rothfels, K. H.,** Chromosomal variability and speciation in black flies, in *Insect Cytogenetics*, Blackman, R. L., Hewitt, G. M., and Ashburner, M., Eds., Blackwell, Oxford, 1980, 207.
193. **Smith, P. a. and Corces, V. G.,** *Drosophila* transposable elements: mechanisms of mutagenesis and interactions with the host genome, *Adv. Genet.*, 29, 229, 1991.
194. **Mason, G. F.,** Sex chromosome polymorphism in the *Simulium tuberosum* complex (Lundström) (Diptera: Simulidae), *Can. J. Zool.*, 62, 647, 1984.
195. **Feraday, R. M., Leonhardt, K. G., and Brockhouse, C. L.,** The role of sex chromosomes in black fly evolution, *Genome*, 32, 538, 1989.
196. **Thompson, P. E. and Bowen, J. S.,** Interactions of differential primary sex factors in *Chironomus tentans*, *Genetics*, 70, 491, 1972.
197. **Feraday, R. M.,** Weak male-determining genes and female heterogamety in *Chironomus tentans*, *Can. J. Genet. Cytol.*, 26, 748, 1984.
198. **Martin, J. and Lee, B. T. O.,** Are there female heterogametic strains of *Chironomus tentans* Fabricius?, *Can. J. Genet. Cytol.*, 26, 743, 1984.
199. **Haig, D.,** The evolution of unusual chromosome systems in sciarid flies: intragenomic conflict and the sex ratio, *J. Evol. Biol.*, 6, 249, 1993.
200. **Stuart, J. J. and Hatchett, J. H.,** Cytogenetics of the hessian fly. I. Mitotic karyotype analysis and polytene chromosome correlations, *J. Hered.*, 79, 184, 1988.
201. **Stuart, J. J. and Hatchett, J. H.,** Cytogenetics of the hessian fly. II. Inheritance and behaviour of somatic and germ-line-limited chromosomes, *J. Hered.*, 79, 190, 1988.
202. **Geyer-Duszyñ, I.,** Experimental research and chromosome elimination in Cecidomyidae (Diptera), *Chromosoma*, 11, 499, 1959.
203. **Kozlova, L. V.,** [On monogeny of gall midges, *Aphidoletes aphidimyza* Rond. (Diptera: Cecidomyidae)], *Nauchn. Tr. Leningr. Ord. Skh. Inst.*, No. 374, 11, 1979.
204. **Went, D. F. and Camenzind, R.,** Sex determination in the dipteran insect *Heteropeza pygmaea*, *Genetica*, 52/53, 373, 1980.

205. **Went, D. F. and Camenzind, R.,** Haemolymph-dependent sex determination in a paedogenetic gall midge, *Naturwissenschaften*, 64, 276, 1977.
206. **Boyes, J. W.,** The chromosomes of Rhagionidae, Stratiomyidae and Xylomyidae (Diptera), *Can. J. Genet. Cytol.*, 15, 255, 1973.
207. **Metz, C. W.,** Chromosome studies in the Diptera. IV. Incomplete synapsis in *Dasyllis grossa*. *Biol. Bull. (Wood's Hole, Mass.)*, 43, 253, 1922.
208. **Mainz, F.,** The genetics of *Megaselia scalaris* Loew (Phoridae): a new type of sex determination in Diptera, *Am. Nat.*, 98, 415, 1964.
209. **Willhoeft, U. and Traut, W.,** Molecular differentiation of the homomorphic sex chromosomes in *Megaselia scalaris* (Diptera) detected by random DNA probes, *Chromosoma*, 99, 237, 1990.
210. **Traut, W. and Willhoeft, U.,** A jumping sex determining factor in the fly *Megaselia scalaris*, *Chromosoma*, 99, 407, 1990.
211. **Dübendorfer, A., Hilfiker-Kleiner, D., and Nöthiger, R.,** Sex determination mechanisms in dipteran insects: the case of *Musca domestica*, *Semin. Dev. Biol.*, 3, 349, 1992.
212. **Franco, M. G., Rubini, P. G., and Vecchi, M.,** Sex-determinants and their distribution in various populations of *Musca domestica* L. of western Europe, *Genet. Res.*, 40, 279, 1982.
213. **Tomita, T. and Wada, Y.,** Multifactorial sex determination in natural populations of the housefly (*Musca domestica*) in Japan, *Jpn. J. Genet.*, 64, 373, 1989.
214. **Jayakar, S. D.,** Some two-locus models for the evolution of sex-determining mechanisms, *Theor. Popul. Biol.*, 32, 188, 1987.
215. **Shono, T. and Scott, J. G.,** Autosomal sex-associated pyrethroid resistance in a strain of housefly (Diptera: Muscidae) with a male-determining factor on chromosome three, *J. Econ. Entomol.*, 83, 686, 1990.
216. **Denholm, I., Franco, M. G., Rubini, P. G., and Vecchi, M.,** Identification of a male determinant on the X chromosome of a housefly (*Musca domestica* L.) population in south-east England, *Genet. Res.*, 42, 311, 1983.
217. **Denholm, I., Franco, M. G., Rubini, P. G., and Vecchi, M.,** Geographical variation in house fly (*Musca domestica* L.) sex determinants within the British Isles, *Genet. Res.*, 47, 19, 1986.
218. **Kerr, R. W.,** Inheritance of DDT resistance in a laboratory colony of the house fly, *Musca domestica*, *Aust. J. Biol. Sci.*, 23, 377, 1970.
219. **Loeschke, V., Nielsen, B. O., and Andersen, D.,** Chromosomal variation, segregation and sex determination in *Hydrotaea meridionalis* (Diptera: Muscidae), *Hereditas*, 118, 229, 1993.
220. **Samoylov, Yu. B.,** [Genetic control for the cabbage root fly II. Localization of male determining factor in the cabbage root fly *Delia brassicae* Bouche], *Genetika*, 21, 1810, 1985.
221. **Inoue, H. and Hiroyishi, I.,** A maternal effect sex transformation mutant of the housefly *Musca domestica*, *Genetics*, 12, 469, 1986.
222. **Maddern, R. H. and Bedo, D. G.,** Properties of the sex chromosomes of *Lucilia cuprina* deduced from radiation studies, *Genetica*, 63, 203, 1984.
223. **Bedo, D. G. and Foster, G. G.,** Cytogenetic mapping of the male-determining region of *Lucilia cuprina* (Diptera: Calliphoridae), *Chromosome*, 92, 344, 1985.
224. **Ribbert, D.,** Die Polytänchromosomen der Borstenbildungszellen von *Calliphora erythrocephala*, *Chromosoma*, 21, 296, 1967.
225. **Ullerich, F.-H.,** Analysis of sex determination in the monogenic blowfly *Chrysomya rufifacies* by pole cell transplantation, *Mol. Gen. Genet.*, 193, 479, 1984.
226. **Clausen, S. and Ullerich, F.-H.,** Sequence homology between a polytene band in the genetic sex chromosomes of *Chrysomya rufifacies* and the *daughterless* gene of *Drosophila melanogaster*, *Naturwissenschaften*, 77, 137, 1990.
227. **Lifschitz, E. and Cladera, J. L.,** *Ceratitis capitata:* cytogenetics and sex determination, in *Fruit Flies, Their Biology, Natural Enemies and Control (World Crop Pests 3B)*, Robinson, A. S. and Harper, G., Eds., Elsevier, Amsterdam, 1989, 3.
228. **Solferini, V. N. and Morgante, J. S.,** $X_1X_1X_2X_2$:X_1X_2Y mechanism of sex determination in *Anastrepha bistrigata* and *A. serpentina* (Diptera: Tephritidae), *Rev. Bras. Genet.*, 13, 201, 1990.
229. **Grewal, J. S. and Kapoor, V. C.,** Karyotypes of some fruitfly species (Tephritidae) of India, in *Fruit Flies of Economic Importance (CEC/IOBC Int. Symp., Rome, 7–10 April 1987)*, Cavallaro, R., Ed., A. A. Balkema, Rotterdam, 1989, 237.
230. **Bush, G. L.,** Female heterogamety in the family Tephritidae (Acalyptrata, Diptera), *Am. Nat.*, 100, 119, 1966.
231. **Anleitner, J. E. and Haymer, D. S.,** Y-Enriched and Y specific DNA sequences from the genome of the Mediterranean fruit fly, *Ceratitis capitata*, *Chromosoma*, 101, 271, 1992.
232. **Hodgkin, J.,** *Drosophila* sex determination: a cascade of regulated splicing, *Cell*, 56, 905, 1989.
233. **Slee, R. and Bownes, M.,** Sex determination in *Drosophila melanogaster*, *Q. Rev. Biol.*, 65, 175, 1990.
234. **Torres, M. and Sanchez, L.,** The segmentation gene *runt* is needed to activate *Sex-lethal*, a gene that controls sex determination and dosage compensation in *Drosophila*, *Genet. Res.*, 59, 189, 1992.
235. **Steinmann-Zwicky, M.,** How do germ cells choose their sex? *Drosophila* as a paradigm, *BioEssays*, 14, 513, 1992.

236. **Cline, T. W.,** The *Drosophila* sex determination signal: how do flies count to two?, *Trends Genet.*, 9, 385, 1993.
237. **Steinemann, M. and Steinemann, S.,** Evolutionary changes in the organization of the major LCP gene cluster during sex chromosomal differentiation in the sibling species *Drosophila persimilis, D. pseudobscura* and *D. miranda, Chromosoma*, 99, 424, 1990.
238. **Ganguly, R., Swanson, K. D., Ray, K., and Krishnan, R.,** A *BamHI* repeat element is predominantly associated with the degenerating NEO-Y chromosome of *Drosophila miranda* but absent in *Drosophila melanogaster* genome, *Proc. Natl. Acad. Sci. U.S.A.*, 89, 1340, 1992.
239. **Steinemann, M., Steinemann, S., and Lottspeich, F.,** How Y chromosomes become genetically inert, *Proc. Natl. Acad. Sci. U.S.A.*, 90, 5737, 1993.
240. **Ohno, S.,** Evolution of sex chromosomes in mammals, *Annu. Rev. Genet.*, 3, 495, 1969.
241. **Jablonka, E. and Lamb, M. J.,** The evolution of heteromorphic sex chromosomes, *Biol. Rev. Camb. Philos. Soc.*, 65, 249, 1990.
242. **Charlesworth, D. and Charlesworth, B.,** Sex differences in fitness and selection for certain fusions between sex chromosomes and autosomes, *Genet. Res.*, 35, 205, 1980.

Chapter 4

HORMONES AND REPRODUCTION

Jim Hardie

CONTENTS

I. INTRODUCTION

This chapter concerns the endocrine control of reproduction in female and male insects. The major gonadotropic hormones are juvenile hormones (JHs) and ecdysteroids, the same hormones that control metamorphosis and moulting. This parsimony in hormone complement occurs because the windows of sensitivity for development and reproduction are, to an extent, isolated from each other and the temporal distribution of hormone receptors differs between different tissues. Thus, during larval and pupal stages, JHs and ecdysteroids are responsible for development, while in the adult (and in certain instances preadult) stages they take on gonadotropic functions. Juvenile hormones are synthesized and released from the corpora allata which are present in all insect stages. The corpora allata are possibly the only site of synthesis, although recently male accessory glands have been reported to produce JH.[1] In contrast, during larval and pupal stadia, ecdysteroids are synthesized and released, for the main part, from the prothoracic glands, which, with certain exceptions (e.g., in Apterygota which alternately moult and reproduce as adults and in solitary locusts), atrophy at the final moult. In adult females, ecdysteroid synthesis occurs in the ovaries, or more specifically in the follicle cells of the ovaries, and in males they are synthesized by the testes, while other

0-8493-6695-X/95/$0.00+$.50

possible sites for production have been suggested.[2] With today's more sensitive techniques for hormone isolation and identification together with molecular biological procedures, the control of JH and ecdysteroid titers and their actions at the gene level are becoming more understood.[3] Recent reviews of the hormonal control of insect reproduction include references 4 through 7.

II. HORMONES AND FEMALES

The endocrine control of reproduction in female insects varies with species; this is, perhaps, not surprising as the reproductive strategies also differ (e.g., oviparous, ovoviparous, viviparous, sexual, parthenogenetic). The present chapter will concentrate on presenting some of the different endocrine strategies with reference to recent investigations of some well-researched insects. Chapter 1 describes the events of oogenesis with some details on endocrine effects on female accessory glands.

A. OOGENESIS

1. Cockroaches

In all species of cockroaches investigated, including oviparous, ovoviparous, and viviparous species, JH has been shown to play a major gonadotropic role. Mated females of the oviparous American cockroach, *Periplaneta americana,* undergo the cyclical production of oothecae (egg cases containing up to 16 eggs) every 4 to 5 days. Removal of the corpora allata from immature females prevents any previtellogenic ovarian development, while in mature females it prevents further ootheca formation by inhibiting vitellogenin synthesis and uptake.[8] Reimplantation of corpora allata or treatment with JH restores the reproductive cycle.[9,10] The precise corpora allata control of the cycle of ootheca production has been the subject of many studies. *In vitro* culture of corpora allata under conditions that allow the synthesis of JH (JH III in this species) has shown a corresponding cycle of JH synthesis and release.[11] More recently, it has been shown that *in vivo* titers of JH III show concomitant cyclical rises, with peak titers occurring during vitellogenesis and low titers at the point where oothecae are deposited.[12] Surgical techniques, more refined than complete allatectomy, have been implemented and show that after unilateral allatectomy (removing one of the pair of corpora allata), ootheca production continues, but at a slightly lower rate, while severing all the nervous connections to the corpora allata had a similar effect.[13] The corpora allata effects on ootheca production could, therefore, be accommodated by the loss of one corpus allatum and, more important, the cyclical activity appeared to be driven by hemolymph-borne factors. However, the same study showed that the cyclical production of oothecae could be reinstated in allatectomized females by treatment with a potent JH analogue which also speeded up the ootheca cycle in intact insects. The study concluded that the ootheca production cycle was not driven by the cycle of JH synthesis and release, but that JH served only to control the speed of an endogenous reproductive cycle: lower JH levels slowed the cycle while higher levels speeded it up; in the absence of JH the cycle stopped.[13] Control of JH synthesis *in vivo* may well be effected by peptide factors such as allatostatins (which inhibit JH synthesis by the corpora allata), recently identified in the brain of the viviparous cockroach, *Diploptera punctata,* and shown to be effective in *P. americana.*[14]

Ecdysteroids are produced by the ovaries of adult cockroaches,[15] but their role has not been fully elucidated. It has been proposed that the co-occurrence of ecdysteroid and choriogenesis might indicate a role at this stage in egg development. They may have an inhibitory efect on JH secretion, and their continued presence in the ootheca indicates that they supply the requirements of the embryos.[16,17]

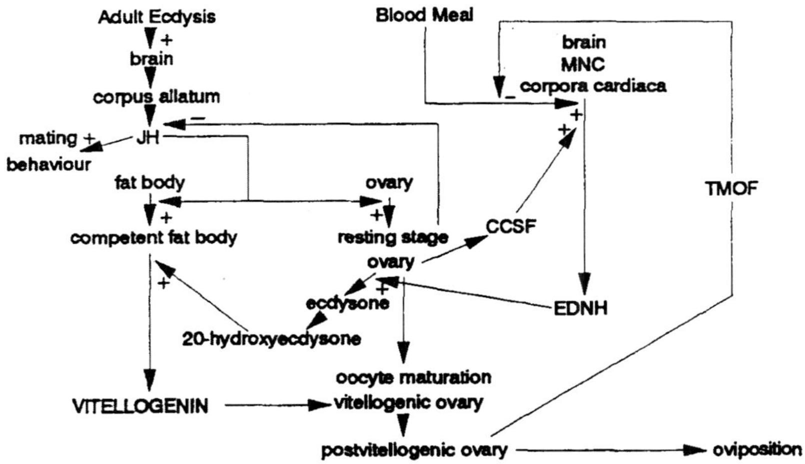

FIGURE 1. Factors regulating ovarian development in the mosquito, *Aedes aegypti*. CCSF = corpus cardiacum-stimulating factor; MNC = medial neurosecretory cells; EDNH = egg development neurosecretory hormone; TMOF = trypsin-modulating oostatic factor. (Modified from Hagedorn, H. *Comprehensive Insect Physiology, Biochemistry and Pharmacology*, Kerkut, G. and Gilbert, L., Eds. Pergamon, Oxford, 1985 page 165.)

2. Dipterans

In a number of Diptera, ecdysteroids replace JH as the major gonadotropic hormone stimulating vitellogenin synthesis/uptake.

a. Mosquitoes

The hormonal control of egg maturation in *Aedes aegypti*, an anautogenous mosquito (requiring a blood meal to develop the first and subsequent egg batches) has been studied intensively.[5] A summary of the endocrine interactions is shown in Figure 1. Adult emergence triggers the release of JH from the corpora allata.[18] JH then acts on the ovary, which contains the primary follicles, to stimulate growth to around twice the original size and form a "resting stage" ovary which is complete within 3 days of emergence. This previtellogenic development of the ovary also involves the production of endocytic organelles by the oocytes, which then become competent for protein uptake.[19] The "resting stage" ovary now exerts an inhibition of JH secretion,[20] but does not develop further until a blood meal has been taken. This initial rise in JH titer also stimulates mating behavior[21] and induces a competence in the fat body to respond to ecdysteroids. The fat body remains unresponsive if allatectomy is performed at adult eclosion,[22] and *in vitro* experiments confirm that JH promotes the competence required for an ecdysteroid response.[23]

After a blood meal, egg development neurosecretory hormone (EDNH)[24] is released from the corpora cardiaca, having been produced in the protocerebrum by the medial neurosecretory cells, and acts on the "resting stage" ovary to induce the production of ecdysone.[25] This EDNH release, also, involves a corpus cardiacum-stimulating factor (CCSF) which is produced by the resting ovary.[26,27] Ecdysone is secreted into the hemolymph and converted to 20-hydroxyecdysone, which then acts on the competent fat body to stimulate vitellogenin synthesis and release. Additionally, the 20-hydroxyecdysone induces the separation of new follicles, the secondary follicles, from the germaria.[28] Control of the endocytotic uptake of vitellogenin by the oocytes has yet to be elucidated, but it is possible that both 20-hydroxyecdysone and an unidentified brain hormone effect vitellogenin uptake.[29] After a short delay, blood meals also induce a 24- to 36-h pulse of elevated JH titer which coincides with decreased levels of

JH-esterase activity.[18] This meal-induced JH pulse stimulates the new follicles to develop to the resting stage and a renewed competence of the fat body to respond to ecdysteroids. The next blood meal reinitiates the egg maturation cycle. Further blood meals are usually taken only after egg batches are laid. If taken before oviposition, the mature postvitellogenic oocytes produce an oostatic hormone, providing an inhibitory feedback which prevents further ovarian development and vitellogenesis.[30] This oostatic hormone has recently been characterized as a decapeptide which inhibits synthesis of a trypsin-like enzyme in the midgut of female *A. aegypti*; it thus prevents the blood-meal digestion and, indirectly, vitellogenesis.[31]

Aedes atropalpus is an autogenous mosquito species which does not require a blood meal to produce the first egg batch. Nevertheless, hormonal control of the ovarian cycle appears similar to the anautogenous *A. aegypti* except that emergence is the stimulus for EDNH release and development of the first egg batch.[5] Indeed, peptides have now been isolated that induce ecdysteroid release from ovaries (i.e., EDNH-like) of both species.[32]

b. Flies

Both JH and ecdysteroids are again implicated in the hormonal control of egg maturation in the higher Diptera (Cyclorrhapha), and finer details are still being revealed. The major JH produced by the corpus allatum of adult Cyclorrhaphan flies (*Drosophila melanogaster, Calliphora vomitoria*) has recently been identified as the bisepoxide of JH III.[33,34]

In an anautogenous colony (requiring a protein meal before egg development) of the black blowfly, *Phormia regina,* it appears that JH is required for the uptake of vitellogenin by the oocyte.[35] Treatment of sugar-fed adult females with JH III or a JH analogue resulted in a proportion of follicles with oocytes containing an opaque material. Immunological procedures showed that this opaque material was not vitellogenin, and it was concluded that protein uptake by the oocytes had been stimulated, but, in the absence of vitellogenin, the opaque material comprised other hemolymph proteins. Sugar-fed flies also retained low levels of ecdysteroids, but after a protein meal, ecdysteroids increased, the ovaries being the major source, and vitellogenin uptake by the oocytes followed.[36] Previous experiments had shown that treatment with precocene II (a pro-allatocidin compound which prevents JH synthesis by the corpus allatum) inhibited oocyte development beyond the previtellogenic stage, but did not prevent vitellogenin synthesis or release into the hemolymph, albeit levels were lower than controls.[37] Recently it has been shown that a peptide factor from the midgut is released in liver-fed females, which stimulates neurosecretory cells in the brain to initiate the neuroendocrine cascade leading to oogenesis.[38] A summary of the proposed endocrine interaction for this fly species is presented in Figure 2 and includes a possible factor released from the brain which induces synthesis and release of ecdysteroids from the ovaries (modified from References 36 and 38).

It has been shown in other flies that both the fat body and the ovaries produce vitellogenin.[39,40] In *Drosophila,* genes for three vitellogenins (termed yolk proteins 1, 2, and 3) are expressed in the fat body and follicular cells of the ovary but the regulation of expression differs between the tissues. Experiments with ligated abdomens showed that ecdysteroids and JHs could induce vitellogenin formation but, in addition, JH promoted vitellogenin uptake by the oocytes. Further studies on the relative amounts of the three vitellogenins indicated that JH stimulated normal synthesis and uptake in the ovary but abnormal synthesis by the fat body, while ecdysteroids had no effect on the ovary but induced normal synthesis by the fat body.[5] This simplistic model may have to be modified in view of more recent work which indicates the presence of other factors mediating nutritional effects on vitellogenin production and uptake.[39,41] By comparison, in the housefly *Musca domestica,* ecdysteroids and a JH analogue have been shown to induce vitellogenin mRNA in both the fat body and the ovaries. However, the JH analogue was the least potent of the two and proved more effective on the ovary than on the fat body.[40]

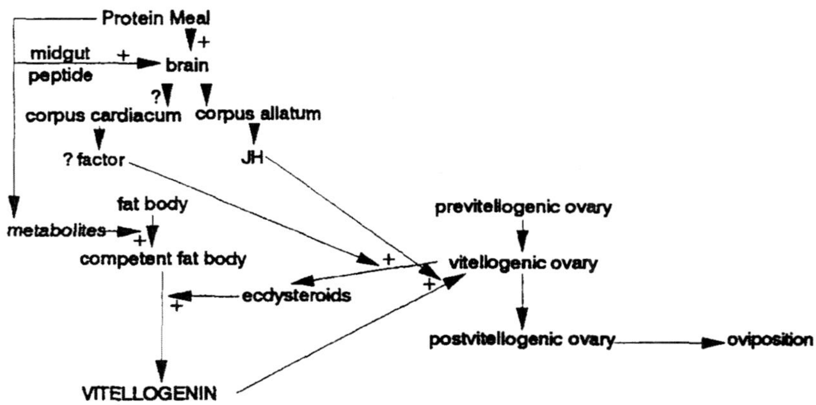

FIGURE 2. Factors regulating ovarian development in the black blowfly, *Phormia regina*. (Modified from Yin et al.[36,38])

3. Hemipterans

The blood-sucking bug *Rhodnius prolixus* was used by Wigglesworth to identify JH as a hormone controlling both metamorphosis and reproduction.[42] He demonstrated that the removal of the corpus allatum prevented vitellogenesis in the ovaries of adult females. This insect has continued to be utilized for the examination of reproductive endocrinology, particularly by Davey and colleagues, and our knowledge of humoral events controlling ovarian development and oviposition has become increasingly detailed. JH probably stimulates vitellogenin synthesis in the fat body, but the study of vitellogenin uptake by the oocytes has taken research precedence.[43] In mated females, the JH has a three-fold action in the stimulation of vitellogenin uptake by the terminal oocytes. Vitellogenins are large protein molecules, and their access to the oocyte cell membrane is facilitated by JH-induced changes in shape of the surrounding follicle cells and the resulting enlargement in the intercellular spaces (increased "patency"; see Chapter 1).[44] However, the presence of JH is also necessary during follicle cell development in order that cells become competent to respond to later increases in JH titer.[45] The uptake of vitellogenin by the oocyte is a calcium-dependent receptor-mediated response, and the receptor binding of vitellogenin is also enhanced by JH.[46] Recent studies also provide evidence of a neural inhibition of corpus allatum activity coming from the brain.[47]

In addition to the gonadotropic effects of JH, *Rhodnius* females possess abdominal neurosecretory organs which produce an antigonadotropic hormone (oostatic hormone).[48,49] Ovaries containing mature oocytes in the pedicels stimulate the release of the antigonadotropin, which counters the effect of JH on the vitellogenin uptake by the oocytes. The changes in patency induced by JH are inhibited in the presence of antigonadotropin.[50]

Ten neurosecretory cells have been identified in the pars intercerebralis of the brain, which produce a peptidergic myotropic factor that initiates ovulation (movement of the fully developed oocyte from the ovary) and subsequently oviposition.[51,52] Unmated females retain eggs in the ovarian pedicel for long periods, resulting in a much lower oviposition rate than mated females, and it is the presence of a spermathecal factor in mated insects that facilitates the release of the oviposition hormone.[53] However, it appears to be the production of ecdysteroids by the ovary, 5 days after a blood meal, that induces the release of the myotropic ovulation hormone from axon terminals in the corpora cardiaca.[54-56] The ecdysteroids do not act directly on the neurosecretory cells but indirectly via aminergic neurons[57] while more recent investigations indicate that there is also a circadian influence over these endocrine events.[58]

4. Homopterans

Aphids form one of the few groups of insects that commonly alternate between parthenogenetic and sexual reproduction (cyclical parthenogenesis).[59] While there are species (and/or clones) that will reproduce only by parthenogenesis, there are none that reproduce solely by sexual means. In a number of species, both the mode of reproduction and ovarian development in the parthenogenetic females is is controlled by the same hormone, JH. Photoperiodic cues are most frequently found to determine the reproductive type: long days result in the formation of parthenogenetic females (virginoparae), while short days promote the development of the sexual forms (oviparae and males).[60]

Work in the late 1970s indicated that the medial neurosecretory cells of the protocerebrum (called the group I cells in aphids)[61] were involved in the determination of female embryos as virginoparae or oviparae.[62] Ablation of these cells by radiocautery, in long-day reared vetch aphids, *Megoura viciae*, induced the production of "short-day" oviparae rather than the expected virginoparae. The observations indicated that a factor produced by these protocerebral cells promoted the appearance of virginoparae, and it was proposed that the factor acted directly upon the developing embryos in the ovarioles. The factor was termed "virginoparin" by Steel.[63] In contemporary studies, topical application of natural juvenile hormones or JH analogues was found to modify embryonic development such that female embryos that were photoperiodically determined as sexual females (oviparae) were redirected to develop as parthenogenetic females (virginoparae).[64-66] In addition to the switch induced in ovary type, there was a concommitant induction of winged or partially winged forms, many of which were sterile, but judicious application of JH produced seemingly normal virginoparous females that were capable of normal reproduction.[65-67] Thus, "virginoparin" and JH had the same effect on the development of female embryos and promoted their development as virginoparae; they were "long-day" hormones.

It was later shown that "virginoparin" and JH were part of the endocrine pathway for the induction of virginoparae.[68] It was shown that corpus allatum volume (this is a single fused gland in the aphid) differed, from day 10 onwards, between short-day- and long-day-reared females that were producing oviparae or virginoparae, respectively. When the group I cells were ablated in long-day insects, and the progeny switched from virginoparae to oviparae, there was a concomittant change in the corpus allatum volume such that it became equal in size to the short-day-reared insect corpora allata. Aphids that underwent sham cautery remained similar to untreated controls; the corpora allata were the same size, and they continued to produce virginoparae. It appears that the group I cells regulate JH synthesis via the corpus allatum. Assessment of JH titers in the aphids showed only JH III present and, although the levels were low, there were higher JH titers in the long-day- than the short-day reared aphids.[69]

There are other subtleties to aphid reproduction in that the parthenogenetic females undergo precocious ovarian development which results in the telescoping of generations.[70] The two ovaries of adult virginoparae (mother) comprise a number of ovarioles containing embryos (daughters) in various stages of development. The most mature of these embryos already contain ovaries with developing embryos (granddaughters). Parthenogenetic aphids are also viviparous and give birth to live young which, at birth, already have ovaries with ovarioles that contain one or two developing embryos. In the black bean aphid *Aphis fabae*, oocyte development and embryogenesis (of the granddaughter generation) within the embryonic (daughter generation) ovaries begins during the mother's fourth instar.[71] Decapitation, which included the removal of the corpora allata, drastically reduced the rate of oocyte differentiation, but this could be restored by JH application. In addition, the growth of the terminal (daughter) embryos is reduced almost completely by decapitation, but again is restored by JH treatment. However, this stimulation of embryogenesis is not reflected in a shortening of the prereproductive period or in the initial rate of reproduction, possibly because other endocrine factors are involved in parturition.[72]

It appears that JH has both a role in determining the parthenogenetic aphid morph and in the regulation of ovarian development in that morph. Additionally, as in other insects, it regulates metamorphosis.[73] However, in most insects the metamorphic role of JH terminates at the final moult, when JH often takes on a reproductive role. Perhaps uniquely in aphids, adult form is determined by low JH titers in the third (penultimate) instar larva and JH can then take on a reproductive role, precociously, in the last larval instar. Control of ovarian development in the sexual females has not been researched, but it will differ. Oocyte differentiation in the germarium occurs later, in the second larval instar (rather than prenatally as in parthenogenetic forms), with vitellogenesis following throughout the later instars, but embryogenesis requires mating, fertilization, and a period of diapause development. Presumptive subunits of vitellogenin are also produced by parthenogenetic females, the ovaries of which produce embryos, not vitellogenic eggs.[74]

B. REGULATION OF SEX PHEROMONE RELEASE

The efficiency of mate location by male insects is increased dramatically by the production of species-specific, volatile cues that can be detected at a distance and used directionally as well as on contact.[75] The production/release of these sex pheromones is intermittent, and recent studies indicate a precise endocrinological control. Barth[76] used behavioral techniques to demonstrate that allatectomy prevented sex pheromone release in some cockroaches, but in other species had no effect. This study formed the basis for the suggestion of neuroendocrine regulation of "calling" (pheromone release) behavior in species with a long adult life span and which mated repeatedly, but with periods of unreceptivity, while there would be no such control in species with a short, nonfeeding adult stage where mating occurs at emergence. Evidence is accumulating that JH may be crucial for both pheromone synthesis and release in the German cockroach, *Blattella germanica,* which mates repeatedly,[77,78] and the brown-banded cockroach, *Supella longipalpa,* which mates only once but lives for several months.[79] JH also appears to regulate sex pheromone production in the female true armyworm, *Pseudaletia unipunctata,* which mate after a period of migration, as well as in male responsiveness to females.[80,81] At least in the case of the female moth, the action of JH is indirect via a brain factor, possibly via pheromone-biosynthesis-activating peptide (PBAN) (see below).

In a number of lepidopterans, a PBAN has been located in the brain–suboesophageal ganglion complex, first in the corn earworm, *Helicoverpa (Heliothis) zea.*[82] The peptide has now been shown to comprise a 33-amino acid sequence in *H. zea,*[83] while Kitamura et al.[84,85] have identified two other PBANs, of similar size, in the silkworm, *Bombyx mori.* An 18-amino acid peptide from the common armyworm, *Pseudaletia* (= *Leucania*) *separata,* showed similar pheromone synthesis effects.[86] The latter peptide shares a common Phe-Ser-Pro-Arg-Leu-NH_2 terminal sequence with the identified PBAN molecules, but also induces cuticular melanization and acts as a weak diapause hormone.[86] The PBAN family of peptides requires further elucidation, and cross-reactivity between different bioassay systems raises questions about the primary actions *in vivo.*[86,87] It is also of interest that the PBAN from *H. zea* stimulates pheromone biosynthesis in the Hessian fly, *Mayetiola destructor* (a cecidomyiid dipteran), and PBAN-like activity has been reported in the cockroach, *B. germanica,* and the locust, *Locusta migratoria.*[88,89] In the housefly, *Musca domestica,* however, ecdysteroids are important not only for vitellogenesis (see above) but also for pheromone production.[90] In this species, sex pheromone production can be artificially induced in males by ecdysteroid treatment, while JH has no effect on pheromone production by either sex.

III. HORMONES AND MALES

The physiological control of spermatogenesis in males has been less well studied than has oogenesis in females. This is partly because mature sperm are often present at eclosion of adult males, and spermatogenesis has occurred under the hormonal conditions required for metamorphosis of the somatic cells.[91] Indeed, experiments on *R. prolixus* indicated precisely this and that spermatogenesis continued autonomously, at a basal rate, in the adult under conditions of low JH and ecdysteroids.[92] Spermatogenesis involves the mitotic cell divisions necessary for the development of fully formed spermatocytes, followed by meiotic division and differentiation of the sperm (spermiogenesis; see Chapter 2). The hormonal regulation of events is not fully understood, and evidence is often contradictory. A number of contradictions undoubtedly occur as some studies focus upon the resumption of testis development after a period of interrupted development (which can include spermatocyte autolysis) during diapause, while others tackle normal development under nondiapausing conditions where spermatogenesis is often arrested during meiotic prophase in early larval insects, only being resumed in late larval (or later) stages.[93] There is no reason to assume that endocrine events leading to renewed development are identical. In lepidopterans and some other species, two types of sperm are produced (see also Chapter 2) — eupyrene (nucleate) sperm which fertilize the egg and apyrene (nonnucleate) sperm whose function is not known.[94] Earlier work tended to look for effects of JH and the results are again contradictory,[4,6,94] but more recent investigations have tended to concentrate on the effects of ecdysteroids.

The testes of male insects, like the ovaries of females, have been shown to produce ecdysteroids. Although their contribution to overall titers is minimal, actions at a more local level could be important.[95-98] A brain ecdysiotropic factor is also reported to induce ecdysteroid secretion by testes.[99]

A. SPERMATOGENESIS

Spermatocysts from diapausing male silkmoth pupae can be stimulated to develop by ecdysteroids *in vitro,* but this effect is indirect and primes the cysts to respond to a large molecular weight chemical (macromolecular factor), found in pupal hemolymph, by increasing the sheath permeability.[100,101] Similar observations have been made on the codling moth, *Cydia pomonella,* but the response was irregular and incomplete,[102] and ultrastructural investigations of ecdysteroid effects on the testicular sheath did not indicate an increased permeability, but an increased metabolic activity and glycogen accumulation.[103] In the latter insect, JH appears to counter the effects of ecdysteroid on spermatogenesis, and topically applied JH analogue induced spermatogenesis arrest in nondiapause-destined larvae, while allatectomy renewed spermatogenesis in diapause-destined larvae.[104] In the cabbage armyworm, *Mamestra brassicae,* ecdysteroid levels are 4.4 times higher in testes of nondiapausing than diapausing pupae. Co-culture of testis and spermatocysts, both from diapausing pupae, resulted in spermiogenesis and correlated with raised ecdysteroids in culture medium, but proteins are also released from the cultured testes and may play a part in spermiogenesis.[105]

In nondiapausing tobacco budworm larvae, *Heliothis virescens,* the initiation of meiosis in young eupyrene spermatocysts occurred *in vitro,* apparently without ecdysteroids, but requiring fetal calf serum (possibly providing a macromolecular-like factor) and testis sheath.[106] The testis sheath factor was not an ecdysteroid nor was it species specific, as reciprocal stimulation occurred with gypsy moth (*Lymantria dispar*) testis sheath.[106] However, in common armyworm, *P. (=Leucania) separata,* spermatogenesis may be autonomous, since no factors have been found to affect spermiogenesis.[107,108]

In nondiapause-destined tobacco hornworm, *Manduca sexta,* eupyrene spermatocytes remain in meiotic prophase from early larval stages to the last larva or pupa. Coincident with the postwandering ecdysteroid peak, they recommence development to meiotic metaphase.[93] This effect can be mimicked in isolated abdomens by 20-hydroxyecdysone injections but not *in vitro,* indicating the ecdysteroid effect may be indirect. In the testes of diapausing pupae, ecdysteroid titers remain low, while in nondiapausing pupae they rise to a peak around day 10, returning to low levels before adult eclosion.[98] During diapause, mitosis and spermatocyte development continues, but apyrene spermatogenesis is arrested with the lysis of premeiotic apyrene spermatocytes, and no apyrene spermatids are found. On the other hand, eupyrene meiosis procedes, but is followed by lysis after differentiation of the eupyrene spermatids. It appears that the lack of ecdysteroids in diapausing pupae allows partial spermatogenesis but leads to resorption of the products at different stages of apyrene and eupyrene spermatogenesis. In the European corn borer, *Ostrinia nubilalis*, apyrene spermiogenesis is directly stimulated by ecdysone and 20-hydroxyecdysone *in vitro*.[109]

B. SPERM RELEASE AND MATURATION

In Lepidoptera, the release of clonal sperm bundles from the testes into the upper vas deferens occurs in the pharate adult, the release being a cyclical, gated event under circadian control.[110,111] In the Mediterranean flour moth *Ephestia kühniella*, release of eupyrene, but not apyrene, sperm bundles from the testes was inhibited by injection of 20-hydroxyecdysone.[112] More recent studies on the gypsy moth *L. dispar* indicate that the release of mature eupyrene sperm is delayed between day 5 of pupation (when the reproductive tract and sperm are fully differentiated) and day 9 by high titers of circulating ecdysteroids.[113] Infusion or injection of 20-hydroxyecdysone further delayed the onset of sperm release in a dose-dependent fashion but did not interfere with the rhythm or gating of release. In addition, exogenous ecdysteroids did not inhibit the release of sperm once the rhythm had been initiated.[113] Thus it appears that the the decline in endogenous ecdysteroid levels on day 7 after pupation are the cue for sperm release to begin, but they do not control the cyclical release. The disparity between the ecdysteroid inhibition of sperm release, after cyclical release had started, in *E. kühniella* but not *L. dispar*, may be due to differences in doses used.[113]

In the male silkmoth *B. mori*, apyrene sperm are present as individual cells in the seminal vesicles while eupyrene sperm remain in bundles. Prior to spermatophore formation in the ejaculatory duct, the apyrene sperm become mobile while the eupyrene sperm undergo maturation, separation to single cells, and activation in the spermatophore. These events are evoked by an endopeptidase, initiatorin, secreted by cells in the ejaculatory duct.[114] The precise regulation of this enzymic activation remains to be elucidated.

C. GONADAL DEVELOPMENT

Testis fusion occurs in the last larval instar of lepidopterans, but not in isolated abdomens with low ecdysteroid titers. Testis fusion has been found to require ecdysteroids.[97,115] In addition, the development of larval spermducts into the seminal vesicles and upper vas deferentia in the pupa requires ecdysteroids.[116-119] However, this may be an indirect effect via factors produced by the fat body and testis sheath, as isolated spermducts fail to develop in the presence of ecdysteroids.[120] The role of ecdysteroids in regulating the development of male accessory glands has recently been reviewed by Happ[7] (see also Chapter 2).

IV. SUMMARY

The examples of hormonal regulation of reproduction in insects demonstrate the wide variation in control mechanisms. This variation occurs even between species with similar life strategies, e.g., blood-feeding mosquitoes and *Rhodnius*. The same two major hormones appear to be involved in all insect species so far investigated, but their roles differ (e.g., ecdysteroids stimulates vitellogenin synthesis in Diptera but the release of a myotropin in *Rhodnius*, while JH stimulates vitellogenin synthesis and uptake in many other insects). The idea of a unified mechanism of endocrinological control in females or males may not be feasible. However, some of the apparent differences in hormonal control of oogenesis between Dipteran species may be due to variation in endogenous hormone titers and the timing of experiments.[121] Current studies of insect reproduction are tending to concentrate on the role of neurosecretory peptides and the molecular biology of hormone synthesis and action.

ACKNOWLEDGMENTS

I thank the Biotechnology and Biological Sciences Research Council for financial support and Stuart Reynolds and Rob Storer for comments on an earlier draft.

REFERENCES

1. **Borovsky, D., Carlson, D.A., Hancock, R.G., Rembold, H., and van Handel, E.,** *De novo* synthesis of juvenile hormone III and I by the accessory glands of the male mosquito, *Insect Biochem. Molec. Biol.*, 24, 437, 1994.
2. **Delbecque, J.-P., Weidner, K., and Hoffmann, K.H.,** Alternative sites for ecdysteroid production in insects, *Invertebr. Reprod. Dev.*, 18, 29, 1990.
3. **Wyatt, G.R.,** Gene regulation in insect development, *Invertebr. Reprod. Dev.*, 20, 1, 1991.
4. **Koeppe, J.K., Fuchs, M., Chen, T.T., Hunt, L.-M., Kovalick, G.E., and Briers, T.,** The role of juvenile hormone in reproduction, in *Comprehensive Insect Physiology, Biochemistry and Pharmacology*, Kerkut, G.A. and Gilbert, L.I., Eds., Pergamon, Oxford, Vol. 8, 1985, 165.
5. **Hagedorn, H.H.,** The role of ecdysteroids in reproduction, in *Comprehensive Insect Physiology, Biochemistry and Pharmacology*, Kerkut, G.A. and Gilbert, L.I., Eds., Pergamon, Oxford, Vol. 8, 1985, 205.
6. **Raabe, M.,** Insect reproduction, *Adv. Insect Physiol.*, 19, 29, 1986.
7. **Happ, G.M.,** Maturation of the male reproductive system and its endocrine regulation, *Annu. Rev. Entomol.*, 37, 303, 1992.
8. **Bell, W.J.,** Dual role of juvenile hormone in the control of yolk formation in *Periplaneta americana*, *J. Insect Physiol.*, 15, 1279, 1969.
9. **Chen, D.H., Robbins, W.E., and Monroe, R.E.,** The gonadotropic action of Cecropia extracts in allatectomised American cockroaches, *Experientia*, 18, 577, 1962.
10. **Girardie, A.,** Étude biometrique de la croissance ovarienne après ablation et implantation de corpora allata chez *Periplaneta americana*, *J. Insect Physiol.*, 8, 199, 1962.
11. **Weaver, R.J., Pratt, G.E., and Finney, J.R.,** Cyclic activity of the corpus allatum related to gonadotrophic cycles in adult female *Periplaneta americana*, *Experientia*, 31, 597, 1975.
12. **Edwards, J.P., Chambers, J., Short, J.E., Price, N.R., Weaver, R.J., Abraham, L., and Walter, C.M.,** Endogenous juvenile hormone III levels and *in vitro* rates of hormone biosynthesis by corpora allata during the reproductive cycle of adult female *Periplaneta americana*, in *Chromatography and Isolation of Insect Hormones and Pheromones*, McCaffery A.R. and Wilson, I.D., Eds. Plenum Press, New York, 1990, 3.
13. **Weaver, R.J. and Edwards, J.P.,** The role of the corpora allata and associated nerves in the regulation of ovarian cycles in the oviparous cockroach *Periplaneta americana*, *J. Insect Physiol.*, 36, 51, 1990.
14. **Weaver, R.J.,** Profile of the responsiveness of corpora allata from virgin female *Periplaneta americana* to an allatostatin fom *Diploptera punctata*, *J. Insect Physiol.*, 37, 111, 1991.
15. **Weaver, R.J., Strambi, A., and Strambi, C.,** The significance of free ecdysteroids in the haemolymph of adult cockroaches, *J. Insect Physiol.*, 30, 705, 1984.
16. **Chiang, A.-S., Burns, E.L., and Schal, C.,** Ovarian regulation of cyclic changes in size and activity of corpus allatum cells in *Blattella germanica*, *J. Insect Physiol.*, 37, 907, 1991.
17. **Pascual, N., Cerdá, X., Benito, B., Tomás, J., Piulachs, M.D., and Bellés, X.,** Ovarian ecdysteroid levels and basal oocyte development during maturation in the cockroach *Blattella germanica* (L.), *J. Insect Physiol.*, 38, 339, 1992.
18. **Shapiro, A.B., Wheelock, G.B., Hagedorn, H.H., Baker, F.C., Tsai, L.W., and Schooley, D.A.,** Juvenile hormone and juvenile hormone esterase in adult females of the mosquito *Aedes aegypti*, *J. Insect Physiol.*, 32, 867, 1986.
19. **Raikhel, A.S. and Lea, A.O.,** Hormone-mediated formation of the endocytotic complex in mosquito oocytes, *Gen. Comp. Endocrinol.*, 57, 422, 1985.
20. **Rossignol, P.A., Feinsod, F.M., and Speilman, A.,** Inhibitory regulation of corpus allatum activity in mosquitoes, *J. Insect Physiol.*, 27, 651, 1981.
21. **Gwadz, R.W., Lounibos, L.P., and Graig, G.B.,** Precocious sexual receptivity induced by a juvenile hormone analog in females of the yellow fever mosquito, *Aedes aegypti*, *Gen. Comp. Endocrinol.*, 16, 47, 1971.
22. **Flanagan, T.R. and Hagedorn, H.H.,** Vitellogenin synthesis in the mosquito: the role of juvenile hormone in the responsiveness to ecdysone, *Physiol. Entomol.*, 2, 173, 1977.
23. **Ma, M., Zhang, J.-Z., Gong, H., and Gwadz, R.,** Permissive action of juvenile hormone on vitellogenin production by the mosquito, *Aedes aegypti*, *J. Insect Physiol.*, 34, 593, 1988.
24. **Lea, A.O.,** The medial neurosecretory cells and egg maturation in mosquitoes, *J. Insect Physiol.*, 13, 419, 1967.
25. **Hagedorn, H.H., Shapiro, J.P., and Hanaoka, K.,** Ovarian ecdysone secretion is controlled by a brain hormone in an adult mosquito, *Nature*, 282, 92, 1979.
26. **Borovsky, D.,** Release of the egg development neurosecretory hormone in *Aedes aegypti* and *Aedes taeniorhynchus* induced by an ovarian factor, *J. Insect Physiol.*, 28, 311, 1982.
27. **Lea, A.O. and Van Handel, E.,** A neurosecretory hormone-releasing factor from ovaries of mosquitoes fed blood, *J. Insect Physiol.*, 28, 503, 1982.

28. **Beckermeyer, E.F. and Lea, A.O.,** Induction of follicle separation in the mosquito by physiological amounts of ecdysterone, *Science*, 209, 819, 1980.
29. **Koller, C.N. and Raikhel, A.S.,** Initiation of vitellogenin uptake and protein synthesis in the mosquito (*Aedes aegypti*) ovary in response to a blood meal, *J. Insect Physiol.*, 37, 703, 1991.
30. **Else, J.G. and Judson, C.L.,** Enforced egg-retention and its effects on vitellogenesis in the mosquito *Aedes aegypti*, *J. Med. Entomol.*, 9, 527, 1972.
31. **Borovsky, D., Carlson, D.A., Griffin, P.R., Shabanowitz, J., and Hunt, D.F.,** Mass spectrometry and characterisation of *Aedes aegypti* trypsin modulating oostatic factor (TMOF) and its analogues, *Insect Biochem. Molec. Biol.*, 23, 703, 1993.
32. **Matsumoto, S., Brown, M.R., Suzuki, A., and Lea, A.O.,** Isolation and characterization of ovarian ecdysteroidogenic hormones from the mosquito, *Aedes aegypti*, *Insect Biochem.*, 19, 651, 1989.
33. **Altaratz, M., Segal, D., Richard, D.S., Gilbert, L.I., and Applebaum, S.W.,** Juvenile hormone production by wild type and apterous[4] mutant *Drosophila melanogaster* corpora allata *in vitro*, in *Insect Neurochemistry and Neurophysiology 1989*, Borkovek, A.B. and Masler, E.P., Eds., Humana Press, Clifton, NJ, 1990, 333.
34. **Duve, H., Thorpe, A., Yagi, K.J., Yu, C.G., and Tobe, S.S.,** Factors affecting the biosynthesis and release of juvenile hormone bisepoxide in the adult blowfly *Calliphora vomitoria*, *J. Insect Physiol.*, 38, 575, 1992.
35. **Stoffolano, J.G., Li, M.-F., Zou, B.-X., and Yin, C-M.,** Vitellogenin uptake, not synthesis, is dependent on juvenile hormone in adults of *Phormia regina* (Meigen), *J. Insect Physiol.*, 38, 839, 1992.
36. **Yin, C.-M., Zou, B.-X., Yi, S.-X., and Stoffolano, J.G.,** Ecdysteroid activity during oogenesis in the black blowfly, *Phormia regina* (Meigen), *J. Insect Physiol.*, 36, 375, 1990.
37. **Yin, C.-M., Zou, B.-X., and Stoffolano, J.G.,** Precocene II treatment inhibits terminal oocyte development but not vitellogenin synthesis and release in *Phormia regina* (Meigen), *J. Insect Physiol.*, 35, 465, 1989.
38. **Yin, C.-M., Zou, B.-X., Yi, M.-F. and Stoffolano, J.G.,** Discovery of a midgut peptide hormone which activates the endocrine cascade leading to oogenesis in *Phormia regina* (Meigen), *J. Insect Physiol.*, 40, 283, 1994.
39. **Bownes, M. and Reid, G.,** The role of the ovary and nutritional signals in the regulation of fat body yolk protein gene expression in *Drosophila melanogaster*, *J. Insect Physiol.*, 36, 471, 1990.
40. **Agui, N., Shimadu, T., Izumi, S., and Tomino, S.,** Hormonal control of vitellogenin mRNA levels in the male and female housefly, *Musca domestica*, *J. Insect Physiol.*, 37, 383, 1991.
41. **Bownes, M.,** The role of juvenile hormone, ecdysone and the ovary in the control of *Drosophila* vitellogenesis, *J. Insect Physiol.*, 35, 409, 1989.
42. **Wigglesworth, V.B.,** The function of the corpus allatum in growth and reproduction of *Rhodnius prolixus*, *Q. J. Microsc. Sci.*, 79, 91, 1936.
43. **Davey, K.G.,** Hormonal control of vitellogenin uptake in *Rhodnius prolixus* Stål. *Am. Zool.*, 21, 701, 1981.
44. **Pratt, G.E. and Davey, K.G.,** The corpus allatum and oogenesis in *Rhodnius prolixus* (Stål.). III. The effects of allatectomy, *J. Exp. Biol.*, 56, 201, 1972.
45. **Abu-Hakima, R. and Davey, K.G.,** Two actions of juvenile hormone on the follicle cells of *Rhodnius prolixus* Stål., *Can. J. Zool.*, 53, 1187, 1975.
46. **Wang, Z. and Davey, K.G.,** Characterization of yolk protein and its receptor on the oocyte membrane in *Rhodnius prolixus*, *Insect Biochem. Molec. Biol.*, 22, 757, 1992.
47. **Davey, K.G. and Chiang, R.G.,** The effect of severing the doral vessel on egg production in *Rhodnius prolixus*, *Arch. Insect Biochem. Physiol.*, 11, 139, 1989.
48. **Heubner, E. and Davey, K.G.,** An antigonadotropin from the ovaries of the insect *Rhodnius prolixus* Stål., *Can. J. Zool.*, 51, 113, 1973.
49. **Davey, K.G. and Kuster, J.E.,** The source of an antigonadotropin in the female of *Rhodnius prolixus* Stål., *Can. J. Zool.*, 59, 761, 1981.
50. **Davey, K.G. and Heubner, E.,** The response of follicle cells of *Rhodnius prolixus* to juvenile hormone and antigonadotropin *in vitro*, *Can. J. Zool.*, 52, 1407, 1974.
51. **Kriger, F.L. and Davey, K.G.,** Identified neurosecretory cells in the brain of female *Rhodnius prolixus* contain a myotropic peptide, *Can. J. Zool.*, 62, 1720, 1984.
52. **Davey, K.G. and Kriger, F.L.,** Variations during the gonotrophic cycle in the titer of the myotropic ovulation hormone and the response of the ovarian muscles in *Rhodnius prolixus*, *Gen. Comp. Endocrinol.*, 58, 452, 1985.
53. **Davey, K.G.,** Copulation and egg production in *Rhodnius prolixus:* the role of the spermathecae, *J. Exp. Biol.*, 42, 373, 1965.
54. **Ruegg, R.P., Kriger, F.L., Davey, K.G., and Steel, C.G.H.,** Ovarian ecdysone elicits release of a myotropic ovulation hormone in *Rhodnius* (Insects: Hemiptera), *Int. J. Invertebr. Reprod.*, 3, 357, 1981.
55. **Ruegg, R.P., Orchard, I., and Davey, K.G.,** 20-Hydroxy-ecdysone as a modulator of electrical activity in neurosecreory cells of *Rhodnius prolixus*, *J. Insect Physiol.*, 28, 243, 1982.
56. **Davey, K.G. and Prasher, A.K.,** Increased synthesis of proteins is closely coupled to release in the myotropic neurosecretory cells of *Rhodnius prolixus*, *Insect Biochem.*, 20, 215, 1990.

57. **Orchard, I., Ruegg, R.P., and Davey, K.G.,** The role of central aminergic neurons in the action of 20-hydroxy-ecdysone on neurosecretory cells of *Rhodnius prolixus, J. Insect Physiol.*, 29, 387, 1983.
58. **Ampleford, E.J. and Davey, K.G.,** Egg laying in the insect *Rhodnius prolixus* is timed in a circadian fashion, *J. Insect Physiol.*, 35, 183, 1989.
59. **Dixon, A.F.G.,** *Aphid Ecology*, Blackie, Glasgow, 1985.
60. **Lees, A.D.,** The control of polymorphism in aphids, *Adv. Insect Physiol.*, 3, 207, 1966.
61. **Johnson, B.,** A histological study of neurosecretion in aphids, *J. Insect Physiol.*, 9, 727, 1963.
62. **Steel, C.G.H. and Lees, A.D.,** The role of neurosecretion in the photoperiodic control of polymorphism in the aphid *Megoura viciae, J. Exp. Biol.*, 67, 117, 1977.
63. **Steel, C.G.H.,** Neurosecretory control of polymorphism in aphids, in *Phase and Caste Determination in Insects, Endocrine aspects*, Lüscher, M., Ed., Pergamon, Oxford, 1976, 117.
64. **Mittler, T.E., Nassar, S.G., and Staal, G.B.,** Wing development and parthenogenesis induced in progenies of kinoprene-treated gynoparae of *Aphis fabae* and *Myzus persicae, J. Insect Physiol.*, 22, 1717, 1976.
65. **Hardie, J.,** Juvenile hormone and photoperiodically controlled polymorphism in *Aphis fabae:* prenatal effects on presumptive oviparae, *J. Insect Physiol.*, 27, 257, 1981.
66. **Hardie, J. and Lees, A.D.,** The induction of normal and teratoid viviparae by juvenile hormone and kinoprene in two aphid species, *Physiol. Entomol.*, 10, 65, 1985.
67. **Corbitt, T.S. and Hardie, J.,** Juvenile hormone effects on polymorphism in the pea aphid, *Acyrthosiphon pisum, Entomol. Exp. Appl.*, 38, 131, 1985.
68. **Hardie, J.,** The corpus allatum, neurosecretion and photoperiodically controlled polymorphism in an aphid, *J. Insect Physiol.*, 33, 201, 1987.
69. **Hardie, J., Baker, F.C., Jamieson, G.C., Lees, A.D., and Schooley, D.A.,** The identification of an aphid juvenile hormone, and its titre in relation to photoperiod, *Physiol. Entomol.*, 10, 297, 1985.
70. **Hardie, J. and Lees, A.D.,** Endocrine control of polymorphism and polyphenism, in *Comprehensive Insect Physiology, Biochemistry and Pharmacology*, Kerkut, G.A. and Gilbert, L.I., Eds., Pergamon, Oxford, Vol. 8, 1985, 441.
71. **Hardie, J.,** Juvenile hormone stimulation of oocyte development and embryogenesis in the parthenogenetic ovaries of an aphid, *Aphis fabae, Int. J. Invertebr. Reprod. Dev.*, 11, 189, 1987.
72. **Hardie, J., Mallory, A.C.L., and Quashie-Williams, C.A.,** Juvenile hormone and host-plant colonization by the black bean aphid, *Aphis fabae, Physiol. Entomol.*, 15, 331, 1990.
73. **Lees, A.D.,** The development of juvenile hormone sensitivity in alatae of the aphid *Megoura viciae, J. Insect Physiol.*, 26, 143, 1980.
74. **Ishikawa, H. and Matsuka, M.,** Hemolymph proteins of chestnut aphid, *Lachnus tropicalis, Comp. Biochem. Physiol.*, 74B, 521, 1983.
75. **Tamaki, Y.,** Sex pheromones, in *Comprehensive Insect Physiology, Biochemistry and Pharmacology*, Kerkut, G.A. and Gilbert, L.I., Eds., Pergamon, Oxford, Vol. 9, 1985, 145.
76. **Barth, R.H.,** Insect mating behavior: endocrine control of a chemical communication system, *Science*, 149, 882, 1965.
77. **Schal, C., Burns, E.L., and Blomquist, G.J.,** Endocrine regulation of female contact sex pheromone production in the German cockroach, *Blattella germanica, Physiol. Entomol.*, 15, 81, 1990.
78. **Schal, C., Burns, E.L., Gadot, M., Chase, J., and Blomquist G.J.,** Biochemistry and regulation of pheromone production in *Blattella germanica* (L.) (Dictyoptera, Blattellidae), *Insect Biochem.*, 21, 73, 1991.
79. **Smith, A.F. and Schal, C.,** Corpus allatum control of sex pheromone production and calling in the female brown-banded cockroach, *Supella longipalpa* (F.) (Dictyoptera: Blattellidae), *J. Insect Physiol.*, 36, 251, 1990.
80. **Cusson, J. and McNeil, J.N.,** Involvement of juvenile hormone in the regulation of pheromone release activities in a moth, *Science*, 243, 210, 1989.
81. **Cusson, J., Yagi, K., Tobe, S.S., and McNeil, J.N.,** Identification of release products of corpora allata of male and female armyworm moths, *Pseudaletia unipuncta, J. Insect Physiol.*, 39, 775, 1993.
82. **Raina, K.A. and Klun, J.A.,** Brain factor control of sex pheromone production in the female corn earworm moth, *Science*, 225, 531, 1984.
83. **Raina, K.A., Jaffe, H., Kempe, T.G., Blacher, R.W., Fales, H.M., Riley, C.T., Klun, J.A., Ridgway, R.L., and Hayes, D.K.,** Identification of a neuropeptide hormone that regulates sex pheromone production in female moths, *Science*, 244, 796, 1989.
84. **Kitamura, A., Nagasawa, H., Kataoka, H., Ando, T. and Suzuki, A.,** Amino acid sequence of pheromone-biosynthesis-activating neuropeptide-II (PBAN-II) of the silkworm, *Bombyx mori, Agric. Biol. Chem.*, 54, 2495, 1990.
85. **Kitamura, A., Nagasawa, H., Kataoka, H., Inoue, T., Matsumoto, S., Ando, T., and Suzuki, A.,** Amino acid sequence of pheromone-biosynthesis-activating neuropeptide (PBAN) of the silkworm, *Bombyx mori, Biochem. Biophys. Res. Commun.*, 163, 520, 1989.

86. **Matsumoto, R., Yamashita, O., Fonagy, A., Kurihara, M., Uchiumi, K., Niagamine, T., and Mitsui, T.**, Functional diversity of a pheromonatropic neuropeptide: induction of cuticular melanization and embryonic diapause in lepidopteran insects by *Pseudaletia* pheromonotropin, *J. Insect Physiol.*, 38, 847, 1992.
87. **Lafont, R.**, Reverse endocrinology, or "hormones" seeking functions, *Insect Biochem.*, 21, 697, 1991.
88. **Foster, S.P., Bergh, J.C., Rose, S., and Harris, M.O.**, Aspects of pheromone biosynthesis in the Hessian fly, *Mayetiola destructor* (Say), *J. Insect Physiol.*, 37, 899, 1991.
89. **Raina, K.A. and Menn, J.J.**, Pheromone biosynthesis activating neuropeptide: from discovery to current status, *Arch. Insect Biochem. Physiol.*, 22, 141, 1993.
90. **Blomquist, G.J., Adams, T.S., Halarnkar, P.P., Gu, P., Mackay, M.E., and Brown, L.A.**, Ecdysteroid induction of sex pheromone biosynthesis in the housefly, *Musca domestica* — are other factors involved?, *J. Insect Physiol.*, 38, 309, 1992.
91. **Dumser, B.J.**, The regulation of spermatogenesis in insects, *Annu. Rev. Entomol.*, 25, 341, 1980.
92. **Dumser, B.J. and Davey, K.G.**, The *Rhodnius* testis: hormonal effects on germ cell division, *Can. J. Zool.*, 53, 1682, 1975.
93. **Friedländer, M. and Reynolds, S.E.**, Meiotic metaphases are induced by 20-hydroxyecdysone during spermatogenesis of the tobacco hornworm, *Manduca sexta*, *J. Insect Physiol.*, 34, 1013, 1988.
94. **Leviatan, R. and Freidländer, M.**, The eupyrene/apyrene dichotomous spermatogenesis of Lepidoptera. I. The relationship with postembryonic development and the role of the decline in juvenile hormone titre towards pupation, *Dev. Biol.*, 68, 515, 1979.
95. **Koolman, J., Scheller, K., and Bodenstein, D.**, Ecdysteroids in the adult male blowfly *Calliphora vicina*, *Experientia.*, 35, 134, 1979.
96. **Loeb, M.J., Woods, C.W., Brandt, E.P., and Borkovec, A.**, Larval testes of the tobacco budworm: a new source of insect ecdysteroids, *Science*, 218, 896, 1982.
97. **Loeb, M.J., Brandt, E.P., and Birnbaum, M.J.**, Ecdysteroid production by testes of the tobacco budworm, *Heliothis virescens*, from last larval instar to adult, *J. Insect Physiol.*, 30, 375, 1984.
98. **Friedländer, M. and Reynolds, S.E.**, Intratesticular ecdysteroid titres and the arrest of sperm production during pupal diapause in the tobacco hornworm, *Manduca sexta*, *J. Insect Physiol.*, 38, 693, 1992.
99. **Loeb, M.J., Brandt, E.P., Woods, C.W., and Borkovec, A.**, An ecdysiotropic factor from brains of *Heliothis virescens* induces testes to produce a immuno-detectable ecdysteroids *in vitro*, *J. Exp. Zool.*, 243, 275, 1987.
100. **Kambysellis, M. and Williams, C.M.**, *In vitro* development of insect tissues. I. A macromolecular factor prerequisite for silkworm spermatogenesis, *Biol. Bull.*, 141, 527, 1971.
101. **Kambysellis, M. and Williams, C.M.**, *In vitro* development of insect tissues. I. The role of ecdysone in the spermatogenesis of silkworms, *Biol. Bull.*, 141, 541, 1971.
102. **Friedländer, M. and Benz, G.**, Control of spermatogenesis resumption in post-diapausing larvae of the codling moth, *J. Insect Physiol.*, 28, 349, 1982.
103. **Friedländer, M.**, 20-Hydroxyecdysone induces glycogen accumulation within the testicular sheath during *in vitro* spermatogenesis renewal in diapausing codling moths (*Cydia pomonella*), *J. Insect Physiol.* 35, 29, 1989.
104. **Friedländer, M.**, Juvenile hormone and regulation of dichotomous spermatogenesis during larval diapause of the codling moth, *J. Insect Physiol.*, 28, 1009, 1982.
105. **Shimizu, T., Moribayashi, A., and Agui, N.**, *In vitro* analyses of spermiogenesis and testicular ecdysteroids in the cabbage armyworm, *Mamestra brassicae* L. (Lepidoptera: Noctuidae), *Appl. Entomol. Zool.*, 20, 56, 1985.
106. **Giebultowicz, J.M., Loeb, M.J., and Borkovec, A.B.**, *In vitro* spermatogenesis in lepidopteran larvae: role of the testis sheath, *J. Invertebr. Reprod. Dev.*, 11, 211, 1987.
107. **Shimizu, T., Yagi, S., and Kuramochi, K.**, Regulation of spermiogenesis in the common armyworm, *Leucania separata* (Lepidoptera: Noctuidae), *Appl. Entomol. Zool.*, 23, 156, 1988.
108. **Shimizu, T., Yagi, S., and Agui, N.**, The relationship of testicular and hemolymph ecdysteroid titer to spermiogenesis in the common armyworm, *Leucania separata*, *Entomol. Exp. Appl.*, 50, 195, 1989.
109. **Gelman, D.B., Woods, C.W., and Borkovec, A.B.**, Effects of ecdysone and 20-hydroxyecdysone on apyrene spermiogenesis in the European corn borer, *Ostrinia nubilalis*, *J. Insect Physiol.*, 34, 733, 1988.
110. **Riemann, J.G., Thorson, B.J., and Ruud, R.L.**, Daily cycle of release of sperm from the testes of the Mediterranian flour moth, *J. Insect Physiol.*, 20, 195, 1974.
111. **Giebultowicz, J.M., Riemann, J.G., Raina, A.K., and Ridgway, R.L.**, Circadian system controlling the release of sperm in insect testes, *Science*, 245, 1098, 1989.
112. **Thorson, B.J. and Riemann, J.G.**, Effect of 20-hydroxyecdysone on sperm release from the testes of the Mediterranian flour moth, *Anagasta kuehniella* (Zeller), *J. Insect Physiol.*, 28, 1013, 1982.
113. **Giebultowicz, J.M., Feldlaufer, M., and Gelman, D.B.**, Role of ecdysteroids in the regulation of sperm release from the testis of the gypsy moth, *Lymantria dispar*, *J. Insect Physiol.*, 36, 567, 1990.
114. **Osanai, M., Kasuga, H., and Aikagi, T.**, Induction of sperm motility of apyrene sperm and dissociation of eupyrene sperm bundles of the silkworm, *Bombyx mori* by initiatorin and trypsin, *Invertebr. Reprod. Dev.*, 15, 97, 1989.

115. **Nowock, J.,** Induction of imaginal differentiation by ecdysone in the testes of *Ephestia kuehniella*, *J. Insect Physiol.*, 18, 1699, 1972.
116. **Szollosi, A. and Landureau, J.-C.,** Imaginal cell differentiation in the spermduct of *Samia cynthia* (Lepidoptera). Responses *in vitro* to ecdysone and ecdysterone, *Biol. Cell.*, 28, 23, 1977.
117. **Shinbo, H. and Happ, G.M.,** Effects of ecdysteroids on the growth of the post-testicular reproductive organs in the silkworm, *Bombyx mori*, *J. Insect Physiol.*, 35, 855, 1989.
118. **Shimizu, T.,** Development of spermduct and seminal vesicle during pharate adult of cabbage armyworm, *Mamestra brassicae*, *Invertebr. Reprod. Dev.*, 15, 221, 1989.
119. **Loeb, M.J.,** Growth and development of spermducts of the tobacco budworm, *Heliothis virescens*, *in vivo* and *in vitro*, *Invertebr. Reprod. Dev.*, 19, 97, 1991.
120. **Loeb, M.J.,** Development of isolated spermducts from *Heliothis virescens* (Lepidoptera) *in vitro*, *Invertebr. Reprod. Dev.*, 20, 67, 1991.
121. **Kelly, T.J., Adams, T.S., Schwartz, M.B., Birnbaum, M.J., Rubenstein, E.C., and Imberski, R.B.,** Juvenile hormone and ovarian maturation in the diptera: a review of recent results, *Insect Biochem.*, 17, 1089, 1987.

Chapter 5

FATAL ATTRACTION: THE DISRUPTION OF MATING AND FERTILIZATION FOR INSECT CONTROL

Richard Wall

CONTENTS

I. INTRODUCTION

The range of powerful insecticides now available provides the means of controlling the vast majority of insect pests. However, complete reliance on these chemicals brings with it a variety of associated problems. These include the development of resistance by the target pest species, a variety of unintentional effects on nontarget organisms, the presence of pesticide residues in food, and high costs both of development and deployment.[1] As a result, there is growing concern that in the future the judicious use of chemical insecticides must, at the very least, be supplemented by the development of a range of noninsecticidal techniques for pest control.[2,3]

0-8493-6695-X/95/$0.00+$.50

In the search for effective and acceptable alternatives to chemical control, considerable attention has been directed toward insect reproduction. Mate location and identification, copulation, insemination and fertilization, external physical and chemical stimuli, and the internal physiological mechanisms that regulate each step present an enormous range of features which are available for disruption or impairment. Disruption of component parts of the process of reproduction allows the insect's own reproductive behavioral repertoire to be turned against itself, to bring about genetic death. These autocidal control techniques have several advantages over simple killing systems, including being highly species specific so that they have minimal impact on the nontarget fauna.

The techniques so far developed for control by manipulation of insect reproduction are those which produce a reduction in fertility. This has been effected principally by the release of males which have been rendered sexually sterile, altered genetically, or otherwise treated to disrupt reproduction in the natural population. In the case of sterile males, the released individuals compete for matings in the wild population, and the eggs of wild females fertilized by their sperm fail to hatch. The release of males with heritable genetic deficiencies[4] is a step in sophistication beyond the use of individuals that have been sterilized by the induction of dominant lethal mutations with radiation because it has the advantage that the resultant effects may persist in the population for several generations.[5] Finally, the development of systems that attract and impair the reproduction of individuals in the field overcomes the need to mass rear and release sterilized or genetically modified insects. Such devices are known as autosterilizing systems and are based on the integration of a chemosterilant or a compound which can produce a similar end result through transovarial effects on eggs or larvae with an attractant into a single control device which can be placed in the field.

This chapter presents an overview of these techniques and the principles behind them. It focuses particularly on Diptera of medical and veterinary importance and seeks, wherever possible, to give detailed examples of practical control operations in the field.

II. STERILE INSECT TECHNIQUE

The release of sterilized male insects to suppress a wild population was first proposed in the late 1930s.[6] It is now referred to as the sterile insect technique or SIT. Irradiated sterile males are released into a wild population in numbers sufficient to allow them to obtain a large proportion of the matings with fertile females. Eggs fertilized by the sperm of the irradiated males fail to hatch. As a result, with continued release of sterile males, the population of wild insects is eventually driven to extinction.

A. THE PRINCIPLES OF CONTROL BY SIT

The effectiveness of the release of sterile males depends critically on the ratio of wild females to sterile males released. At the simplest level, in a population of 100 fertile females and an equal number of fertile males, if 900 sterile males are released to give a total male population of 1000, on average only 100 × (100/1000) = 10 fertile matings will occur. If each fertile female produces only a single female offspring, a further release of the same number of sterile males would make it unlikely that any fertile matings would occur in the second generation.[7]

The effectiveness of the suppression is influenced by the competitiveness of the released males, immigration of already mated females into the release area, and the recovery capacity or seasonal changes in the target population.[5] These factors can be incorporated into a simple equation[8] which can be used to explore the effects of sterile male release:

$$W_n + 1 = (W_n \times \left[\frac{W_n}{W_n + (R \times C)}\right] + M) \times D \tag{1}$$

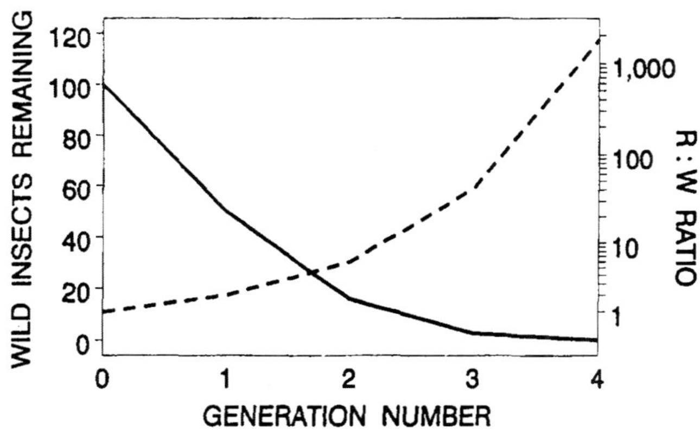

FIGURE 1. The number of wild insects of either sex remaining in a population subjected to the release of a constant number of sterile males, equivalent to the initial number of fertile females (solid line). Also, the change in ratio of released males (R) to wild (W) insects of either sex (dashed line). There is no immigration (M = 0). The wild population would have remained constant if there had been no intervention, and density-dependent factors do not operate to give it any tendency to recover (D = 1). The released sterile males are assumed to be fully sterile and fully competitive with wild males (C = 1).

W is the number of wild individuals of either sex in the population each generation n, assuming an equal sex ratio. Sterilization is considered to be complete, and R is the number of completely sterile males released. C is the probability of a released male mating with a wild female, relative to that of a wild male doing so, where a value of 1 represents full competitiveness. M is the number of fertile, mated, immigrant females. These are assumed to be refractory to further mating. D simulates the tendency of the target population to increase or decrease due to seasonal or density dependent factors. When D = 1 and R = 0, each adult female produces an average of one female offspring and with no immigration (M = 0) the population size remains constant. Sterile females released are not included in the equation since they are not expected to affect the dynamics of the population.

Using the equation to examine the effects of altering the various parameters shows that with a constant rate of release of sterile males (R), the population declines towards eradication because as the number of wild individuals falls, the $R:W_n$ ratio increases (Figure 1). However, the rate at which eradication of the wild population is achieved would be retarded if the competitiveness of released males was lower than that of wild males (Figure 2).

The rate of eradication is also strongly affected by the immigration of fertile females since, even if the released males sterilize all the indigenous females, this can never push the reproducing population below the number of inseminated female immigrants per generation. The rate of reduction due to sterile male release may also be affected in the short term by underlying increases or decreases in the target population, but more important, may be permanently prevented if density-dependent regulatory factors are strong enough to compensate for the maximum sterility which can be achieved. Finally, in practical terms, the initial size of the wild population is of importance, since low density populations require the release of smaller numbers of sterile males to achieve a given rate of reduction.

The major advantages of the technique are that it is species specific, so that in most cases it presents no environmental hazards. In addition, it may be an effective means of "mopping up" the residual population left after insecticidal methods have achieved the maximum possible effect, since even at low population densities, the behavioral mechanisms that bring the sexes together for mating mean that released male flies are likely to be able to locate the remaining fertile females.

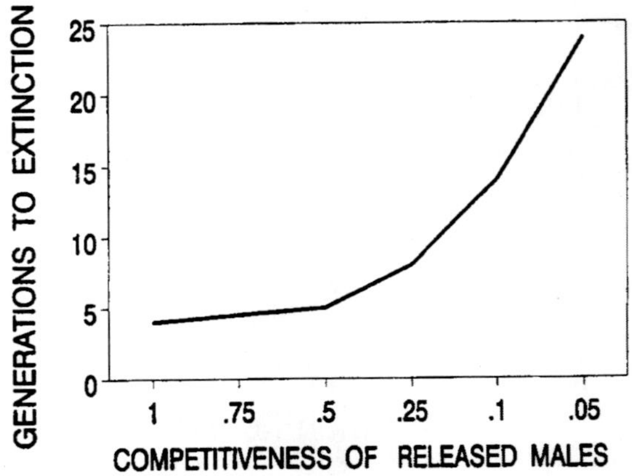

FIGURE 2. The number of generations required to reduce a wild population to less than 1, by the release of a constant number of sterile males, equivalent to the initial number of fertile females. Released males differ in their competitiveness with wild males (C). There is no immigration and D in equation 1 is unity.

B. PRACTICAL CONTROL BY SIT

As the foregoing discussion shows, there are numerous practical imperatives which must be achieved for the effective use of SIT. Most critically, to ensure success, the release area must be isolated as far as possible to protect against immigration. The fact that this is seldom achievable makes effective control difficult. For example, suppression of populations of the blood-sucking pest the stable fly *Stomoxys calcitrans*, with sterilized males was attempted in central Florida.[9] Sterilization was brought about by exposure of pupae to 2 krad from a ^{60}Co source. Adults were allowed to emerge in the laboratory and released in areas of high cattle density. Each day sterile flies were released, initially at equal and, eventually, at double the estimated initial wild population. Despite daily losses of about 35% of the released males through mortality and emigration and the fact that the released males were only about half as competitive as wild males, after one generation, over 85% sterility was found in the native population. With continued release, sterility rapidly approached 100%, while the field population fell in size by 97%. Although sterility remained high in the release area for a number of weeks after the last release of sterile males, subsequent reinfestation was rapid.[9]

The usefulness of the sterile insect technique for the control or eradication of tsetse flies (*Glossina* spp.), the vectors of African trypanosomiasis, has been demonstrated by several experiments and field projects.[10-12] However, the extensive rearing facilities and logistic support required to breed and release large numbers of insects make application of SIT expensive. This is particularly so when more than one species or strain of a single species need to be controlled. For example, control of *Glossina palpalis palpalis* was achieved in central Nigeria, through the release of more than 1.5 million sterile males in an area of 1500 km^2.[13] Males were released at a ratio of at least 10 sterile to each fertile wild male, for at least 3 generations. Control was assisted by the fact that *G. p. palpalis* populations are concentrated in areas of riverine habitat, thus allowing the release of sterile males to be focused in these areas. The release area was maintained free of *G. p. palpalis* for 3 years by deploying insecticide trap and target barriers around the cleared area. However, the problem of animal trypanosomiasis was not eliminated because of the presence of the sympatric species *Glossina tachinoides*. Additionally, when dealing with pests that are harmful to man or animals such as tsetse flies, the release of hundreds of thousands of potential vectors, although sterilized, could increase substantially the transmission of disease. Therefore, the wild population must be reduced significantly with insecticides or traps prior to the release.

The competitiveness of the males reared for release is often difficult to maintain under mass rearing regimens in the laboratory. For example, attempts to control the mosquito *Anopheles quadrimaculatus* in Florida involved the release of 433,600 irradiated males over a period of

14 months.[14] No reduction in mosquito numbers and no reduction in fertility of wild females resulted. It was concluded that the failure to suppress the wild population was the result of the fact that the laboratory strain used was too inbred to be competitive with wild males. Similar lack of success was recorded in an attempt to control the mosquito *Aedes aegypti*, lack of competitiveness on this occasion being attributed to too high a radiation dose.[15]

One further disadvantage of sterile male release and all other sterilizing systems over simply killing the target pest with insecticides is the delay in suppressing the population. If a pest has already reached economic damage levels, considerable further losses would occur before the sterilized flies has any affect on the wild population.

In addition to the above examples, control by release of sterile males has been attempted for the suppression of populations of numerous other species of pest Diptera, such as the Mediterranean fruit fly *Ceratitis capitata*,[16] horn flies *Haematobia irritans*,[17] the blowfly *Lucilia sericata*,[18] and the heel fly *Hypoderma lineatum*.[19] However, in many cases, while trials have given promising results, the technique has been found too expensive to be used on a larger scale in the field, or little more than temporary, local population suppression has been achieved. The major exception to this pattern, however, has been the use of the sterile insect technique for the eradication of the New World screwworm fly *Cochliomyia hominivorax* from North America and North Africa.

1. Screwworm Eradication in North America

The screwworm fly *C. hominivorax* is an obligate primary parasite of warm-blooded animals, including humans.[20] Adult females lay batches of up to 350–450 eggs on the edges of open wounds or in body orifices. On hatching, within 24 hours the maggots start to feed, burrowing into the living tissue.[21] The resulting extensive wound may rapidly lead to the death of the struck animal.[22]

The natural range of the fly extends from the southern states of the U.S. through Central America and the Caribbean Islands to northern Chile, Argentina, and Uruguay. Each summer the fly used to spread north and west into more temperate zones from its overwintering areas near the U.S./Mexican border. The fly was of greatest significance as a pest of livestock, necessitating the continued costs of vigilance, treatment and control. In the epidemic year of 1935 in Texas, there were approximately 230,000 cases in livestock and 55 in humans.[23]

In 1954 the first full-scale operation to eradicate *C. hominivorax* was mounted on the island of Curaçao, 50 miles off the coast of Venezuela. Eradication was achieved in 6 months by swamping the wild population with about 800 sterile males per square mile per week with screwworm files sterilized with X-rays.[24] Attention was then turned to removal of the screwworm fly from mainland U.S. The fly was eradicated first of all from the Florida peninsula in 1957 through 1959, assisted by the unusually cold winters of 1957 and 1958. Pupae 5 days old were exposed to ^{60}Co radiation and both sexes were released as adults at a peak rate of about 500 km^2/week.[25]

In subsequent stages, the fly was progressively pushed south from areas of the U.S. Flies were produced at a large rearing factory in Texas capable of producing 200 million flies per week. These were sterilized with ^{137}Cs at a minimum dose of 5500 rads and released from the air as adults. Eventually a 3100-km long barrier zone was established against reinvasion, from Texas to California along the Mexico border.[24,26]

Between 1972 and 1976, however, the number of infestations of *C. hominivorax* recorded in the U.S. increased dramatically.[25] This may have been the result of the intense selection pressure generated by the massive scale of the release of irradiated screwworm flies, leading to rapid evolution of a strain of wild flies that no longer mated readily with the released males. In addition, factory rearing may have favored the selection of a domesticated strain unable to compete with wild males.[27] The outbreak in Texas in 1976 was estimated to have cost consumers and producers between U.S. $280–370 million. Nevertheless, effective control of *C. hominivorax* in the U.S. was reimposed in 1979, and no screwworm infestations have been

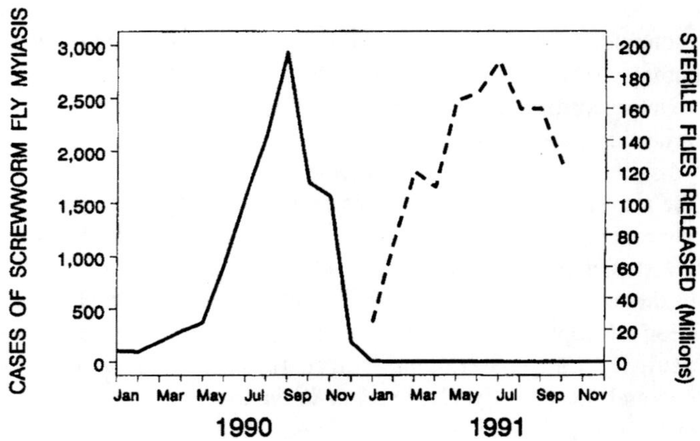

FIGURE 3. The number of confirmed cases of New World screwworm fly infestation recorded each month in Libya in 1990 and 1991 (solid line); the number of sterile screwworm flies released each month in 1991 (dashed line). (Modified from FAO *New World Screwworm Newsletter* 1991, 1.)

detected since 1984.[28,29] The U.S. Department of Agriculture program has subsequently been directed against the fly in Mexico, Puerto Rico, Vieques, and the Virgin Islands,[28,29] and in 1991 Mexico was declared officially free of screwworm fly.

The effects of weather on the outcome of the SIT control campaigns has been the subject of recent debate. Weather has been recognized to affect screwworm fly populations; higher temperatures and humidity increase the survival of this subtropical species and affect underlying rates of population increase.[24] It is notable that the successful screwworm SIT control campaigns in the U.S. coincided with the particularly cold winters of 1957, 1958, and 1962. It has been suggested that a contributory factor to the breakdown of screwworm control in the U.S. between 1972 and 1976 may have been the favorable warm winters in those years. These may have allowed flies to invade the control area and overwinter, or may have enabled undetected residual populations to expand rapidly.[30] Furthermore, it was suggested that given the presence of undetected residual populations, new outbreaks may well occur when climatic conditions again favor the pest. The existence of residual populations has been disputed, however, on the grounds that they could not remain undetected, given the fecundity and mobility of *C. hominivorax* and the clinical severity of screwworm infestations. It was further suggested that there were no grounds for supposing that climatic factors are critical to the success of screwworm eradication, especially in subtropical areas such as southern Mexico.[26,31] Nevertheless, by influencing the initial size of the target population and the underlying rate of population increase, climatic factors will inevitably affect the ease with which a population can be suppressed by SIT.

2. Screwworm Eradication in North Africa

In 1988 *C. hominivorax* screwworm flies were discovered in an area 10 km south of Tripoli in Libya.[32] They probably arrived with a shipment of contaminated livestock from South America. This was the first known established population of this species outside the Americas. The fly quickly spread to infest about 25,000 km^2. In 1989 there were about 150 cases of myiasis by *C. hominivorax*, but by 1990, a total of 12,068 confirmed cases of screwworm fly myiasis were recorded and, at its peak, almost 3000 cases were seen in the single month of September 1990 (Figure 3). It was estimated that if unchecked the infestation could cost the Libyian livestock industry about U.S. $30 million per year and the North African region approximately U.S. $280 million per year.[33] Considerable alarm was expressed about the effects of allowing the fly to remain, since it is well adapted for rapid dispersal[34] and could have spread throughout North Africa and possibly into more favorable sub-Saharan habitats.[35]

As a result, an international program to eradicate the fly was established, coordinated by the Food and Agriculture Organization of the United Nations.

C. hominivorax were reared at the Tuxtla factory in Mexico, where they were sterilized by exposure to gamma radiation as late stage pupae. These were then transported in refrigerated vehicles to Mexico City, and then air-freighted to Libya. On arrival, the boxes were stored in environmentally controlled trailers and a sugar-based food was supplied to sustain the emerging flies. When approximately 80% of the adults had emerged, boxes of 1600 pupae were dispersed at about three to ten boxes per minute, from light aircraft flying at 240 km/h and at an altitude of 500 m. The aircraft flew along predetermined flight paths about 4 km apart. The boxes opened in midair or on impact with the ground, releasing the newly emerged sterile adults.[33]

The screwworm fly population in Libya was able to overwinter, though it showed a substantial fluctuation in response to weather (Figure 3). After a trial program in December 1990, full-scale releases began in February 1991, at the time when the screwworm fly population was at its seasonal minimum. Flies were released at initial rates of 3.5 million per week, escalating to peak rates in July 1991 of 40 million per week, at densities ranging from 500–1200 per km^2. By May 1991 an area of 41,000 km^2 was being treated each week. The release of sterile males was backed up by extensive control operations on the ground, including surveillance teams and quarantine stations to prevent movement of infested animals outside the infested area. In 1990 a total of 12,068 confirmed cases of *C. hominivorax* infestation has been recorded, but in 1991 only 6 cases were seen, the last occurring in April (Figure 3). By November, after 6 months with no further cases, the release of sterile flies was terminated and eradication announced.[36] The final cost of the program was estimated to be approximately U.S. $64 million.

The outstanding success of this sterile male release operation can be attributed in part to the ready availability of large numbers of screwworm flies and access to the already developed U.S. and Mexican expertise. In addition, the success of the operation was due to the absence of immigration, the confinement of the outbreak population in North Africa by the ocean to the north and desert to the south, and to the fact that in Libya, *C. hominivorax* was probably close to the edge of its climatic range, a marked trough in abundance occurring at unfavorable times of year.

III. GENETIC CONTROL

A. INTRODUCTION

Advances in SIT stimulated interest in the development of alternative mechanisms by which insects could be rendered effectively sterile. This interest has resulted in the identification of a wide range of genetic mechanisms that can be used to manipulate the genetic composition of natural insect populations.[4] These can broadly be divided into two categories. The first include mechanisms that have an effect for one generation only after release, and include dominant lethal mutations such as those induced by radiation or chemosterilants, cytoplasmic incompatibility, and hybrid sterility. The second includes inherited partial sterility, recessive lethal genes, and chromosome translocations. These mechanisms lead to inheritable distortions which, therefore, continue to exert an effect on the population for some generations after release. The advantages of sterility effects in the target population, prolonged over several generations, have concentrated attention on the second category of genetic mechanism, and translocations have, to date, appeared to be the most useful of these.

B. GENETIC SEXING SYSTEMS

1. General

Chromosome breakage is a common result of the irradiation of cells. If the broken sections of two nonhomologous chromosomes rejoin to the wrong partners, the result is called reciprocal translocation. Meiosis in individuals heterozygous for such translocations may then result in the production of gametes with an unbalanced, lethal, genetic complement. When these fertilize normal gametes, nonviable embryos are produced. The heterozygote is referred to as semisterile. For a single translocation, half of the progeny of a mating between a heterozygous and normal individuals are usually affected. A homozygote, which inherits the same translocation from both parents, may have impaired fertility or viability. Nevertheless, identification and selection of such homozygotes may allow them to be reared in sufficient numbers in the laboratory for their release in to the field. Every mating between an individual homozygous for the translocation and a normal individual will then result in the production of heterozygotes, which at meiosis will produce a proportion of nonviable gametes. When the semisterile heterozygote mates, only a proportion of its offspring will mature. The rearing and release of insects carrying semisterilizing autosomal translocations as a means of pest control was first proposed by Serebrovsky.[37]

The three main types of translocations that can occur are between an autosome (non-sex chromosome) and the Y chromosome, an autosome and X chromosome, or between two autosomes. In the first case, semisterility is inherited through the male. In the second, it can be inherited through both the male and female, but only female homozygotes are possible. In the third case, the translocation can be inherited through both sexes and both can be homozygotes.

The artificial linkage of deleterious selectable genes to sex, using Y-autosome translocations is known as a genetic sexing system. The development of genetic sexing systems may be of considerable value in the separation of the sexes in laboratory cultures, allowing males to be isolated prior to release in conventional SIT programs.[38] Hence, they may be particularly important in mosquitoes where males do not bite and mass release of biting females would be unacceptable. Genetic sexing systems have been proposed for a number of species of Diptera, including the housefly *Musca domestica*,[39] the tsetse flies *G. austeni* and *G. morsitans*,[40,41] the mosquitoes *Culex tritaeniorhynchus*[42,43] and *A. albimanus*[44] and the Mediterranean fruit fly *C. capitata*,[45-47] although functioning genetic sexing systems have been established for relatively few.[48] The use of Y-autosome translocations for control has been evaluated most comprehensively in Australia for the sheep blowfly *Lucilia cuprina*.[49-52]

2. Genetic Control of the Australian Sheep Blowfly

The blowfly *L. cuprina* is the most important myiasis pest of sheep in Australia.[53] Annual costs of production losses, prevention, and treatment were estimated to be approximately $149 million in 1985 and, in the absence of effective preventative measures, the sheep industry would be nonviable over much of the present grazing land in Australia. The primary means of combating flystrike in Australia has been with insecticides. However, concerns over growing insecticide resistance resulted in the initiation of research into SIT and genetic control.

It was found that exposure of *L. cuprina* to radiation can induce a range of mutations including translocations where one or more sections of the Y chromosome are swapped with sections of one or more autosomes. Males possessing the translocation can be identified and selected in the laboratory. Possession of the Y-linked translocation confers partial sterility, and a proportion of the eggs produced in a cross between a modified male and a wild female will fail to hatch.[49,54] Among the individuals that do hatch, the deleterious translocation is carried by the male progeny but not the females.

However, in a further refinement of the technique, it was found to be possible to translocate wild type alleles of autosomal mutations affecting eye color onto the Y chromosome. The recessive mutations are carried on the nontranslocation chromosome set by the genetically modified males and in heterozygotic females, but are not expressed because of the presence of the dominant wild type alleles. In female homozygotes, the mutations are expressed as white or yellow eye color. These females lack the light-filtering pigments that give the normal blowfly eye its reddish-brown color, so the homozygous mutant flies are effectively blind in

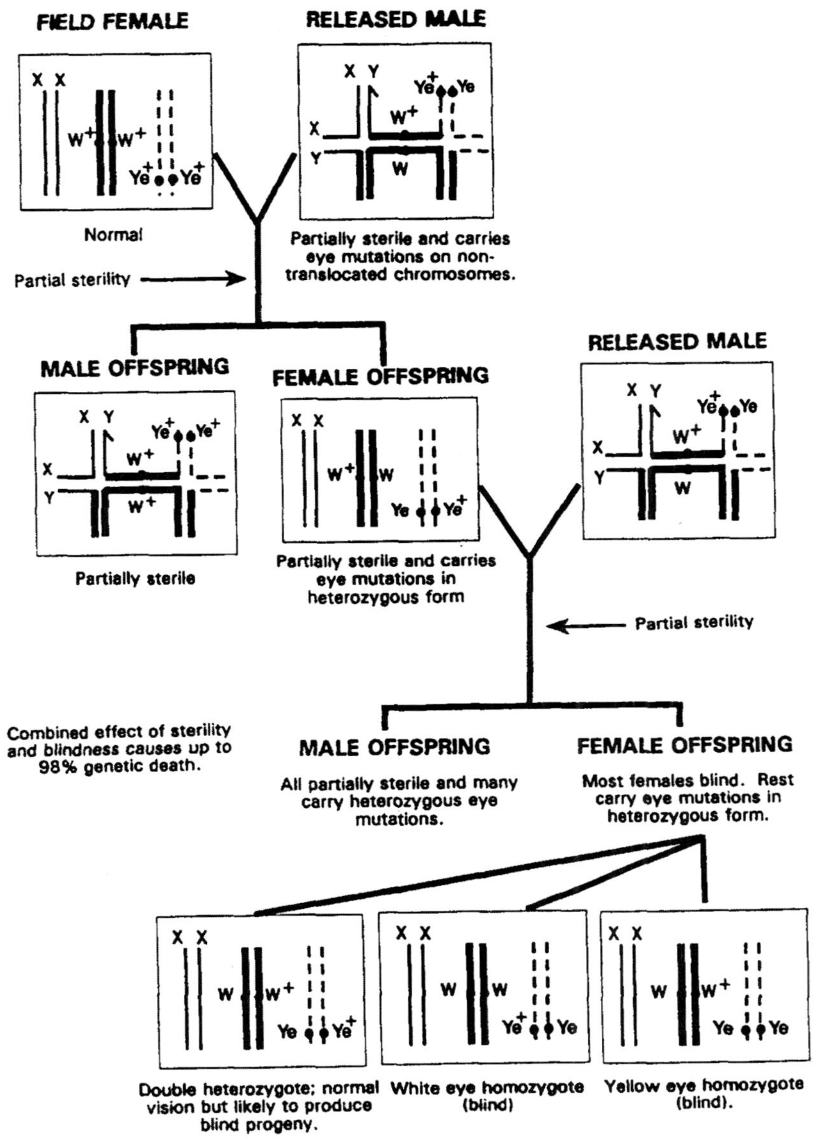

FIGURE 4. Transfer of genes (white eye W^+, yellow eye Ye^+, and the white and yellow eye mutations W and Ye) and translocated chromosomes from genetically modified male *Lucilia cuprina* released in the field.

daylight. When heterozygous females mate with the genetically altered males, a large percentage of their female progeny are homozygous for one or more of the eye pigment mutations. The blind females can readily be reared in the laboratory, but die rapidly in the field. The system is termed a female-killing (FK) system or the genetically impaired female technique (GIFT).[5]

Hence, release of the modified strains causes genetic death, partly from semisterility caused by the chromosome rearrangement and partly from death of the female descendants of the released males due to homozygosity for the mutations (Figure 4). An additional development (not shown in Figure 4) is that the released males also carry inversions to try to ensure maintenance of the linkage of the genes to the translocation and also to contribute an additional degree of potential sterility from females which inherit the inversions in heterozygous form. Computer simulations indicated that a theoretical maximum death rate of 98% per generation could be achieved by release of males possessing the chromosome abnormalities and the eye color mutations.[55] Furthermore, genetic death from semisterility and homozygosity should persist in the wild population for several generations after cessation of releases, giving this control system a considerable advantage over conventional SIT in which suppression ceases when release stops.[52,55] At lower release rates GIFT would be expected to result in more rapid suppression of the target population than SIT.

In field trials on Flinders Island, which is about 40 km^2 and 27 km off the Australian coast, 34,000 modified male *L. cuprina* were released per km^2 each week between August 1985 and May 1986. The induced rate of genetic death peaked at 87%, 6 months after the trial began and the blowfly population declined from 345 females per hectare in October 1985 to less than 1 female per hectare in May 1986, when releases were terminated. The population remained at below 4 females per hectare for the following 10 months.[52,56]

However, population suppression with the release of genetically modified males also suffers from many of the problems associated with conventional SIT, particularly the immigration of females from outside the release area and the noncompetitiveness of released males.[57] While the Flinders Island trial showed that it is possible to reduce an isolated *Lucillia* population by this method, further trials in the much larger Furneaux Island group encountered significant problems. Difficulties were experienced in rearing the 15 million modified flies required per week to swamp the wild population, and problems also arose from the breakdown of the genetic strains under large-scale rearing, due to the spontaneous recombination of genes.[58] The instability of genetic-sexing systems has also been recorded in a wide variety of other insect species.[59] As a result of these considerations in combination with the expense of maintaining large-scale rearing facilities, to date removal of *L. cuprina* from large areas of the Australian mainland has not been attempted.

IV. AUTOSTERILIZATION

Many of the problems associated with the need to mass rear insects for the release of sterilized or genetically modified males could be overcome by the use of devices which attract and disrupt the fertility of wild flies in the field. Such devices are known as autosterilizing systems. These devices may either sterilize flies directly, using a chemosterilant, or produce an equivalent end result via transovarial effects on eggs, larvae, or pupae. The principles that apply to the use of autosterilization systems are quite different from conventional killing systems or SIT. However, the potential for effective and efficient suppression of pest populations by autosterilization is in some respects much greater, although it does not have the advantage of the sterile:fertile male ratio improving as the wild population declines, as it does with SIT.

A. THE PRINCIPLES OF CONTROL BY AUTOSTERILIZATION

A device that simply attracts and kills greater numbers of males is unlikely to have any impact on the target pest population, since in most insect populations females mate infrequently, males are capable of inseminating several females, and the operational sex ratio, therefore, is heavily biased towards males. As a result, even a very small proportion of surviving males are usually capable of finding and inseminating all the available females. This is a critical problem for traps baited with chemicals, such as pheromones, or sound, that attract only males. Attracting, sterilizing, and then releasing males only would be more effective than attracting and killing these males, since the sterile males would mate with fertile wild females. The effects would then be similar to conventional SIT. Autosterilizing males also has the benefits of reduced costs and complexity when compared with mass-rearing and release. However, it would be difficult to achieve high sterile-to-fertile male ratios and, in most circumstances, attracting and sterilizing males only would not bring about suppression of the population as effectively as simply attracting and killing both sexes with insecticide.

Attracting and sterilizing females only would have no advantage over killing them at the same rate with insecticide. The attraction and sterilization of both sexes, however, would be

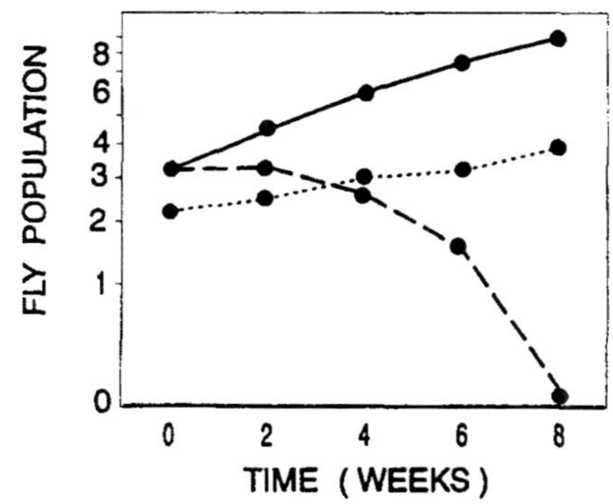

FIGURE 5. The change in abundance (log number) in a simulated housefly population that is untreated (solid line) or treated with insecticides that kill 90% of each generation (thin dashed line) or sterilants that sterilize 90% of each generation (thick dashed line). (Modified from Borkovec, A. *Advances in Pest Control Research*, Vol. 7, Interscience, New York.)

the most effective tactic.[60,61] For example, if a housefly population was treated with insecticides that kill 90% per generation, too few individuals would be killed to overcome the high potential for increase (Figure 5). However, if the population was exposed to material that sterilized 90% of both sexes of each generation, because sterile females are effectively genetically dead and there are nine sterile to each fertile male attempting to obtain a mating with each remaining fertile female, on average only 1% of the matings are between fertile females and fertile males and produce viable offspring (Figure 5). Hence, sterilization would be expected to eradicate the population in four generations.[62]

The value of sterilization of both sexes over killing depends on the proportion of the population that can be attracted and sterilized each generation.[61] At low rates of kill or sterilization there is little to choose between the two techniques, but as a greater and greater number of individuals are affected, the relative advantage of sterilization of both sexes over killing increases steeply (Figure 6). This is because the proportion of fertile matings is the product of the proportions of males and females *not* entering traps.[63] As a result, the availability of highly attractive baits and traps is critical to the development of practical autosterilizing systems.

Autosterilizing systems have a number of further theoretical advantages over insecticidal treatment or mass-rearing and release. Differential survivorship of fertile and sterile males and

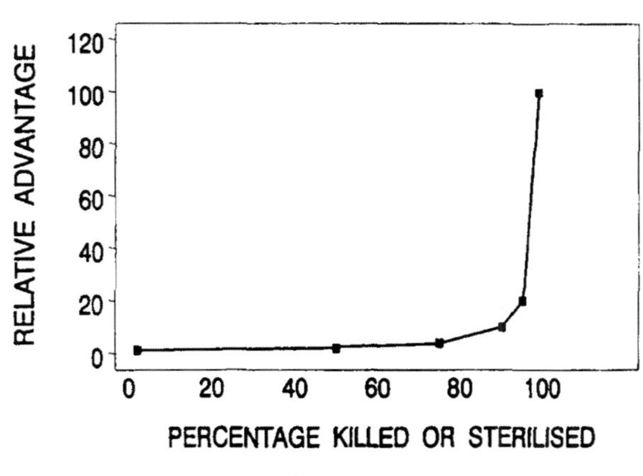

FIGURE 6. Relative superiority of sterilization over killing for control, plotted against the percentage sterilized or killed. Superiority is calculated as the ratio of the proportion of the population reproducing after treatment with insecticides to the proportion of the population reproducing after treatment with sterilant. (Modified from Knipling, E. *U.S.D.A. Agriculture Handbook*, No. 512, U.S. Department of Agriculture, Washington, D.C., 1979, 659.)

residual fertility of males visiting the baits make much less difference to the outcome of autosterilizing systems than they do to a conventional sterile release program, because much of the effect comes from the sterilizing of the trapped females. In addition, with sterilization of wild insects in the field, no additional insects are being added to the system.[60] Using autosterilizing devices would also reduce the strength of selection for behavioral resistance to attractants compared with killing traps, since attracted sterilized flies would mate with many of the unattracted individuals.

The action of food-bated traps containing insecticide would be expected to be density independent with respect to pest mortality. In contrast, the relative increase in efficiency of sterilization of both sexes over killing would increase at lower pest densities, because of the continued ability of individuals to find mates. The operation of density-dependent population regulation also would add to the advantage of autosterilization over insecticide use.[64] In populations which experience high levels of density-dependence, a smaller absolute number of individuals will need to be attracted and sterilized because of the lower density at which the population equilibrates, although the percentage of the population that would need to be attracted and sterilized per day would remain the same.[61] However, as with other control methods based on the induction of sterility, immigration into the control area would be especially detrimental to the use of autosterilization.[65]

B. ATTRACTANTS AND STERILANTS

As discussed earlier, the relative advantages of using an autosterilizing system depend on the nature and potency of the attractant, since this determines how many and whether only one or both sexes can be attracted. Attraction can be achieved either with olfactory behavior-controlling chemicals (semiochemicals) or visual or auditory cues. Among the semiochemicals, sex pheromones, particularly those produced by Lepidoptera, have received most attention.[66,67] However, most of the pheromones identified are produced by females for the stimulation of males. Additionally, among the Diptera most pheromones, at best, have been found to have only relatively low volatility. The hydrocarbons concerned often serve as contact sex-recognition chemicals, arrestants, and mating stimuli for the male and induce male copulatory activity when used in physiological amounts.[68-71] Pheromones produced by males for female attraction have been identified in few flies, with the exceptions of fruit flies of the genera *Dacus*[72] and *Drosophila*. Since attraction of males alone is of only limited use in an autosterilizing system, nonpheromonal semiochemicals, the kairomones, particularly those used by insects to locate host animals and plants for feeding or oviposition and which can attract both sexes or predominantly females, may be of considerably greater value as attractants.[73,74]

Vision has been shown to be of considerable importance in the behavior of many species of insect.[75,76] A wide range of features, such as shape, size, color, and movement have been found to contribute to the attraction of a number of species of flies to baited traps. However, of these, the responses to color have been shown to be of particular importance.[77-79] In general, attraction to blue, black, and red hues have been identified as most attractive in hematophagous Diptera, such as tabanids, blackfly, and mosquitoes.[76] In contrast, diurnal herbivorous insects have been found generally to be attracted to yellow hues with reflectivities between 500–580 nm.[75] The development of attractive baits for insects has been the subject of extensive work, which has been recently comprehensively reviewed by Muirhead-Thompson.[80]

The theoretical advantages of autosterilization have prompted the search for chemical sterilants that are effective and environmentally acceptable. Ideal sterilants should sterilize both sexes of a target species or, alternatively, separate male and female sterilants should be developed and used simultaneously. Sterilization should be lifelong and not impair mating competitiveness. The materials used should affect only the target pest, either by being biochemically species specific or by using a selective physical method of application.

A vast number of chemicals have been investigated.[81] However, many of the most potent materials, such as the alkylating agents[82] or the thioaziridine sterilant bisazir,[83,84] are potent not only against insects but have such high mammalian toxicity as to be unsafe for deployment in the field. A potent but entirely insect-specific sterilant, which acted against both sexes, would represent a major advance in the development of autosterilizing systems.

Considerable interest has been given to the use of pathogenic viruses as pesticides.[85] However, as yet, few appear to have shown practical promise as sterilants.

The search for alternative compounds has resulted in intensive studies of hormonal disruption of fertility or reproductive behavior using insect growth regulators (IGRs).[86,87] Amongst these the juvenile hormones and chitin synthesis inhibitors are of particular interest. The juvenile hormones, their synthetic mimics, and the chitin synthesis inhibitors have long been known to have a range of effects that might be exploited for control purposes, including disruption of normal embryogenesis, molt inhibition and disruption interference with diapause, stimulation of precocious egg development, and, most commonly, the derangement of metamorphosis. Although they have the advantages of being highly species specific and having low persistence in the environment, as yet only a small number have found commercial application for insect population suppression. As used at present, the mode of action of the IGRs currently available means that they generally act as "third generation pesticides"[88] in terms of population suppression, rather than in the manner of sterilants, but as a result of their considerable potential for further development, they are currently the focus of intensive research.[89-92]

C. AUTOSTERILIZING SYSTEMS

1. General

Some early successes in the development of practical autosterilizing systems incorporating chemosterilants were made for use for the control of stable flies,[93] houseflies,[94] and mosquitoes.[95] A combination of chemosterilant and sweetened bait was used to control houseflies *M. domestica* in poultry houses.[96,97] However, their development and use in the field has been limited by the high mammalian toxicity of the chemosterilants currently available and particularly by the relatively low percentage of the target population that can be attracted by the baits available.

For control of an insect population with baits, the number that need to be treated per day is dependent on the potential for increase of that population.[98] With an insect such as a blowfly or screwworm fly, where a single female may produce batches of 200 to 400 eggs several times during its lifetime, approximately 20 to 40% of the females in the population need to be killed per day to bring about its elimination. If both males and females could be sterilized, approximately 10–20% of each sex would need to be affected per day. This level of attraction, however, is seldom achievable with the baits currently available.

In contrast, the relatively low rate of reproduction of tsetse flies makes them good candidates for control by trapping and autosterilization. Imposed increases in daily mortality of only 2–4% are required to bring about their eventual eradication,[99] and baits available at present make this level of attraction a practical reality.[100]

2. Control of Tsetse Flies by Autosterilization

Initial experiments showed that it was possible to sterilize tsetse flies in the field with the chemosterilant metepa,[101] and that both males and females could be attracted using host odors. This was shown to be capable of suppressing an island population of two species of tsetse.[102] However, metepa was considered to be too toxic for use in large-scale trials in the field. More recently it has been shown that the unusual mode of reproduction of tsetse flies favors the use of a juvenile hormone mimic, pyriproxyfen, as a practical chemosterilant.[103,104]

The tsetse fly has an advanced form of adenotrophic viviparity. After mating, a single egg is ovulated and retained in the oviduct where it hatches and undergoes three larval instars

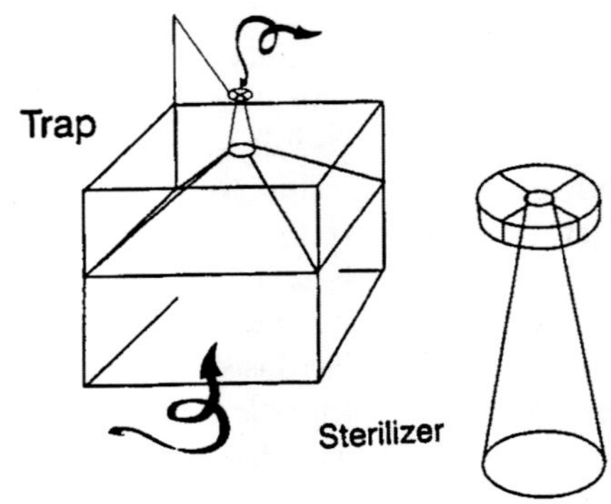

FIGURE 7. A tsetse F2-trap[77] with sterilizer attached and detail of the sterilizer, which consists of a wire frame covered with mutton cloth dipped in pyriproxyfen solution. The frame is inserted into a plastic casing and attached to the apex of the trap. Flies enter at the base of the trap and move up through the trap towards the light, where they are focused by a mesh cone into the sterilizer tower. They leave via the sterilizer, after contacting the pyriproxyfen-treated cloth. (Adapted from Hargrove, J. and Langley, P. *Bull Entomol. Res.*, 80, 397, 1990.)

feeding on a nutritive secretion produced by the uterine glands of the female. Larviposition occurs when the fully mature larva is ready to pupate.[105] Following suitable formulation in an oil, the compound is absorbed through the adult female cuticle on tarsal contact. In the female, pyriproxyfen is transferred to the larva "in utero." The third instar larva appears to be normal, but after pupariation, metamorphosis is disrupted and the pupa dies.

Only very small amounts of pyriproxyfen are required, and a single treatment will ensure that the female is effectively sterilized for life. Laboratory studies have shown that males making contact with a treated surface can transfer sterilizing doses to females when they mate.[104]

A field trial conducted in Zimbabwe showed that pyriproxyfen can be used in traps for tsetse control.[106] Tsetse entering traps, baited with host-odor olfactory attractants, were forced to brush against material dosed with pyriproxyfen to effect their escape (Figure 7). Emergence rates from puparia of *G.m. morsitans* and *G. pallidipes* fell to 30% and 2.7%, respectively. The decline in birth rate was estimated as sufficient to cause a population reduction to 10^{-6} of its original level if the autosterilization program was maintained for 1 year.

3. Development of an Autosterilizing System for Housefly Control

In addition to attracting sufficiently large numbers to the bait, a further practical obstacle to the disruption of reproduction for autosterilization with IGRs is inducing flies to pick up an effective dose of the compound. For example, significant egg hatch inhibition in the housefly, *M. domestica,* following contact with surfaces treated with a 20% suspension concentrate of the chitin synthesis inhibitor triflumuron, requires tarsal contact for at least one hour.[107] Such concentrations and contact times are not realistic for a working autosterilizing device. Nevertheless, recent studies have shown that it is possible to induce female houseflies to pick up an effective dose by presenting triflumuron on sugar-baited targets.[108] Direct effects on females may be a result of ingestion of triflumuron combined with tarsal contact allowing absorption of the chemical across the cuticle. In addition, male houseflies exposed to sugar and triflumuron treated targets are also capable of affecting the reproductive performance of unexposed females. The effects of triflumuron on males may be through direct impairment of spermatogenesis or insemination ability, or exposed males may simply pick up sufficient triflumuron on their bodies to allow them to pass on an effective dose to normal females during mating. Experimental studies have shown that after exposure of both sexes to triflumuron on sugar-baited targets, egg hatch may be reduced to less than 4% and, for those larvae that do eclose, less than 3% pupariate, giving a cumulative mortality of 98.8%.[109] If maintained, this level of mortality would be sufficient to suppress a field population of houseflies. This autosterilizing system is currently being tested for housefly control in poultry houses with encouraging results.

V. DISRUPTION OF BEHAVIOR

Identification and isolation of the complex interaction of olfactory, visual, physiological, and tactile cues used by insects to locate their mates and complete successful mating and insemination may allow them to be used to provide inappropriate stimuli to manipulate and disrupt reproductive behavior. For example, in many species of Diptera, males form persistent swarms near a prominent marker. The males may pursue, court, or capture and copulate with females passing briefly through the swarm.[110] It has been demonstrated that blood-seeking females of the mosquito *Cx. tritaeniohynchus* can be attracted to artificial swarm markers and killed.[111]

Following initial copulation, females of most species of insect become unreceptive to further mating attempts by males for substantial periods, if not for life. This is believed to be caused, at least in the short term, by the transfer of a receptivity-inhibiting substance, passed in the form of an accessory gland secretion during copulation.[112,113] Identification of these chemicals and discovery of a means of contaminating virgin females in the field could be used to switch off their receptivity, effectively rendering them sterile.

In tsetse flies (*Glossina* spp.), the contact sex pheromone of the female induces the initial stage of male copulatory behavior. This includes arrestment of locomotion, orientation over the surface of the female, extension of the hypopygium, and engagement of the genitalia.[114,115] It has been shown that the presence of synthetic tsetse sex pheromone on female decoys sewn onto screens in the field will induce copulatory behavior in males that make contact with them.[116-120] However, it has not been possible to make any practical use of this behavior, using either insecticide or chemosterilant added to the decoys, since insufficient males can be attracted. In contrast, pheromones have been used effectively, as attractants and mating disruptants for male Lepidoptera, in a number of field trials.[121-123]

There is clearly considerable scope for further work in the development of techniques that disrupt the mating behavior of insects by providing inappropriate stimuli or misdirecting behavior. However, it should be noted that techniques that simply block or disrupt some stage of the reproductive process act as direct alternatives to insecticides and have no inherent superiority as in some respects do SIT, genetic control, and autosterilization.

VI. CONCLUSIONS

The efficacy of the release of sterile or genetically altered strains for insect population suppression is considerable. The techniques are most appropriate for use against pest populations that exist at naturally low densities, or are subject to substantial seasonal population density fluctuations. They may also be used to prevent the establishment of immigrant populations in new areas or as a follow-up technique against populations that have already been reduced by other methods. However, the costs associated with the scale on which these techniques must be used render them impractical and uneconomic in most instances and against most pest species. As a result, in the 50 years since the techniques were first proposed, they have found relatively limited practical application. Only where it is feasible to eradicate an isolated population of a particularly virulent pest, or to eradicate it from an area which will subsequently be protected by a barrier zone, will these techniques be cost effective. Screwworm eradication in North America and Mexico and more recently in North Africa are the classic examples.

In contrast, the use of autosterilizing devices baited with pheromone or kairomone attractants and a sterilizing agent seem to have substantially greater potential for widespread application in field control programs. Autosterilization has a number of advantages over the other autocidal control techniques. The sterilization of both sexes at a specific rate would be expected to achieve more efficient suppression than the release of sterile or genetically altered males at the same rate or killing both sexes at the same rate with insecticides.[60,61] Autosterilizing systems provide a means of achieving many of the benefits of the release of sterile or genetically altered males without the need to mass rear and release. However, considerable costs may be incurred in the deployment and maintenance of sterilizing devices, particularly in areas of terrain with difficult access. Nevertheless, they are, theoretically, highly cost effective systems of autocidal pest control.

Autosterilizing systems may be particularly valuable where the aims of the control program are less comprehensive than total eradication. In practical terms, autosterilization would be most effective as a control technique when used against insects with low birth rates and high adult survivorship and against isolated populations. The latter may be most easily found in relatively closed systems such as poultry houses, barns, or grain storage silos. They also may be important components of integrated pest management systems. For example, an autosterilizing system could be used in combination with the release of sterile or genetically impaired males. It could also be used in conjunction with the release of insect biocontrol agents such as predators or parasitoids, so long as the autosterilizing device either did not attract these or the chemosterilant used was inactive against them. The use of conventional insecticides or traps is unhelpful with these techniques because they kill sterilized and fertile insects equally.

Advances in the development of autosterilizing systems for a range of pest insects are highly dependent on the identification of potent attractants and on the development of powerful but entirely insect-specific sterilants, both of which should affect males and females. The isolation of a range of chemical kairomones and visual attractants shows considerable promise.[80] While relatively few attractants yet appear to be able to lure sufficiently large numbers to overcome the high intrinsic rates of increase of populations of many species of pest Diptera, strategic timing of the deployment of an autosterilizing device so that it has maximum impact on the target populations may contribute to its efficient use.[124,125] Strategic deployment requires a detailed understanding of the seasonal population dynamics of the target species.

In the development of autosterilizing systems, less progress has been made in the search for effective sterilants. A great deal of interest has been given to the use of IGRs. However, the most common effects of juvenile hormones and juvenile hormone mimics are the disruption of various aspects of metamorphosis. They do not generally affect the egg production of higher Diptera. The development of successful autosterilizing systems for tsetse control has been possible, using a juvenile hormone mimic, but it is the unusual viviparous habit of the tsetse which permits disruption of metamorphosis through treatment of the adult female. A similar mode of application is unlikely to be effective, therefore, for other flies. Greater success for Diptera may be achieved using the chitin synthesis inhibitors, such as triflumuron.[107-109] The considerable benefits of specificity and environmental safety that would be gained through using an IGR or other biochemical sterilant, should direct intensive research towards this area. While it might initially appear almost impossible to affect male fertility using IGRs, the aim of affecting males could be achieved by the development of delivery systems which attract and allow males to pick up sufficient material to enable them to transfer effective doses to females during subsequent mating. The males thereby would be rendered effectively sterile. The use of sex pheromones, as behavioral arrestants for males, may have an important role in this context.

The manipulation of the reproductive physiology and behavior of insects to bring about their destruction has the potential for highly effective suppression of pest populations. However, the development of autocidal techniques for pest control is complex. Detailed information relating to the ecology, genetics, behavior, and physiology needs to be obtained for every pest species to be controlled. Basic information is also needed concerning the seasonal fluctuations, mobility, rates of increase, and, in particular, the extent to which the attempts to suppress a target population are likely to be negated by density-dependent compensation. The growing recognition of the need to develop more specific, effective, and environmentally sensitive pest control techniques will increase the urgency of these areas of research.

ACKNOWLEDGMENTS

I am grateful to Dr. M. J. R. Hall and an anonymous referee for their helpful comments on this chapter. The financial support of a Royal Society University Research Fellowship is gratefully acknowledged.

REFERENCES

1. **Curtis, C. F. and Lines, J. D.,** Insecticides in the management of insect vectors of tropical disease, *Insect Sci. Applic.*, 8, 709, 1987.
2. **Worthington, E. B.,** Insecticides, *Nature*, 234, 55, 1971.
3. **Strong, L. and Wall, R.,** The chemical control of livestock parasites: problems and alternatives, *Parasitol. Today*, 6, 291, 1990.
4. **Dame, D. A.,** Control of insects of veterinary importance by genetic techniques, *Prev. Vet. Med.*, 2, 515, 1984.
5. **Foster, G. G., Weller, G. L., James, W. J., Paschalidis, K. M., and McKenzie, L. J.,** Advances in sheep blowfly genetic control in Australia, in *IAEA/FAO Int. Symp. Management Insect Pests: Nuclear and Related Molecular and Genetic Techniques*, International Atomic Energy Agency/Food and Agricultural Organization, Vienna, 1992.
6. **Bushland, R. C. and Hopkins, D. E.,** Experiments with screw-worm flies sterilized by X-rays, *J. Econ. Entomol.*, 44, 725, 1951.
7. **Knipling, E. F.,** The Basic Principles of Insect Population Suppression and Management, U.S.D.A. Agriculture Handbook No. 512, U.S. Department of Agriculture, Washington, D.C., 1979, 659.
8. **Curtis, C. F.,** Genetic control: growth industry or lead balloon?, *Biol. J. Linn. Soc.*, 2, 359, 1985.
9. **La Brecque, G. C., Meifert, D. W., and Weidhaas, D. E.,** Potential of the sterile-male technique for the control or eradication of stable flies, *Stomoxys calcitrans* Linnaeus, in *Sterility Principle for Insect Control*, IAEA/FAO, Vienna, 1974, 449.
10. **Dame, D. A. and Schmidt, C. H.,** The sterile-male release technique against tsetse flies, *Glossina* spp., *Bull. Entomol. Soc. Am.*, 16, 24, 1970.
11. **Williamson, D. L., Dame, D. A., Gates, D. B., Cobb, P. E., Bakuli, B., and Warner, P. V.,** Integration of insect sterility and insecticides for control of *Glossina morsitans morsitans* Westwood (Diptera: Glossinidae) in Tanzania. V. The impact of sequential releases of sterilized flies, *Bull. Entomol. Res.*, 73, 391, 1983.
12. **Takken, W., Olandunmade, M. A., Dengwat, L., Feldmann, H. H., Onah, J. A., Tenabe, S. O., and Hamann, H. J.,** The eradication of *Glossina palpalis palpalis* (Robineau-Desvoidy) (Diptera: Glossinidae) using traps, insecticide-impregnated targets and the sterile insect technique in central Nigeria, *Bull. Entomol. Res.*, 76, 275, 1986.
13. **Olandunmade, M. A., Feldmann, U., Takken, W., Tenabe, S. O., Hamann, H. J., Onah, J., Dengwat, L., Van Der Vloedt, A. M. V., and Gingrich, R. E.,** Eradication of *Glossina palpalis palpalis* (Robineau-Desvoidy) (Diptera: Glossinidae) from agropastoral land in central Nigeria by means of the sterile insect technique, in *Sterile Insect Technique for Tsetse Control and Eradication*, IAEA/FAO, Vienna, 1990, 5.
14. **Weidhaas, D. E., Schmidt, E. L., and Seabrook, E. L.,** Field studies on the release of sterile males for the control of *Anopheles quadrimaculatus*, *Mosq. News*, 22, 283, 1962.
15. **Morlan, H. B., McCray, E. M., and Kilpatrick, J. W.,** Field tests with sexually sterile males for control of *Aedes aegypti*, *Mosq. News*, 22, 295, 1962.
16. **Mellado, L.,** La tecnica de machos esteriles en el control de la mosca Mediterrano. Programas realizados en Espana, in *Sterility Principle for Insect Control or Eradication*, IAEA/FAO, Vienna, 1971, 49.
17. **Kunz, S. E., Graham, M. R., Hogan, B. F., and Eschle, J. L.,** Effect of releases of sterile horn flies into a native population of horn flies, *Environ. Entomol.*, 3, 159, 1974.
18. **MacLeod, J. and Donnelly, J.,** Failure to reduce an isolated blowfly population by the sterile males release method, *Entomol. Exp. Appl.*, 4, 101, 1961.
19. **Kunz, S. E., Drummond, R. O., and Weintraub, J.,** A pilot test to study the use of the sterile insect technique for eradication of cattle grubs, *Prev. Vet. Med.*, 2, 523, 1984.

20. **James, M. T.,** *The Flies That Cause Myiasis in Man,* U.S. Dep. Agric. Misc. Publ., No. 631, 1947, 175.
21. **Thomas, D. B. and Mangan, R. L.,** Oviposition and wound visiting behaviour of the screwworm fly, *Cochliomyia hominivorax* (Diptera: Calliphoridae), *Ann. Entomol. Soc. Am.,* 82, 526, 1989.
22. **Humphrey, J. D., Spradbery, J. P., and Tozer, R. S.,** *Chrysomya bezziana*: pathology of old world screw-worm fly infestations in cattle, *Exp. Parasitol.,* 49, 381, 1980.
23. **Dove, W. E.,** Myiasis of man, *J. Econ. Entomol.,* 30, 29, 1935.
24. **Bushland, R. C.,** Screwworm research and eradication, *Bull. Entomol. Soc. Am.,* 21, 23, 1975.
25. **Baumhover, A. H., Husman, C. N., Skipper, C. C., and New, W. D.,** *J. Econ. Entomol.,* 52, 1202, 1959.
26. **Knipling, E. F.,** The eradication of the screwworm fly, *Sci. Am.,* 203, 54, 1960.
27. **Bush, G. L., Neck, R. W., and Kitto, G. B.,** Screwworm eradication: inadvertent selection for noncompetitive ecotypes during mass rearing, *Science,* 193, 491, 1976.
28. **Graham, O. H.,** Symposium on eradication of the screwworm from the United States and Mexico, *Misc. Publ. Entomol. Soc. Am.,* 62, 1985, 1.
29. **Krafsur, E. S., Whitten, C. J., and Novy, J. E.,** Screwworm eradication in North and Central America, *Parasitol. Today,* 3, 131, 1987.
30. **Readshaw, J. L.,** Screwworm eradication a grand delusion?, *Nature,* 320, 407, 1986.
31. **Krafsur, E. S., Townson, H., Davidson, G., and Curtis, C. F.,** Screwworm eradication is what it seems, *Nature,* 323, 495, 1986.
32. **Gabaj, M. M., Wyatt, N. P., Pont, A. C., Beesley, W. N., Awam, M. A. Q., Gusbi, A. M., and Benhaj, K. M.,** The screwworm fly in Libya: a threat to the livestock industry of the Old-World, *Vet. Rec.,* 125, 347, 1989.
33. **Lindquist, D. A., Abusowa, M., and Hall, M. J. R.,** The New World screwworm fly in Libya: a review of irs introduction and eradication, *Med. Vet. Entomol.,* 6, 2, 1992.
34. **Hightower, B. G., Adams, A. L., and Alley, D. A.,** Dispersal of released irradiated laboratory-reared screwworm flies, *J. Econ. Entomol.,* 58, 373, 1965.
35. **Hall, M. J. R. and Beesley, W. N.,** The New World screwworm fly in North Africa, *Pestic. Outlook,* 1, 34, 1990.
36. **FAO,** *New World Screwworm Newsletter,* FAO/Screwworm Emergency Centre for North Africa (SECNA), Rome, 1991, 1.
37. **Serebrovsky, A. S.,** On the possibility of a new method for the control of insect pests, *Zool. Zh.,* 19, 618, 1940 [Russian].
38. **Busch-Peterson, E.,** Genetic sex separation in sterile insect technique pest management programmes, with special reference to the medfly *Ceratitis capitata* (Weid.), in *Fruit Flies of Economic Importance,* Cavalloro, R., Ed., CEC/IOBC Int. Symp. 7–10 April, 1987, Rome, Italy, 1989, 225.
39. **Wagoner, D. E., Nickel, A. C., and Johnson, O. A.,** Chromosomal translocation heterozygotes in the house fly, *J. Hered.,* 60, 301, 1969.
40. **Curtis, C. F., Southern, D. I., Pell., P. E., and Craig-Cameron, T. A.,** Chromosome translocations in *Glossina austeni, Genet. Res. Camb.,* 20, 101, 1972.
41. **Curtis, C. F., Langley, P. A., Mews, A. R., Offori, E. D., Southern, D. I., and Pell, P. E.,** Sex ratio distortion and semi-sterility in the progeny of irradiated *Glossina morsitans, Genet. Res. Camb.,* 21, 153, 1973.
42. **Baker, R. H., Reisen, W. K., Sakai, R. K., Hayes, C. G., Aslamkhan, M., Saifuddin, U. T., Mahmood, F., Perveen, A., and Javed, S.,** Field assessment of mating competitiveness of male *Culex tritaeniorhynchus* carrying a chromosomal aberration, *Ann. Entomol. Soc. Am.,* 72, 751, 1979.
43. **Baker, R. H., Reisen, W. K., Sakai, R. K., Rathor, H. R., Raana, K., Azra, K., and Niaz, S.,** Mating behaviour and competitiveness of males carrying a complex chromosomal aberration, *Ann. Entomol. Soc. Am.,* 73, 581, 1980.
44. **Suguna, S. G., Seawright, J. A., Joslyn, D. J., and Rabbani, M. G.,** Homozygous pericentric inversions and other studies of inversions on chromosome 3 in the mosquito *Anopheles albimanus, Can. J. Genet. Cytol.,* 23, 57, 1981.
45. **Rossler, Y.,** Recombination in males and females of the Mediterranean fruit fly (Diptera: Tephritidae) with and without chromosomal aberrations, *Ann. Entomol. Soc. Am.,* 75, 619, 1979.
46. **Robinson, A. S. and Van Heemert, C.,** *Ceratitis capitata* — a suitable case for genetic sexing, *Genetica,* 58, 229, 1982.
47. **Saul, S. H.,** Genetic sexing in the Mediterranean fruit fly *Ceratitis capitata* (Weidemann) (Diptera: Tephritidae): conditional lethal translocations that preferentially eliminate females, *Ann. Entomol. Soc. Am.,* 77, 280, 1984.
48. **Kaiser, P. E., Seawright, J. A., Dame, D. A., and Joslyn, D. J.,** Development of a genetic sexing system for *Anopheles albimanus, Econ. Entomol.,* 71, 766, 1978.
49. **Whitten, M. J.,** Insect control by genetic manipulation of natural populations, *Science,* 171, 682, 1971.
50. **Whitten, M. J., Foster, G. G., Vogt, W. G., Kitching, R. L., Woodburn, T. L., and Konovalov, C.,** Current status of genetic control of the Australian sheep blowfly, *Lucilia cuprina* Weidemann (Diptera: Calliphoridae), *Proc. XV Int. Congr. Entomol.,* Washington, D.C., 1976, 129.

51. **Foster, G. G., Vogt, W. G., and Woodburn, T. L.,** Genetic analysis of field trials of sex-linked translocation strains for genetic control of the Australian sheep blowfly *Lucilia cuprina* Weidemann, *Aust. J. Biol. Sci.*, 38, 275, 1985.
52. **Foster, G. G.,** The sheep blowfly genetic control programme in Australia, *Insect and Pest Control Newsletter*, FAO/IAEA, 43, 23, 1989.
53. **Foster, G. G., Kitching, R. L., Vogt, W. G., and Whitten, M. J.,** Sheep blowfly and its control in the pastoral ecosystem of Australia, *Proc. Ecol. Soc. Aust.*, 9, 213, 1975.
54. **Whitten, M. J.,** The use of genetically-selected strains for pest replacement or suppression, in *Genetics in Relation to Insect Management*, Hoy, M. A. and McKelvey, J. J., Eds., The Rockefeller Foundation, New York, 1979, 31.
55. **Foster, G. G., Vogt, W. G., Woodburn, T. L., and Smith, P. H.,** Computer simulation of genetic control. Comparison of sterile male and field-female killing systems, *Theor. Appl. Genet.*, 76, 870, 1988.
56. **Davidson, S.,** Sheep blowfly control by genetic sabotage, *Rural Res.*, 145, 19, 1990.
57. **Vogt, W. G., Woodburn, T. L., and Foster, G. G.,** Ecological analysis of field trials conducted to assess the potential of sex-linked translocation strains for genetic control of the Australian sheep blowfly, *Lucilia cuprina* (Weidemann), *Aust. J. Biol. Sci.*, 38, 259, 1985.
58. **Foster, G. G.,** Chromosomal inversions and genetic control revisited: the use of inversions in sexing systems for higher Diptera, *Theor. Appl. Genet.*, 81, 619, 1991.
59. **Busch-Peterson, E. and Baumgartner, H.,** Simulation of instability in genetic-sexing strains: effects of male recombination in association with other biological parameters, *Bull. Entomol. Res.*, 81, 11, 1991.
60. **Langley, P. A. and Weidhaas, D.,** Trapping as means of controlling tsetse, *Glossina* spp. (Diptera: Glossinidae): the relative merits of killing and sterilization, *Bull. Entomol. Res.*, 76, 89, 1986.
61. **Wall, R. and Howard, J.,** Autosterilization for control of the housefly *Musca domestica*, *J. Theor. Biol.*, 117, 431, 1994.
62. **Borkovec, A. B.,** Insect chemosterilants, Vol. 7, *Advances in Pest Control Research*, Interscience, New York, 1966, 143 pp.
63. **Knipling, E. F.,** The Basic Principles of Insect Population Suppression and Management, *U.S.D.A. Agriculture Handbook*, No. 512, U.S. Department of Agriculture, Washington, D.C., 1979, 659.
64. **Barclay, H. J. and Mackauer, M.,** The sterile insect release method for pest control: a density dependent model, *Environ. Entomol.*, 9, 810, 1980.
65. **Barclay, H. J.,** Pheromone trapping models for pest control: effects of mating patterns and immigration, *Res. Popul. Ecol.*, 26, 303, 1984.
66. **Klassen, W., Ridgway, R. L., and Inscoe, M.,** Chemical attractants in integrated pest management programmes, in *Insect Suppression with Controlled Release Pheromone Systems*, Vol. 1, Kydonieus, A. F. and Beroza, M., Eds., CRC Press, Boca Raton, FL, 1982, 13.
67. **Sanders, C. J.,** Monitoring spruce budworm population density with six traps, *Can. Ent.*, 120, 1175, 1988.
68. **Bartell, R. J., Shorey, H. H., and Barton-Browne, L.,** Pheromonal stimulation of the sexual activity of males of the sheep blowfly *Lucilia cuprina* (Calliphoridae) by the female, *Anim. Behav.*, 17, 576, 1969.
69. **Carlson, D. A., Mayer, M. S., Silhacek, D. L., James, J. D., Beroza, M., and Bierl, B. A.,** Sex-pheromone attractant of the house fly: isolation, identification and synthesis, *Science*, 174, 76, 1971.
70. **Carlson, D. A., Langley, P. A., and Huyton, P.,** Sex pheromones of the tsetse fly: isolation identification and synthesis of contact aphrodisiac, *Science*, 201, 750, 1978.
71. **Adams, T. S. and Holt, G. G.,** Effects of pheromone components when applied to different models on male sexual behaviour in the housefly *Musca domestica*, *J. Insect Physiol.*, 33, 9, 1987.
72. **Fletcher, B. S.,** Storage and release of a sex-pheromone by the Queensland fruit fly, *Dacus tryoni* (Diptera: Trypetidae), *Nature*, 219, 631, 1968.
73. **Wall, R. and Langley, P. A.,** From behaviour to control: the development of trap and target techniques for tsetse fly population management, *Agric. Zool. Rev.*, 4, 137, 1991.
74. **Coppedge, J. R., Broce, A. B., Tannahill, F. H., Goodenough, J. L., Snow, J. W., and Crystal, M. M.,** Development of a bait system for suppression of adult screwworms, *J. Econ. Entomol.*, 71, 483, 1978.
75. **Prokopy, R. J. and Owens, E. D.,** Visual detection of plants by herbivorous insects, *Annu. Rev. Entomol.*, 28, 337, 1983.
76. **Allan, S. A., Day, J. F., and Edman, J. D.,** Visual ecology of biting flies, *Annu. Rev. Entomol.*, 32, 297, 1987.
77. **Green, C. H. and Flint, S.,** An analysis of colour effects in the performance of the F2 trap against *Glossina pallidipes* Austen and *G. morsitans morsitans* Westwood (Diptera: Glossinidae), *Bull. Entomol. Res.*, 76, 409, 1986.
78. **Holloway, M. P. T. and Phelps, R. J.,** The responses of *Stomoxys* spp. (Diptera: Nuscidae) to traps and artificial host odours in the field, *Bull. Entomol. Res.*, 81, 51, 1991.
79. **Wall, R., Green, C., French, N., and Morgan, K.,** Development of an attractive target for the sheep blowfly *Lucilia sericata*, *Med. Vet. Entomol.*, 6, 67, 1992.
80. **Muirhead-Thompson, R. C.,** *Trap Responses of Flying Insects. The Influence of Trap Design on Capture Efficiency*, Academic Press, London, 1991, 287.

81. **Fye, R. L. and LaBreque, G. C.,** Bibliography of Arthropod Chemosterilants, ARS S-93, Agricultural Research Service, U.S. Department of Agriculture, 1976, 54 pp.
82. **La Brecque, G. C.,** Studies with three alkylating agents as house fly sterilants, *J. Econ. Entomol.*, 54, 684, 1961.
83. **Langley, P. A. and Carlson, D. A.,** Laboratory evaluation of bisazir as a practical chemosterilant for tsetse control, *Bull. Entomol. Res.*, 76, 583, 1986.
84. **Hall, M. J. R. and Langley, P. A.,** Development of a system for sterilizing tsetse flies, *Glossina* spp., in the field, *Med. Vet. Entomol.*, 1, 201, 1987.
85. **Moore, N. F., King, L. A., and Possee, R. D.,** Viruses of insects, *Insect Sci. Appl.*, 8, 275, 1987.
86. **Schniderman, H. A.,** Insect hormones and insect control, in *Insect Juvenile Hormones Chemistry and Action*, Academic Press, London, 1972, 3.
87. **Ruscoe, C. N. E.,** Insect control by hormones, in *Biology in Pest and Disease Control*, Price Jones, D. and Solomon, M. E., Eds., Blackwell, Oxford, 1974, 147.
88. **Williams, C. M.,** Third-generation insecticides, *Sci. Am.*, 217, 13, 1967.
89. **Friedel, T., Hales, D. F., and Birch, D.,** Cyromazine-induced effects on the larval cuticle of the sheep blowfly, *Lucilia cuprina*: ultrastructural evidence for a possible mode of action, *Pestic. Biochem. Physiol.*, 31, 99, 1988.
90. **Langley, P. A., Felton, T., and Oouchi, H.,** Juvenile hormone mimics as effective sterilants for the tsetse fly *Glossina morsitans morsitans*, *Med. Vet. Entomol.*, 2, 29, 1988.
91. *Advances in Regulation of Insect Reproduction*, Bennettová, B., Gelbic, I., and Soldán, T., Eds., Institute of Entomology, Czechoslovakian Academy of Science, Academia Praha, Prague. 1992, 329.
92. *Insect Juvenile Hormone Research. Fundamental and Applied Approaches. Chemistry Biochemistry and Mode of Action.*, Mauchamp, B., Couillaud, F., and Baehr, J. C., Eds., Institut National de la Recherche Agronomique, Paris, 1992, 296.
93. **La Brecque, G. C., Meifert, D. W., and Rye, J., Jr.,** Experimental control of stable flies *Stomoxys calcitrans* (Diptera: Muscidae) by release of chemosterilized adults, *Can. Entomol.*, 104, 885, 1971.
94. **La Brecque, G. C., Fye, R. L., and Morgan, J., Jr.,** Induction of sterility in adult house flies and stable flies by chemosterilization of pupae, *J. Econ. Entomol.*, 65, 751, 1972.
95. **Ikeshoji, T. and Yap, Y. Y.,** Monitoring and chemosterilization of a mosquito population, *Culex quefasciatus* (Diptera: Culicidae) by sound traps, *Appl. Entomol. Zool.*, 22, 474, 1987.
96. **La Brecque, G. C. and Meifert, D. W.,** Control of house flies (Diptera: Muscidae) in poultry houses with chemosterilants, *J. Med. Entomol.*, 3, 232, 1966.
97. **Meifert, D. W. and La Brecque, G. C.,** Integrated control for the suppression of a population of houseflies, *Musca domestica* L., *J. Med. Entomol.*, 8, 43, 1971.
98. **Weidhaas, D. E. and Haile, D. G.,** A theoretical model to determine the degree of trapping required for insect population control, *Bull. Entomol. Soc. Am.*, 24, 18, 1978.
99. **Hargrove, J. W.,** Tsetse: the limits to population growth, *Med. Vet. Entomol.*, 2, 203, 1988.
100. **Vale, G. A., Bursell, E., and Hargrove, J. W.,** Catching-out the tsetse fly, *Parasitol. Today*, 1, 106, 1985.
101. **House, A. P. R.,** Chemosterilization of *Glossina morsitans morsitans* Westwood and *G. pallidipes* Austen (Diptera: Glossinidae) in the field, *Bull. Entomol. Res.*, 72, 529, 1982.
102. **Vale, G. A., Hargrove, J. W., Cockbill, G. F., and Phelps, R. J.,** Field trials of baits to control populations of *Glossina morsitans morsitans* Westwood and *G. pallidipes* Austen (Diptera: Glossinidae), *Bull. Entomol. Res.*, 76, 179, 1986.
103. **Langley, P. A., Felton, T., and Oouchi, H.,** Juvenile hormone mimics as effective sterilants for the tsetse fly *Glossina morsitans morsitans*, *Med. Vet. Entomol.*, 2, 29, 1988.
104. **Langley, P. A., Felton, T., Stafford, K., and Oouchi, H.,** Formulation of pyriproxyfen, a juvenile hormone mimic, for tsetse control, *Med. Vet. Entomol.*, 4, 127, 1990.
105. **Denlinger, D. A. and Ma, W. C.,** Dynamics of the pregnancy cycle in the tsetse *Glossina morsitans*, *J. Insect Physiol.*, 20, 1015, 1974.
106. **Hargrove, J. W. and Langley, P. A.,** Sterilizing tsetse (Diptera: Glossinidae) in the field: a successful trial, *Bull. Entomol. Res.*, 80, 397, 1990.
107. **Howard, J. and Wall, R.,** The effects of triflumuron, a chitin synthesis inhibitor, on the house fly, *Musca domestica* (Diptera: Muscidae), *Bull. Entomol. Res.*, 85, 71, 1995.
108. **Howard, J. and Wall, R.,** The use of triflumuron on sugar-baited targets for autosterilization of the house fly, *Musca domestica, Ent. Exp. Appl.*, 74, 1995.
109. **Howard, J. and Wall, R.,** Autosterilization of the house fly, *Musca domestica*, using the chitin synthesis inhibitor Triflumuron on sugar-baited targets, *Med. Vet. Entomol.*, 10, 1995.
110. **Downes, J. A.,** The swarming and mating flight of Diptera, *Annu. Rev. Entomol.*, 14, 271, 1969.
111. **Bidlingmeyer, W. L. and Hem, D. G.,** The range of visual attraction and the effect of competitive visual attractants upon mosquito (Diptera: Culicidae) flight, *Bull. Entomol. Res.*, 70, 321, 1980.

112. **Gillott, C.,** *Arthropoda — Insecta. Reproductive Biology of Invertebrates,* Vol. 3, *Accessory Sex Glands,* Adiyodi, G. G. and Adiyodi, R. D., Oxford, New Dehli, 1988, 319.
113. **Smith, P. H., Gillott, C., Barton Browne, L., and van Gerwen, A. C. M.,** The mating induced refractoriness of *Lucilia cuprina* females: manipulating the male contribution, *Physiol. Entomol.,* 15, 469, 1990.
114. **Huyton, P. M., Langley, P. A., Carlson, D. A., and Coates, T. W.,** The role of sex pheromones in initiation of copulatory behaviour by male tsetse flies, *Glossina morsitans morsitans, Physiol. Entomol.,* 5, 243, 1980.
115. **Huyton, P. M., Langley, P. A., Carlson, D. A., and Schwarz, M.,** Specificity of contact sex pheromones in tsetse flies, *Glossina* spp., *Physiol. Entomol.,* 5, 253, 1980.
116. **Langley, P. A., Coates, T. W., Carlson, D. A., Vale, G. A., and Marshall, J.,** Prospects for autosterilization of tsetse flies, *Glossina* spp. (Diptera: Glossinidae), using sex pheromone and bisazir in the field, *Bull. Entomol. Res.,* 72, 319, 1982.
117. **Hall, M. J. R.,** The orientation of tsetse flies to pheromone-baited decoy 'females' in the field, *Bull. Entomol. Res.,* 77, 487, 1987.
118. **Hall, M. J. R.,** Characterisation of the sexual responses of male tsetse flies, *Glossina morsitans morsitans* to pheromone-baited decoy 'females' in the field, *Physiol. Entomol.,* 13, 49, 1988.
119. **Hall, M. J. R. and Langley, P. A.,** The responses of individual males in an isolated population of *Glossina morsitans morsitans* Westwood (Diptera: Glossinidae) to pheromone-baited decoy 'females', *Bull. Entomol. Res.,* 79, 319, 1989.
120. **Wall, R.,** Sexual responses of males of *Glossina morsitans morsitans* and *G. pallidipes* to traps and targets in the field, *Bull. Entomol. Res.,* 79, 335, 1989.
121. **Critchley, E. R., Campion, D. G., McVeigh, L. J., Hunter-Jones, P., Hall, D. R., Cork, A., Nesbit, B. F., Marrs, G. J., Jutsum, A. R., Hosny, M. M., and El-Syed, A. N.,** Control of pink bollworm, *Pectinophora gossypiella* Saunders (Lepidoptera: Gelechiidae) in Egypt by mating disruption using an aerially applied microencapsulated pheromone formulation, *Bull. Entomol. Res.,* 73, 289, 1983.
122. **Webb, R. E., Tatman, K. M., Leonhardt, B. A., Plimmer, J. R., Boyd, V. K., Bystrak, P. G., Schwalbe, C. P., and Douglass, L. W.,** Effects of aerial application of racemic disparlure released on male trap catch and female mating success of Gypsy Moth (Lepidoptera: Lymantriidae), *J. Econ. Entomol.,* 81, 268, 1990.
123. **Webb, R. E., Leonhardt, B. A., Plimmer, J. R., Tatman, K. M., Boyd, V. K., Cohen, D. L., Schwalbe, C. P., and Douglass, L. W.,** Effects of racemic disparlure released from grids of plastic ropes on mating success of Gypsy Moth (Lepidoptera: Lymantriidae) as influenced by dose and by population density, *J. Econ. Entomol.,* 83, 910, 1990.
124. **Wall, R., French, N., and Morgan, K. L.,** Sheep blowfly population control: development of a simulation model and analysis of management strategies, *J. Appl. Ecol.,* 30, 743, 1993.
125. **Wall, R., French, N., and Morgan, K. L.,** Predicting the abundance of the blowfly *Lucilia sericata* Meigen (Diptera: Calliphoridae), *Bull. Entomol. Res.,* 83, 431, 1993.

Chapter 6

PARTHENOGENESIS IN INSECTS WITH PARTICULAR REFERENCE TO THE ECOLOGICAL ASPECTS OF CYCLICAL PARTHENOGENESIS IN APHIDS

A. F. G. Dixon

CONTENTS

I. INTRODUCTION

The term "parthenogenesis" was coined by Richard Owen[78] in 1849, who defined it as the successive production of procreating individuals from a single ovum. He also viewed the increase in numbers by parthenogenesis as similar to growth, a view which was supported by Huxley[48] in 1858, and much later by Janzen,[52] who stressed that in apomictic parthenogens natural selection acts at the level of a clone, which he designated the "evolutionary individual." That is, a clone produced by parthenogenesis is equivalent to the body of a sexually reproducing organism although split up into a number of physically separate units. Etymologically, the word parthenogenesis means "reproduction by a virgin."

Gametic reproduction either involves genetic mixing, as in normal sexual reproduction and automictic parthenogenesis, or no genetic mixing, as in apomictic parthenogenesis. The distinction between apomictic and automictic parthenogenesis is blurred because there are some forms of automictic parthenogenesis (functional apomictics) in which genetic mixing is suppressed.[5,105] Parthenogenesis is widespread in the Insecta, occurring in Coleoptera, Collembola, Diptera, Embioptera, Ephemeroptera, Hemiptera, Isoptera, Lepidoptera, Orthoptera,

0-8493-6695-X/95/$0.00+$.50

Psocoptera, Strepsiptera, Trichoptera, and Thysanoptera.[88] Indeed, parthenogenesis was first experimentally confirmed by Bonnet,[14] working on aphids in the 18th century, and virgin birth was observed in aphids by Leeuwenhoek in the 17th century. However, in birds and mammals parthenogenesis is completely absent, and it is extremely rare in molluscs. A major challenge is to develop a theory to account for the phylogenetic distribution of parthenogenesis. The high incidence among insects may be because they are relatively immobile compared with larger organisms and therefore less able to seek refuge from adverse climatological and biological conditions. Under these conditions, parthenogenesis may have been at a selective advantage because it conserves general purpose genotypes.[62] However, the comparable dispersal ability but markedly different incidence of parthenogenesis in aphids and marine molluscs tends to refute this suggestion.

II. EVOLUTION OF PARTHENOGENESIS

Reasons generally cited for the abandonment of sex are that periodic shortages of males in small local populations would favor females that are parthenogenetic; parthenogens are better able to conserve particular genotypes and so more easily exploit extreme environments, or parthenogenesis offers advantages for the colonization and rapid exploitation of ephemeral habitats.[22,25,27,62,85,90] This applies to parthenogenetic ability that is under genetic control, e.g., References 43 and 50. A very different approach is to see parthenogenesis in terms of the advantages for microorganisms or cytoplasmic elements, transmitted solely in the female line, of manipulating the sex ratio in favor of females.[86]

Hymenoptera have diploid females and haploid males, i.e., sex is dependent on whether an egg is fertilized, which is under the control of the mother. Some mothers, however, produce only females without fertilization (thelytoky). Interestingly, the completely parthenogenetic *Encarsia* and *Trichogramma* can be rendered permanently bisexual by treatment with antibiotics. There is no evidence that this is genetically controlled, but there is evidence of extrachromosomal inheritance. The males that develop after treatment with antibiotics produce sperm and mate with females, but are unable to inseminate them. It is suggested that the parthenogenetic females lay unfertilized eggs and that microorganisms cause the restoration of diploidy.[109] Although the female-biased sex ratios of Hymenoptera are explicable in terms of an evolutionary stable strategy of the insect's genome, under certain conditions this strategy and adaptations of the insect's symbionts coincide.[86] The presence of ovarially transmitted microorganisms or cytoplasmic elements capable of specifically killing males appears to be widespread in insects. There are examples known from Coleoptera, Diptera, Hemiptera, Hymenoptera, and Lepidoptera.[47] Therefore, their role in determining sex ratios and, more important for this review, their possible role in the evolution of parthenogenesis, should not be neglected.

Deleterious or nonfunctional alleles accumulate in continuously parthenogenetic strains[50,73] as they have no way of reducing the overall load. That is, mutation load cannot be less than that already existing in the organism's at present least-loaded lines. This is known as Muller's ratchet principle.[6,75] The expectation is that exclusively parthenogenetic reproduction should lead, despite compensating mutation,[94] to rapid extinction on an evolutionary time scale, with the clonal species that do evolve not giving rise to higher taxonomic categories.[87,88] However, as pointed out by Maynard Smith[67] and Williams,[106] there are notable challenging exceptions. The whole of the rotifer order Bdelloidea, with its several families and many genera and species, and the freshwater gastrotrich order Chaetonotoidea are exclusively parthenogenetic.[68] In the case of the diploid parthenogenetic brine shrimp, *Artemia*, the diversity and the 5 million years' existence of the parthenogenetic lineage is thought to have been sustained by automictic recombination.[20]

Apomictic parthenogenesis is thought to have evolved either from automictic parthenogenesis in which the meiotic division is retained[43,90,104] or abruptly in its full form.[105] Because of the imperfect cytogenetic mechanism involved in automictic parthenogenesis, it initially confers a lower fitness than either pure sexuality or apomixis, most strikingly because of poor egg hatch, but also the retention of mating behavior in spite of nearly complete absence of males.[21,58,64,90,92] These disadvantages might explain why automicts are relatively rare and apomicts, which do not incur these costs, flourish.[5,43]

That apomictic parthenogenesis may have evolved abruptly in its full form is supported by both aphids and *Daphnia* having genes that suppress meiosis. A genetic change results in the failure to produce sexual forms and may result in the coexistence of cyclic parthenogenetic and continuously parthenogenetic clones. In response to environmental stimuli, some clones produce sexuals (holocyclic), others some males and parthenogenetic females (androcyclic), and others only parthenogenetic females (anholocyclic). In the case of the aphid *Myzus persicae*, these life cycle traits are inherited monofactorily, with androcycly recessive to holocycly, which allows the exchange of genes between androcyclic and holocyclic clones and the generation of other anholocyclic clones.[8,50] Further studies on the inheritance of these life history traits are needed for a better understanding of the evolution of apomictic parthenogenesis and in particular cyclical parthenogenesis.

Alternatively, as exposing parthenogenetic strains of some species of Hymenoptera to high temperatures also induces the production of males, it has been suggested that this might form the basis for the seasonal control of sexual reproduction in the cyclical parthenogenesis of organisms like aphids. There is some evidence that the maternally transmitted symbionts in aphids have a nutritive role, but they have not yet been shown to have a causative role in gametogenesis or sex determination, although they are known not to enter male and soldier embryos of some but not all aphids.[41,49,86]

III. PARTHENOGENESIS IN APHIDS

Most aphids show cyclical parthenogenesis in which there is several generations of parthenogenetic reproduction between each bout of sexual reproduction. Owen[78] thought sexual reproduction occurred on the exhaustion of the spermatic force associated with fertilization. However, a hundred years earlier, Bonnet[14] had correctly attributed the appearance of sexuals to the effect of long chilly autumnal nights. In parthenogenetic reproduction each ovum is produced mitotically (apomixis), and as a consequence the offspring are identical. This view was challenged by Cognetti,[26] who claimed that there is an exchange of alleles between homologous chromosomes within the prophase nucleus of the oocyte, a process known as endomeiosis, which generates variation within clones. This led to a controversy, which has been reviewed by Blackman[9-11] and Pagliai-Bonvicni and Cognetti.[79] Suffice it to say, exhaustive cytological and morphological studies and even one using DNA fingerprinting,[24] have failed to reveal any significant intraclonal variation.

Four strains of the raspberry aphid, *Amphorophora idaei* Borner, differentially infest cultivars of raspberry. The aphid strains result from segregation at two gene loci and the plant resistance to aphids from different combinations of 14 genes.[15,56] Similarly, in *Acyrthosiphon pisum* (Harris), *Metopolophium dirhodum* (Walker), *M. persicae* (Sulzer), *Schizaphis graminum* (Rhondani), and *Sitobion avenae* (F.), particular genotypes are better adapted to living on certain plant species or cultivars within the aphid's host range.[65,74,80,81,91,100,101] Thus, contrary to Blakeley's[13] claim, there is support for the idea that genetic variation among host plants is important for the retention of sex in aphids. That is, sex and the resultant genetic recombination are at a selective advantage because all the other organisms that aphids interact with are evolving.[93] However, the increasing use of sterile males and lures baited with sex pheromones for capturing males for controlling pests, combined with the widespread planting of genetically uniform crops, would appear to tip selection in favor of parthenogenesis in insect pests.

Although sex in aphids has advantages in heterogeneous environments, nevertheless continuously parthenogenetic species of aphids have rapidly adapted to a wide range of conditions.

This has been achieved by mutation and is particularly well illustrated by the history of *Therioaphis trifolii* forma *maculata* (Monell) in North America. In 10 years, a continuously parthenogenetic clone introduced into New Mexico from the Mediterranean area colonized most of the southern areas of North America and evolved biotypes that were able to thrive on previously resistant cultivars of alfalfa and also became resistant to certain insecticides, and in the northern areas it even started to reproduce sexually and overwinter as eggs. Therefore, the lack of genetic recombination did not prevent the spotted alfalfa aphid from adapting to new conditions.[12,28,77] The success of this parthenogen, as in other pest aphids, comes from infesting crops that tend to be spatially invariant habitats, in which a single clone may gain dominance[31] and there is a mutation rate sufficient to allow it to adapt to changes in its largely man-made and relative homogeneous habitat.

Within the family Aphididae there is only one exception to the notion that exclusively parthenogenetic forms do not give rise to higher taxonomic categories — that is, the Tramini. This tribe consists of 4 genera and 32 species,[39] and it would be of considerable interest to know more about this group's taxonomic status and evolutionary history. In addition, only 1 out of 30 species of aphids is continuously parthenogenetic.[11] The potentially greater rate of increase of a continuously parthenogenetic mutant[76] should rapidly result in the competitive elimination of cyclically parthenogenetic strains. Therefore, there must be a markedly compensating advantage of sex.[61] This, as indicated above, is thought to be the production of rare recombinant genotypes that can survive in the face of new adversity. In retaining cyclical parthenogenesis, aphids have the advantages of both parthenogenesis and sex. The prevalence of cyclically parthenogenetic over continuously parthenogenetic species of aphids is a consequence of the balance between the extinction of asexual populations and the origination by fixation of asexual mutants.[23]

IV. ECOLOGY OF CYCLICAL PARTHENOGENESIS

Once parthenogenetic, certain lines of development become possible, in particular, the telescoping of generations, polyphenism, and host alternation. Aphids have developed these life history traits, which to a large extent accounts for their success.

A. TELESCOPING OF GENERATIONS

The strong relationship between the rates of increase and the size of organisms indicates that aphids have a faster rate of increase than one would expect. They achieve rates of increase that are more characteristic of organisms the size of mites, i.e., one to two orders of magnitude smaller. [32,42] Most organisms first grow and then switch to reproduction, and an important feature that constrains their rate of increase is the time it takes them to develop from birth to maturity.

Aphids, as a consequence of being parthenogenetic, have developed a different strategy. They invest in both somatic and gonadal growth during their larval development with soma growing logistically and the gonads exponentially.[19] At birth aphids already have embryos developing in their gonads, and their most advanced embryos have also started to develop gonads. This telescoping of generations is characteristic of aphids. That is, throughout larval development aphids simultaneously invest in growth of soma and gonads such that on becoming adult most aphids are ready to give birth, i.e., they do not indulge in a bang-bang reproductive strategy.[84] What favors the simultaneous commitment to growth and reproduction during larval development in aphids?

If there is no constraint on the relative rate of development of the gonads, it is best to first grow and then reproduce, i.e., a bang-bang strategy.[84] Kindlmann and Dixon[54] argue for a constraint on the rate of development of the gonads. Fecundity is the conversion of gonadal tissue into offspring containing their own gonads. Gonads are crucial for the future, while soma only assimilates and then dies. The best strategy for maximizing growth rate is to use all the energy gained from the soma to keep the gonadal developmental rate maximal. After moulting to the adult stage, aphids do not grow anymore and all the energy assimilated is utilized for reproduction. Thus there must be an optimal somatic (s_A) to gonadal (g_A) ratio for an adult. If the ratio were bigger there would be a surplus of energy, as the gonads would be unable to utilize all of the energy because of the constraint on the rate of development of the gonads (R). If the ratio were smaller, then there would be insufficient energy to ensure the maximum rate of development of the gonads. Therefore, either energy would be wasted or gonadal development rate would be submaximal during the adult stage if s_A/g_A is not optimal.

Embryos assimilate nutrients over the whole surface of their body.[16] Therefore, their prenatal developmental time (i.e., the period from ovulation to birth) is unlikely to be affected by the size of their gonads. Assuming that the maximum rate of development of the gonads, R, applies to the embryos, then whether an embryo can achieve birth size before its mother reaches maturity depends on when during its mother's development it is ovulated. If the prenatal developmental time is longer than the mother's larval developmental time, then it is advantageous for ovulation to occur and development to commence before the birth of the mother. If not, then the embryo would have to complete its development after its mother becomes adult, which would lead to further increase in the size of the mother's gonads during adulthood and break the optimal balance between s_A/g_A. If ovulation occurs before a mother is born, then she can give birth just after reaching maturity. In the case of aphids, ovulation occurs in the grandmother such that each individual actually consists of individuals of three generations telescoped together. After birth of offspring the mother's gonads are smaller, the optimal balance between s_A/g_A is broken, and there is now surplus energy to support further growth of the gonads, allowing maturation of other embryos and ovulation.

That aphids generally produce their offspring singly at intervals in time (iteroparous) rather than all at once (semeloparous) can be similarly explained. The adult soma provides enough energy to cover the respiration costs of a body of size s_A+g_A and the Rg_A needed for the growth of the gonads. However, if there is a constraint on the rate of development of the gonads, R, then on giving birth, gonad size decreases to g where $g < g_A$ and only requires sufficient energy to support $Rg < Rg_A$. Thus a certain amount of energy is unused until the gonads reach the size g_A again. Therefore, it is advantageous to keep the size of the gonads as close to g_A as possible in order to use most of the energy assimilated by the soma. This is achieved by the anterior germaria ovulating first and ovulation occurring over a prolonged period so that there are differences in the developmental status of embryos within an ovariole and, for example, between the basal embryos in each of the ovarioles.[17] In this way, aphids are able to produce offspring continuously (iteroparously) and maintain the size of their gonads close to g_A.

What evidence is there for a constraint on development? Under congenial conditions aphids generally take approximately a week to develop from birth to maturity. In contrast, many other insects take approximately 3 weeks. However, if one takes into consideration that an aphid starts developing inside its grandmother, then the actual development time is 2.5 times longer than it takes an aphid to develop from birth to maturity, i.e., approximately 3 weeks. Therefore, there appears to be a minimum "time" required for development. Telescoping of generations enabled aphids to overcome this constraint on development and achieve rates of increase more typical of much smaller organisms. Similarly, the oocytes in the gonads of the immature stages of the parthenogenetic cecidomyid fly *Heteropeza pygmaea* may develop parthenogenetically, which results in the larvae reproducing, a phenomenon known as paedogenesis.[103] Although different from the telescoping of generations, paedogenesis in omitting the adult stage also circumvents the constraint on development.

B. POLYPHENISM

An aphid life cycle usually consists of a sequence of parthenogenetic generations and ends in sexual reproduction. Normally, individual aphids are short-lived, and experience only a small part of the total range of environmental and nutritional conditions experienced by the

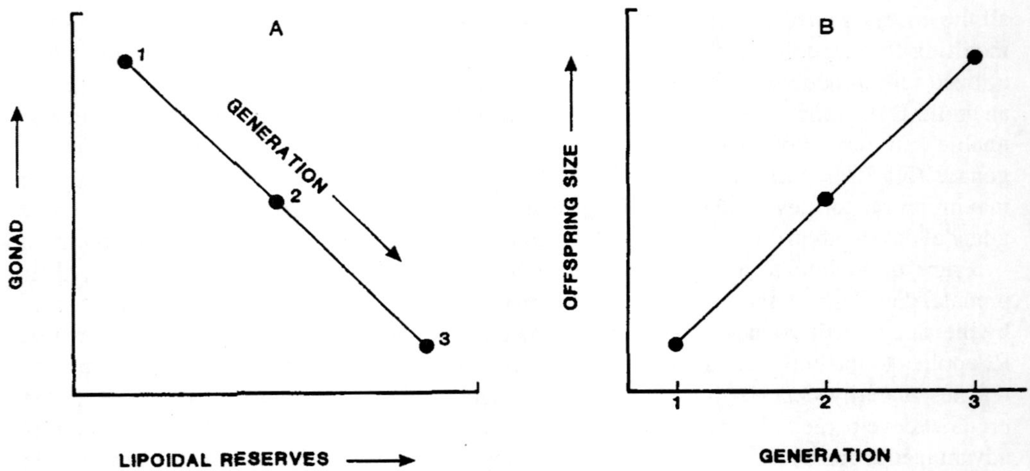

FIGURE 1. Diagrammatic representation of the relationships: (A) between gonadal size and lipoidal reserves in individuals of generations 1 to 3; and (B) the size of the offspring born to individuals in generations 1 to 3.

species during a season. This has favored the evolution of generation-specific reproductive strategies rather than a single but flexible strategy for all morphs of a species.[37,51,55] Individuals of each generation are differentially adapted for survival, reproduction, and migration. However, this division of labor in which individuals specialize in one or other of these functions imposes constraints in terms of resource allocation, physiology, and structure for carrying out other functions. This concept of the changing importance of roles over time emphasizes the functional aspect of polyphenism and largely avoids the use of the confusing array of names for the various morphs.

1. Reproduction

Early in the season, while plants are actively growing, they are a rich source of food for aphids, but generally become less suitable as they cease growing. Associated with this change in host quality is a programmed change within clones of aphids in which individuals of the first generation have proportionally larger gonads and smaller fat reserves than those of later generations.[34,102] First-generation individuals also tend to produce rapidly many small offspring and the later generations fewer but larger offspring (Figure 1). Large offspring are better able to survive on poor quality hosts than are small offspring.[31]

2. Migration

As plants eventually die or are killed, aphids need to disperse to ensure their long-term survival. In species that show alary dimorphism, minor movements on a plant and between adjacent plants are mainly undertaken by unwinged individuals, with the winged individuals migrating over greater distances.

Tactile stimulation associated with crowding[60,83] deterioration in host quality,[35] a combination of host quality and crowding,[46,99] day length,[108] or a combination of these factors[66] can induce the development of winged forms. By responding to a number of stimuli rather than one, aphids can possibly achieve a closer and more reliable tracking of environmental conditions. As it takes a generation or more for a change of form, aphids tend to respond to cues that enable them to anticipate the onset of adverse conditions; however, on reaching maturity, they only migrate if conditions are adverse.[33] This enables them to produce the more fecund and faster developing nonmigratory forms[29,107] while the host plant is favorable and, as it becomes unfavourable, to switch to producing migratory forms.

3. Defense/Cleaning

The existence of a "soldier" morph in species of *Colophina, Pemphigus*, and *Pseudoregma* is a particularly good example of the division of labor within a clone, with some individuals specialized for defense and in some cases even cleaning. They can make up 13% of a colony, and are usually very short-lived, may not feed or reproduce, and clean and defend the colony against insect enemies.[1,3,7] Similarly, approximately 50% of the aphids in the large galls produced by *Astegopteryx styracicola* are "bitters" and possibly afford the gall protection against predators like squirrels.[2,4,44] Because a colony is likely to consist mainly of the descendants of a founder aphid (i.e., it is a clone), investing in soldiers, which seize and kill insect enemies or clean honeydew, shed skins or dead aphids from the colony could be advantageous because it increases the fitness of the clone.

4. Estivation/Hibernation

In summer, many plants, especially shrubs and trees, cease growing and, until their leaves become senescent at the onset of autumn, are poor hosts for aphids. Several species of aphids estivate during the summer. The second generation of sycamore aphids, *Drepanosiphum platanoidis*, mature with small gonads and large fat body and may not reproduce for up to 8 weeks.[30,37] The aphid *Periphyllus testudinaceus*, also living on sycamore, estivates as a first instar nymph. Poor nutrition induces this aphid to produce small flattened nymphs whose bodies are covered and fringed with minute plates. These nymphs attach themselves very closely to the surface of leaves and can only be removed with great difficulty by predators.

Many species can overwinter viviparously. In some species there is a special overwintering form, called a hiemalis.[45] For example, about a fifth of the autumnal apterae of the lettuce root aphid (*Pemphigus bursarius*) can survive in the soil without a host plant for 48 weeks at 3°C.[38] They have poorly developed gonads and a well-developed fat body. The environmental cue that induces the development of the hiemalis in this root-feeding aphid is low temperature. The hiemalis remains in hibernation until soil temperatures increase and its host plants resume growing the following spring.[53]

C. COMPLEX LIFE CYCLES

Seasonally, 10% of aphids switch between unrelated host plants. This phenomenon, known as host alteration or heteroecy, has long attracted interest. Bonnet[14] suspected it existed, and it was first described by Walker.[95] The typical life cycle involves migration between a woody host, on which sex occurs and eggs are laid (the primary host) and an herbaceous plant, where reproduction is solely parthenogenetic (the secondary host).

Why aphids host-alternate was first addressed by Mordvilko.[69-72] Clearly, the twice-yearly migratory cycle with associated mortality is a cost, especially in a taxon such as aphids with poor host location abilities. Even though a clone can be represented by many host locators, nevertheless only an average of 1 in 100, or even 1000, are thought to survive each host transfer.[89] Why then do aphids host alternate? Throughout the literature of the last century, there are two themes: (1) as aphids can produce many generations in the course of a season they are able to exploit the complementary growth patterns of herbaceous and woody plants; and (2) although it is advantageous for aphids to shift all life cycle stages to the phylogenetically younger and nutritionally more favorable herbaceous plants, the morphs produced in spring have undergone such extensive morphological and physiological changes to living on woody plants that it is impossible for them to live on other plants.[71] An extensive review of host alternation in aphids[63] and an experimental analysis of host alternation in *Cavariella*[57] give little support to the second hypothesis. It is concluded that such obligate resource switching occurs when changing hosts is predictably more profitable than remaining on the same plant species. The coexistence of plants displaying asynchronous phenologies supplied the potential, and the ability of aphids to produce a number of highly prolific generations in quick succession, which amplify small differences in performance on the different plants, supplied the means of exploiting the potential.

D. FACULTATIVE TRACKING OF THE ENVIRONMENT

Generation-specific strategies have enabled aphid clones to track very closely the predictable seasonal trends in habitat quality. Although at any one time the habitat is generally good or poor, there is nevertheless a great deal of variability. It is probably no coincidence, therefore, that many aphids have a facultative ability to vary their investment in gonads and energy reserves.

In the unwinged individuals of each generation, in many aphids there is a directly proportional trade-off in the investment in gonads and lipoidal reserves, which is independent of size. Individuals with small gonads are better able to resist periods of starvation and seek out high quality feeding sites, whereas those with large gonads are better able to exploit the high quality habitats. Although the range in investments in gonads and lipoidal reserves in each clone appears to be fixed, the proportion with small gonads and large lipoidal reserves and vice versa, is variable, but under maternal control. In harsh habitats mothers produce proportionally more individuals with small gonads and large lipoidal reserves and vice versa, but the range remains the same.[18,96]

Similarly, winged individuals show an intraclonal trade-off between reproductive investment and size of the fat body, which is also independent of size. Winged aphids with small gonads have a greater urge to migrate, may fly longer and more frequently, and are better able to survive starvation than aphids with large gonads. The short-distance migrants consist mainly of individuals with large gonads, and the long-distance migrants should consist mainly of individuals with small gonads. If they settle on a poor host they tend to retain functional wing muscles longer than when they settle on a good host. The degree of differential migratory ability shown by an aphid clone will probably reflect both the within and between year heterogeneity of the environment. In favorable environments, individuals with a high reproductive investment ("risk takers"), will be fitter than those with a low reproductive investment ("risk averse"), and vice versa.[36,59,97,98]

Alary dimorphism and intraclonal tactical diversity in reproductive investment is particularly characteristic of one of the ten subfamilies of the Aphididae, the Aphidinae, and may partly account for their success, with 60% of aphid species belonging to this subfamily.

V. CONCLUSIONS

The high incidence of parthenogenesis in the Insecta compared to many other groups of animals is puzzling. Although ovarially transmitted microorganisms are undoubtedly implicated in the evolution of parthenogenesis in some Hymenoptera, this is not the case for other insect orders where parthenogenesis appears to be under genetic control. Parthenogenesis is thought to be advantageous in extreme habitats because it conserves general-purpose genotypes and in ephemeral habitats because it enhances their colonization and rapid exploitation.

In adopting cyclical parthenogenesis, aphids benefit from both sex and parthenogenesis. The genetic variation generated by recombination during sexual reproduction enables them to keep up with the generation of habitat heterogeneity, and parthenogenesis permits the prolific propagation of the more successful genotypes. Parthenogenesis also enabled aphids to overcome the developmental constraint on the rate of increase by beginning their development inside their grandmother. Short individual life span also made it possible for aphids to develop generation-specific strategies, which along with a facultative ability to vary their investment in gonads and energy reserves enabled them to track very closely seasonal trends in habitat quality. Their high rate of increase amplifies small differences in performance on different species of plants and so makes host plant specialization and host alternation advantageous. Although the "altruistic" behavior of soldier aphids in defending a colony has rightly attracted a lot of attention, it should not be allowed to obscure the fact that this is only a part of a functional polyphenism involving defence, migration, reproduction, and survival, which is likely to have been selected for because it increased fitness at the clonal level. Elton[40] compared phenotypic plasticity to a conjuror's magic kettle, dispensing the beverage the environment demands, which is particularly apt for aphids.

Given that cyclical parthenogenesis is a very effective mechanism for coping with both long- and short-term environmental uncertainty, it is relevant to ask: why is it so rare? The view that cyclical parthenogenesis is advantageous if the individual's life spans a small part of the seasonal range of conditions is supported by the occurrence of cyclical parthenogenesis in cladocera and aphids. Taxa with longer lived individuals lack the ability to develop the temporal correspondence of strategy with habitat,[82] so characteristic of aphids. However, a short life span is common to many other groups of insects that do not show cyclical parthenogenesis.

ACKNOWLEDGMENT

The author is indebted to Dr. Richard Sequeira for reading and commenting on the manuscript.

REFERENCES

1. **Aoki, S.,** *Colophina clematis* (Homoptera, Pemphigidae), an aphid species with 'soldiers', *Kontyû,* 45, 276, 1977.
2. **Aoki, S.,** Further observations on *Astegopteryx styracicola* (Homoptera: Pemphigidae) an aphid species with soldiers biting man, *Kontyû,* 47, 99, 1979.
3. **Aoki, S., Akimoto, S., and Yamane, D.,** Observations on *Pseudoregma alexanderi* (Homoptera, Pemphigidae), an aphid species producing pseudoscorpion-like soldiers on bamboos, *Kontyû,* 49, 355, 1981.
4. **Aoki, S., Yamane, S., and Kiuchi, M.,** On the biters of *Astegopteryx styracicola* (Homoptera, Aphidoidea), *Kontyû,* 45, 563, 1977.
5. **Bell, G.,** *The Masterpiece of Nature: the Evolution and Genetics of Sexuality,* Croom Helm, London, 1982.
6. **Bell, G.,** Recombination and immortality of the germ line, *J. Evol. Biol.,* 1, 67, 1988.
7. **Benton, T. G. and Foster, W. A.,** Altruistic housekeeping in a social aphid, *Proc. R. Soc. Lond. B.,* 247, 199, 1992.
8. **Blackman, R. L.,** Biological approaches to the control of aphids, *Phil. Trans. R. Soc. Lond. B.,* 274, 473, 1976.
9. **Blackman, R. L.,** Early development of the parthenogenetic egg in three species of aphids (Homoptera: Aphididae), *Int. J. Insect Morphol. Embryol.,* 7, 33, 1978.
10. **Blackman, R. L.,** Stability and variation in aphid clonal lineages, *Biol. J. Linn. Soc.,* 11, 259, 1979.
11. **Blackman, R. L.,** Chromosomes and parthenogenesis in aphids, in *Insect Cytogenetics,* Blackman, R. L., Hewitt, G. M., and Ashburner, M., Eds., *Symp. Roy. Entomol. Soc. Lond.,* 10, 133, 1980.
12. **Blackman, R. L.,** Species, sex and parthenogenesis in aphids, in *The Evolving Biosphere,* Forey, P. L., Ed., Cambridge University Press, Cambridge, 1981, 75.
13. **Blakely, N.,** Biotic unpredictability and sexual reproduction: do aphid genotype-host genotype interactions favour aphid sexuality?, *Oecologia (Berlin),* 52, 396, 1982.
14. **Bonnet, C.,** *Traité d'insectologie ou Observations sur les Pucerons,* Chez Durand, Paris, 1745.
15. **Briggs, J. B.,** The distribution, abundance, and genetic relationships of four strains of the rubus aphid (*Amphorophora rubi* (Kalt)) in relation to raspberry breeding, *J. Hort. Sci.,* 40, 109, 1965.
16. **Brough, C. N. and Dixon, A. F. G.,** Follicular sheath (ovarian sheath) structure in virginoparae of the vetch aphid, *Megoura viciae* Buckton (Homoptera: Aphididae), *Int. J. Insect Morphol. Embryol.,* 18, 217, 1989.
17. **Brough, C. N. and Dixon, A. F. G.,** Reproductive investment and the inter-ovariole differences in embryo development and size in virginoparae of the vetch aphid, *Megoura viciae, Entomol. Exp. Appl.,* 52, 215, 1989.

18. **Brough, C. N. and Dixon, A. F. G.,** Intraclonal trade-off between reproductive investment and size of fat body in the vetch aphid, *Megoura viciae* Buckton, *Funct. Ecol.*, 3, 747, 1989.
19. **Brough, C. N., Dixon, A. F. G., and Kindlmann, P.,** Pattern of growth and fat content of somatic and gonadal tissues of virginoparae of the vetch aphid, *Megoura viciae, Entomol. Exp. Appl.*, 56, 269, 1990.
20. **Browne, R. A.,** Population genetics and ecology of *Artemia*: insights into parthenogenetic reproduction, *TREE*. 7, 232, 1992.
21. **Buda, V. G., Mozuraitis, R. L., and Myarchaitis, V. P.,** Parthenogenesis and signal behavior of the butterflies *Lithocolletis emberizaepennella* Bouche (Lepidoptera Gracillariidae), *Dokl. Acad. Nauk SSSR*, 319, 512, 1991.
22. **Bulmer, M. G.,** Cyclical parthenogenesis and the cost of sex, *J. Theoret. Biol.*, 94, 197, 1982.
23. **Charlesworth, B.,** Reply to Bulmer on the cost of sex with cyclical parthenogenesis, *J. Theoret. Biol.*, 94, 223, 1982.
24. **Carvalho, G. R., Maclean, N., Wratten, S. D., Carter, R. E., and Thurston, J. P. O.,** Differentiation of aphid clones using DNA fingerprints from individual aphids, *Proc. Roy. Soc. Lond. B*, 243, 109–114, 1991.
25. **Clark, W. C.,** The ecological implications of parthenogenesis, in *Perspectives in Aphid Biology*, Lowe, A. D., Ed., *Entomol. Soc. New Zeal. Bull.*, No. 2, 103, 1973.
26. **Cognetti, G.,** Citogenetica della partenogenesi negli Afidi, *Arch. Zool. Ital.*, 46, 89, 1961.
27. **Cuellar, O.,** Animal parthenogenesis, *Science*, 197, 837, 1977.
28. **Dickson, R. C.,** Development of the spotted alfalfa aphid population in North America, *Int. Kong. Entomol. (Wien 1960)*, 2, 26, 1962.
29. **Dixon, A. F. G.,** Fecundity of brachypterous and macropterous alatae in *Drepanosiphum dixoni* (Callaphididae, Aphididae), *Entomol. Exp. Appl.*, 15, 335, 1972.
30. **Dixon, A. F. G.,** Seasonal changes in fat content, form, state of gonads and length of adult life in the sycamore aphid *Drepanosiphum platanoids* (Schr.), *Trans. Roy. Entomol. Soc. Lond.*, 127, 87, 1975.
31. **Dixon, A. F. G.,** *Aphid Ecology*, Blackie, Glasgow, 1985.
32. **Dixon, A. F. G.,** Individuals, populations and patterns, in *Individuals, Populations and Patterns in Ecology*, Leather, S. R., Watt, A. D., Mills, N. J., and Walters, K. F A., Eds., Intercept, Andover, 1994, 449.
33. **Dixon, A. F. G., Burns, M. D., and Wangboonkong, S.,** Migration in aphids: response to current adversity, *Nature*, 220, 1337, 1968.
34. **Dixon, A. F. G. and Dharma, T. R.,** Number of ovarioles and fecundity in the black bean aphid, *Aphis fabae*, *Entomol. Exp. Appl.*, 28, 1, 1980.
35. **Dixon, A. F. G. and Glen, D. M.,** Morph determination in the bird cherry-oat aphid, *Rhopalosiphum padi* L., *Ann. Appl. Biol.*, 68, 11, 1971.
36. **Dixon, A. F. G. and Howard, M. T.,** Dispersal in aphids, a problem in resource allocation, in *Insect Flight: Dispersal and Migration*, Danthanarayana, W., Ed., Springer-Verlag, Berlin, 1986, 145.
37. **Dixon, A. F. G. and Wellings, P. W.,** Seasonality and reproduction in aphids, *Int. J. Invertebr. Reprod.*, 5, 83, 1982.
38. **Dunn, J. A.,** The survival in soil of apterae of the lettuce root aphid, *Pemphigus barsarius* (L.), *Ann. Appl. Biol.*, 47, 766, 1959.
39. **Eastop, V. F.,** Worldwide importance of aphids as virus vectors, in *Aphids as Virus Vectors*, Harris, K. F. and Maramorosch, K., Eds., Academic Press, New York, 1977, 4.
40. **Elton, C.,** *Animal Ecology and Evolution*, Oxford University Press, Oxford, 1930.
41. **Fakatsu, T. and Ishikawa, H.,** Soldiers and male of an eusocial aphid *Colophina arma* lack endosymbiont: implication for physiological evolutionary interaction between host and symbiont, *J. Insect Physiol.*, 38, 1033, 1992.
42. **Fenchel, T.,** Intrinsic rate of natural increase: the relationship with body size, *Oecologia (Berlin)*, 14, 317, 1974.
43. **Fuyama, Y.,** Genetics of parthenogenesis in *Drosophila melanogaster*. XI. Characterization of a gynogenetically reproducing strain, *Genetics*, 114, 495, 1986.
44. **Geiler, H.,** Eichhornchen, *Sciurus vulgaris fuscoater* Altum. 1876, als Vertilger von Blattläusen (Aphidae), *Säugentierkundl. Mitt.*, 4, 13, 1956.
45. **Hille Ris Lambers, D.,** Polymorphism in Aphididae, *Annu. Rev. Entomol.*, 11, 47, 1966.
46. **Howard, M. T. and Dixon, A. F. G.,** The effect of plant phenology on the induction of alatae and the development of populations of *Metopolophium dirhodum* (Walker), the rose-grain aphid, on winter wheat, *Ann. Appl. Biol.*, 120, 203, 1992.
47. **Hurst, G. D. D., Majerus, M. E. N., and Walker, L. E.,** Cytoplasmic male killing elements in *Adalia bipunctata* (Linnaeus) (Coleoptera: Coccinellidae), *Heredity*, 69, 84, 1992.
48. **Huxley, T. H.,** On the agamic reproduction and morphology of *Aphis* — Part 1, *Trans. Linn. Soc.*, 22, 193, 1858.
49. **Ishikawa, H.,** Biochemical and molecular aspects of the aphid endocytobiosis, in *Insect Endocytobisis: Morphology, Physiology, Genetics, Evolution*, Schwemmerl, W. and Gassner, G., Eds., CRC Press, Boca Raton, 1989, 123.

50. **Innes, D. J. and Herbert, P. D. N.,** The origin and genetic basis of obligate parthenogenesis in *Daphnia pulex*, *Evolution*, 42, 1024, 1988.
51. **Iwasa, Y.,** Pessimistic plant: optimal growth schedule in stochastic environments, *Theor. Popul. Biol.*, 40, 246, 1991.
52. **Janzen, D. H.,** What are dandelions and aphids? *Am. Nat.*, 111, 586, 1977.
53. **Judge, F. D.,** Overwintering in *Pemphigus bursarius* (L.), *Nature*, 216, 1041, 1967.
54. **Kindlmann, P. and Dixon, A. F. G.,** Developmental constraints in the evolution of reproductive strategies: telescoping of generations in parthenogenetic aphids, *Funct. Ecol.*, 3, 531, 1989.
55. **Kindlmann, P. and Dixon, A. F. G.,** Optimum body size: effects of food quality and temperature, when reproductive growth rate is restricted, with examples from aphids, *J. Evol. Biol.*, 5, 677, 1992.
56. **Knight, R. L. and Alston, F. H.,** Pest resistance in fruit bleeding, in *Biology in Pest and Disease Control*, Price Jones, D. and Solmon, M. E., Eds., Blackwell, Oxford, 1974, 73.
57. **Kundu, R. and Dixon, A. F. G.,** Evolution of complex life cycles in aphids, *J. Anim. Ecol.*, 64, 246, 1995.
58. **Lamb, R. Y. and Willey, R. B.,** Are parthenogenetic and related bisexual insects equal in fertility? *Evolution*, 33, 774, 1979.
59. **Leather, S. R., Wellings, P. W., and Dixon, A. F. G.,** Habitat quality and the reproductive strategies of the migratory morphs of the bird cherry-oat aphid, *Rhopalosiphum padi* (L.) colonizing secondary host plants, *Oecologia (Berlin)*, 59, 302, 1983.
60. **Lees, A. D.,** The production of the apterous and alate forms in the aphid *Megoura viciae* Buckton, with special reference to the role of crowding, *J. Insect Physiol.*, 13, 289, 1967.
61. **Lloyd, D. G.,** Benefits and handicaps of sexual reproduction, *Evol. Biol.*, 13, 69, 1980.
62. **Lynch, M.,** Destabilizing hybridization, general-purpose genotypes and geographic parthenogenesis, *Q. Rev. Biol.*, 59, 259, 1984.
63. **Mackenzi, A. and Dixon, A. F. G.,** An ecological perspective of host alternation in aphids (Homoptera: Aphidinea: Aphididae), *Entomol. Gener.*, 16, 265, 1991.
64. **Marescalchi, O., Pijnacker, L. P., and Scali, V.,** Cytology of parthenogenesis in *Bacillus whitei* and *Bacillus lynceorum* (Insecta Phasmatodea), *Invertebr. Reprod. Develop.*, 20, 75, 1991.
65. **Markkula, M.,** Studies on the pea aphid *Acyrthosiphon pisum* Harris (Hom. Aphididae) with special reference to the difference in the biology of the green and red forms, *Ann. Agric. Fenn.*, 2, 1, 1963.
66. **Matsuka, M. and Mittler, T. E.,** Enhancement of alata production by an aphid, *Myzus persicae*, in response to increases in day length, *Bull. Fac. Agric. Tamagawa Univ. Tokyo*, 18, 1, 1978.
67. **Maynard Smith, J.,** The evolution of recombination, in *The Evolution of Sex*, Michod, R. E. and Levin, B. R., Eds., Sinauer, Sunderland, MA, 1988, 106.
68. **Meglitsch, P. A. and Schram, F. R.,** *Invertebrate Zoology*, 3rd ed., Oxford University Press, Oxford, 1991.
69. **Mordvilko, A. K.,** Zur Biologie und Morphologie der Pflanzenläuse (Fam. Aphididae Pass.), *Horae Soc. Entomol. Ross.*, 31, 1, 1900.
70. **Mordvilko, A. K.,** Zur Biologie und Morphologie der Pflanzenläuse (Fam. Aphididae Pass.), *Horae Soc. Entomol. Ross.*, 33, 162, 1901.
71. **Mordvilko, A. K.,** The evolution of cycles and the origin of heteroecy (migration) in plant-lice, *Ann. Mag. Nat. Hist.*, 2, 570, 1928.
72. **Mordvilko, A.,** On the evolution of aphids, *Arch. Naturgesch. N.S.*, 3, 1, 1934.
73. **Mukai, T.,** The genetic structure of natural populations of *Drosophila melanogaster*. 1. Spontaneous mutation rates of polygenes controlling variation, *Genetics*, 50, 1, 1964.
74. **Müller, F. P.,** Biotypen und Unterarten der "Erbsenlaus" *Acyrthosiphon pisum* Harris, *Z. Pflanzenkr. Pflanzenschutz*, 69, 129, 1962.
75. **Muller, H. J.,** The relation of recombination to mutational advance, *Mutation Res.*, 1, 2, 1964.
76. **Newton, C. and Dixon, A. F. G.,** The cost of switching from asexual to sexual reproduction in an aphid, *Entomol. Exp. Appl.*, 47, 283, 1988.
77. **Nielson, M. W. and Don, H.,** A new virulent biotype of the spotted alfalfa aphid in Arizona, *J. Econ. Entomol.*, 67, 64, 1974.
78. **Owen, R.,** *On Parthenogenesis or the Successive Production of Procreating Individuals from a Single Ovum*, John van Voorst, London, 1849.
79. **Pagliai-Bonvicni, A. M. and Cognetti, G.,** Partenogenesi e variabilita genetica negli afidi, *Mem. Soc. Entomol. Ital.*, 60, 63, 1982.
80. **Puterka, G. J. and Peters, D. C.,** Inheritance of greenbug virulence to Gb2 and Gb3 resistance genes in wheat, *Genome*, 32, 109, 1989.
81. **Puterka, G. J. and Peters, D. C.,** Sexual reproduction and the inheritance of virulence in the greenbug, *Schizaphis graminum* (Rondani), in *Aphid-Plant Genotype Interactions*, Campbell, R. K. and Eikenbary, R. D., Eds., Elsevier, New York, 1990, 289.
82. **Ruvinsky, A. O., Perelygin, A. A., Lobkov, Yu. I., and Belyaeu, D. K.,** Factors organizing and maintaining polymorphism in a cyclic parthenogenetic species: *Daphnia pulex*, *Heredity*, 57, 15, 1986.

83. **Shaw, M. J. P.,** Effects of population density on alienicolae of *Aphis fabae* Scop. 1. The effect of crowding on the production of alatae in the laboratory, *Ann. Appl. Biol.*, 65, 191, 1970.
84. **Sibly, R. M. and Calow, P.,** *Physiological Ecology of Animals. An Evolutionary Approach,* Blackwells Scientific Publications, Oxford, 1986.
85. **Stalker, H. D.,** On the evolution of parthenogenesis in Lonchoptera (Diptera), *Evolution,* 10, 154, 1956.
86. **Stouthamer, R., Luck, R. F., and Hamilton, W. D.,** Antibiotics cause parthenogenetic *Trichogramma* (Hymenoptera/Trichogrammatidae) to revert to sex, *Proc. Natl. Acad. Sci. U.S.A.*, 87, 2424, 1990.
87. **Suomalainen, E., Saura, A., and Lokki, J.,** Evolution of parthenogenetic insects, *Evol. Biol.*, 9, 209, 1976.
88. **Suomalainen, E., Saura, A., and Lokki, J.,** *Cytology and Evolution in Parthenogenesis,* CRC Press, Boca Raton, FL, 1987.
89. **Taylor, L. R.,** Migration and the spatial dynamics of an aphid, *Myzus persicae, J. Anim. Ecol.*, 46, 411, 1977.
90. **Templeton, A. R.,** The prophecies of parthenogenesis, in *Evolution and Genetics of Life Histories,* Dingle, H. and Hegmann, P., Eds., Springer, Hamburg, 1982, 75.
91. **Ueda, N. and Takada, H.,** Differential relative abundance of green-yellow and red forms of *Myzus persicae* (Sulzer) (Homoptera: Aphididae) according to host plant and season, *Appl. Entomol. Zool.*, 12, 124, 1977.
92. **Uyenoyama, M. K.,** On the evolution of parthenogenesis: a genetic representation of the "cost of meiosis," *Evolution,* 38, 87, 1984.
93. **Van Valen, L.,** A new evolutionary law, *Evol. Theory,* 1, 1, 1973.
94. **Wagner, G. P. and Gabriel, W.,** Quantitative variation in finite parthenogenetic populations: what stops Muller's ratchet in the absence of recombination, *Evolution,* 44, 715, 1990.
95. **Walker, F.,** Remarks on the migration of Aphides, *Ann. Mag. Natur. Hist.*, 1, 372, 1848.
96. **Walters, K. F. A., Brough, C., and Dixon, A. F. G.,** Habitat quality and reproductive investment in aphids, *Ecol. Entomol.*, 13, 337, 1988.
97. **Walters, K. F. A. and Dixon, A. F. G.,** Migratory urge and reproductive investment in aphids: variation within clones, *Oecologia (Berlin),* 58, 70, 1983.
98. **Ward, S. A. and Dixon, A. F. G.,** Spreading the risk, and the evolution of mixed strategies: seasonal variation in aphid reproductive biology, *Adv. Invertebr. Reprod.*, 3, 367, 1984.
99. **Watt, A. D. and Dixon, A. F. G.,** The role of cereal growth stages and crowding in the induction of alatae in *Sitobion avenae* and its consequences for population growth, *Ecol. Entomol.*, 6, 441, 1981.
100. **Weber, G.,** On the ecological genetics of *Sitobion avenae* (F.) (Hemiptera, Aphididae). *Z. Angew. Entomol.*, 100, 100, 1985.
101. **Weber, G.,** On the ecological genetics of *Metopolophium dirhodum* (Walker) (Hemiptera, Aphididae), *Z. Angew Entomol.*, 100, 451, 1985.
102. **Wellings, P. W., Leather, S. R., and Dixon, A. F. G.,** Seasonal variation in reproductive potential: a programmed feature of aphid life cycles, *J. Anim. Ecol.*, 49, 975, 1980.
103. **Went, D. F.,** Paedogenesis in the dipteran insect *Heteropeza pygmaea*: an interpretation, *Int. J. Invertebr. Reprod.*, 1, 21, 1979.
104. **Went, D. F.,** Parthenogenetic strategies in insect reproduction, *Adv. Invertebr. Reprod.*, 3, 303, 1984.
105. **White, M. J. D.,** *Animal Cytology and Evolution,* 3rd ed., Cambridge University Press, Cambridge, 1973.
106. **Williams, G. C.,** *Natural Selection: Domains, Levels, and Challenges,* Oxford University Press, New York, 1992.
107. **Wratten, S. D.,** Reproductive strategy of winged and wingless morphs of the aphids *Sitobion avenae* and *Metoplophium dirhodum, Ann. Appl. Biol.*, 86, 319, 1977.
108. **Yagamuchi, H.,** Biological studies on the todo-fir aphid *Cinara todicola* Inouye with special reference to its population dynamics and morph determination, *Bull. Gov. For. Exp. Stn. Tokyo,* 283, 1, 1976.
109. **Zchori-Fein, E., Roush, R. T., and Hunter, M. S.,** Male production induced by antibiotic treatment in *Encarsia formosa* (Hymenoptera: Aphelinidae), an asexual species, *Experentia,* 48, 102, 1992.

Chapter 7

FACTORS AFFECTING FECUNDITY, FERTILITY, OVIPOSITION, AND LARVIPOSITION IN INSECTS

Simon R. Leather

CONTENTS

0-8493-6695-X/95/$0.00+$.50

I. INTRODUCTION

The production of its offspring is arguably the most important event in the life span of an insect. It is affected by a number of factors, many of which have already been discussed in the preceding chapters. This chapter examines critically the role that reproductive load and host plant quality have on the reproductive patterns of herbivorous insects. Most of the examples used will refer to the Lepidoptera and the Hemiptera, as these two groups have been extensively studied and the strategies displayed are representative of those shown by insect groups that deposit eggs and live young, respectively. This does not imply that other insect groups will be ignored; examples from other groups will be discussed to demonstrate how similar the strategies employed by insects are, regardless of phylogenetic classification.

II. FECUNDITY AND FERTILITY

Insect fecundity and fertility, the number of offspring produced and the viability of those offspring, are affected by a number of factors and these are discussed at length below. The number and viability of offspring produced by an individual insect is of great importance to the population dynamics of the species as a whole as well as to the individual concerned. The evolution of reproductive strategies (Chapter 10) and specific oviposition behaviors are results of the factors that affect fecundity and fertility, although to a certain extent, oviposition behavior can influence fecundity and is intrinsically linked with reproductive strategies. There is some confusion in the literature between potential fecundity and achieved (realized) fecundity. Potential fecundity can be defined in a number of ways, but is most often some measure of the reproductive capacity of an insect that is measured before the reproductive event occurs, e.g., number of eggs in the reproductive tract. Achieved fecundity is the actual number of offspring produced by an insect during its life span. The difference between potential and achieved fecundity can be substantial. Achieved fecundity depends upon potential fecundity and various ecological factors, particulary nutrition.[10] Nutrition affects size, which is one of the major factors affecting fecundity (see II.B.1.a.i). The number of ovarioles within an insect has a large bearing on the potential fecundity of insects and it is useful to discuss the likely implications of variability in this characteristic before dealing with any other factor.

A. OVARIOLE NUMBER

Insects are characterized by the possession of two ovaries within which are a number of ovarioles. Ovariole number in insects is not always a constant for either family or species. In some groups, e.g., Lepidoptera, it is nearly always a constant, with eight being the norm,[97] however, in some rare individuals of the butterfly *Heliconius charitonius*, six, seven, and nine ovarioles have been reported.[60] Ovariole number also varies slightly in the butterfly *Colias philodice eriphyle:* individuals with eight, nine, and twelve ovarioles have been observed.[188] No explanation for this phenomenon has been sought and it is so rare as to have no bearing on reproductive strategies. There is considerable variation between species and within species in other groups, e.g., Aphididae and Diptera.[72,139] Within those insect species with a variable number of ovarioles, ovariole number is positively related to fecundity.[56,72,73,132,221]

For some insect groups such as the Diptera, the number of ovarioles is positively related to the weight of the insect.[13,73,132] The number of ovarioles in the mosquito *Aedes punctor* is strongly correlated with wing length (an indicator of size), but wing length is less well correlated with achieved fecundity.[168] In other insects, e.g., *Melanoplus* grasshoppers, ovariole number is variable but is strongly related to size. In addition, the number of ovarioles is equal to the number of eggs laid in a pod, so ovariole number in this case is a good indicator of achieved fecundity.[12a]

Other insects, such as aphids, show no satisfactorily demonstrated significant relationship between weight and ovariole number,[132,211,221] but the more ovarioles an individual possesses, the more fecund that individual is likely to be.[132] Thus, within the same ovariole number class, large individuals are more fecund than smaller ones, but a female with six ovarioles and the same weight as one with ten ovarioles would be less fecund than the latter. Ovariole number in aphids is both a function of generational morph,[139,221] and of host quality. Ovariole number in the majority of the aphid species studied is high at the beginning of the life cycle, namely those generations emerging from eggs or, as in the case of *Rhopalosiphum padi*, in the second generation from the egg. Ovariole number decreases as the season progresses, but shows an increase in the autumnal generations. The morphs associated with the highest host quality available, expanding buds and leaves in the spring and senescing leaf tissue in the autumn, are those with the greatest numbers of ovarioles.[49,53] By having a large flexibility in reproductive potential, these aphids are able to fully exploit this high host quality and rapidly increase their numbers before the leaves mature and host quality decreases or the leaves drop. Aphids exploiting plants with a low host quality tend to have fewer ovarioles and to be less fecund[132,139] (see below).

B. THE ROLE OF NUTRITION

Nutrition, i.e., the quality and amount of food experienced by an insect during larval and sometimes adult life, has a major effect on both fecundity and fertility. In some insects, e.g., within the Lepidoptera, it has been considered that adult nutrition is of little importance to fecundity and that the major influence is from nutrition afforded to the larvae. Other insects, e.g., aphids,[55] are considered to be affected by both adult and juvenile nutrition. The following sections consider how both adult and larval nutrition affect fecundity and fertility and also consider how substances not generally thought to be nutritious can also affect these life history parameters.

1. Larval Nutrition and Adult Size

The importance of larval nutrition cannot be underestimated. Whether the insect feeds as an adult or not, the size that it attains as an adult is dependent on the nutrition experienced by an immature larva. The size (weight) an adult insect achieves is generally a good indicator of its fecundity. It must be noted, however, that although adult size is almost invariably correlated with the individual's potential fecundity, it is not always correlated with achieved fecundity[121] (see 1.a). It is those factors which affect both potential and achieved fecundity that will be discussed below.

If the stored product moth *Cadra cautella* is crowded as a larva, then fecundity can be reduced by as much as 67%.[81] Even more dramatic responses to larval nutrition can be seen in other Lepidoptera. For example, when larvae of *Heliothis zea* were fed on corn, 92% of the eggs that were produced by the subsequent adult females were fertile. However, if reared on cotton squares, adult females were totally infertile.[143]

The sawfly *Diprion pini* is strongly affected by the host plant as a larva; even clones of the same species of host plant have a strong effect on larval performance. For example, when reared as a larvae on four different clones of *Pinus sylvestris*, the fecundity of the subsequent adults was highly affected. Those host plants that allowed good larval growth and survival were associated with large, highly fecund females.[8] Larval feeding is also very important in determining ovarian development and subsequent fecundity of the sawfly *Pristiphora erichsonii*.[91]

In general, if insects are reared under poor conditions of larval nutrition, e.g., crowding or poor-quality food plants, then the adults that arise are smaller and tend to be less fecund than those adults arising from larvae developing under better conditions. For example, larvae of the chrysomelid beetle *Chrysomela knabi*, when reared on a less suitable willow host, e.g., *Salix nigra* under low moisture conditions, were smaller and less fecund as adults than when reared on *S. humulis*.[95] Larval competition in the predatory stoneflies *Megarcys signata* and *Kogatus modestus* results in the production of small adults that are less fecund than the large adults that arise when larvae are reared under low density conditions.[171]

a. Adult Size

The size that an insect achieves as an adult is dependent on a number of factors, many due to larval nutrition (see above). The size of the adult insect has many important life history implications, ranging from effects of adult life span, mating success, fecundity, etc. In the following sections, the effect that adult size has on fecundity and fertility are discussed. Other effects of size on insect reproduction are discussed elsewhere, e.g., Chapter 9.

i. Size and Fecundity

Until recently it was believed that for most arthropods increased weight and/or size resulted in increased fecundity.[79,215] Certainly, this has been reported for many insects, especially aphids. However, this relationship is not as straightforward as might at first appear.[121] Many studies have used potential fecundity (i.e. the number of eggs or offspring within an insect at the adult moult) as the index of fecundity for their estimate of lifetime fecundity. In those insects that have produced all their eggs or embryos by the time adulthood is reached, such as in the pine looper moth *Bupalus piniarius*,[18] and the sycamore aphid *Drepanosiphum platanoidis*,[140] this is indeed an adequate prediction. However, in species where ovulation does not cease after the adult moult and the number of steps between the dependent variable (offspring number) and independent variable (weight) is increased, then this relationship becomes less robust. For example, using embryo or large embryo counts as indicators of fecundity in aphids is a common practice.[19,48,62] In an aphid such as *R. padi*, in which ovulation continues after the adult moult and the initial fecundity estimate can be increased by 135%,[114] then this method would produce a highly erroneous result.

The size/fecundity relationship is affected by a number of factors, many of which can be linked either directly or indirectly with host quality. The effect of host quality on this relationship can act through the larval stage as well as on the reproductive adult stage. Moreover, even if large insects of the same species are more fecund than smaller ones of the same species, confounding factors can occur. For example, large adults of the sawfly *Neodiprion sertifer* produce more eggs than small adults, but fewer of their eggs are fertile.[89] In some aphids (e.g., *Aphis fabae*), the relationship between size and fecundity is indirect and acts mainly through reproductive rate and mortality.[197] That is, small aphids have a faster rate of reproduction than large aphids, but do not live as long and are thus less fecund. In Lepidoptera, even more confusion can arise. For example, adult weight in the pine beauty moth *Panolis flammea* is a good indicator of the potential fecundity.[115] However, this has little bearing on the number of eggs that are actually laid, with the achieved fecundity sometimes being as little as 1% of the potential fecundity.[135] In fact, fewer than 15% of laboratory reared *P. flammea* lay even 70% of their maximum potential number of eggs.[121] In field conditions this is likely to be even lower,[33] and the maximum fecundity estimate for field *P. flammea* is 28% of its maximum reproductive potential.[135] In other Lepidoptera, for example the moth *Epiphyas postvittana*, there is a good correlation between adult weight and achieved fecundity when the moths are fed, but not if they are starved.[83]

As an adjunct to the above it is interesting to consider the effect of male size on female fecundity. For example in *Drosophila melanogaster*, females that mate with small males have a greater fecundity than those that have mated with large males. This apparently contradicts the theory that large males are fitter than small males and enjoy greater mating success (Chapter 9), but it appears that small males copulate longer and transfer more sperm than large males.[177]

ii. Size and Fertility

The size of a female insect can have marked effects on the fertility of the eggs laid. For example, large individuals of the codling moth *Cydia pomonella* lay more fertile eggs than small ones.[46] In the butterfly *Pararge aegeria*, small eggs tend to be infertile, but since small eggs also tend to be laid at the end of the reproductive period, it may be that female age influences egg fertility rather than egg weight *per se*.[224] This area requires further study.

2. Adult Nutrition

Adult insects can be divided into two main categories — those that feed after becoming adult and those that do not feed, or if they do, feed only minimally. The former are usually long lived and obtain the nutrients required for reproduction as adults. The latter are usually short lived and have laid down their reserves for reproduction during larval life, as is the case with many Lepidoptera. There are, of course, exceptions. Aphids, for example, feed during adult life but are generally short lived, and larval (nymphal) experience plays a large part in how they function as adults.[55] Some adult Lepidoptera feed on pollen and even blood[61,115] and not solely on carbohydrates.

Adult feeding has for some time been thought only important in those species of insects that are relatively long lived as an adult or which require a specific food source unobtainable as a larvae to commence reproduction, e.g., the blood meal of a female mosquito. Insect groups such as the Lepidoptera, where larval development is long and reproductive reserves are laid down during the larval period, have been considered to be essentially nonfeeding as adults despite the fact that many species are known to be nectar feeding.[23]

However, significant effects in Lepidoptera have been demonstrated in several studies. For example, carbohydrate intake has been shown to increase longevity and egg production in the butterfly *Colias philodice eurytheme*, the moth *Euxoa messoria*, and the checkerspot butterfly *Euphydryas editha*.[30,160,191] In addition, the pine beauty moth *P. flammea*, when given access to water, is twice as fecund as when deprived of a water source, and if given a carbohydrate source is three times as fecund.[115] *E. messoria* is more than three times more fecund when fed on honey solution than when deprived of water.[30] The potato tuber moth *Phthorimaea operculella* lays almost four times as many eggs as unfed moths when fed on sucrose solution,[67] and the moth *Heliothis virescens* shows a 50% increase in fecundity when fed on a sucrose solution.[143] If *H. virescens* is unable to obtain nectar in the field, its reproductive capacity decreases as the moth ages.[224a] Like *P. flammea*, if fed it is able to maintain its body weight and devote its reserves to reproduction.[115] In the field, the spruce budworm *Choristoneura fumiferana* enhances adult fecundity by feeding on honeydew.[150]

Adult feeding in Lepidoptera is thus important in determining realized fecundity. Unfed adult *E. postvittana* are 30% less fecund than moths that have access to normal levels of adult nutrition, but interestingly the fertility of the eggs is unaffected, suggesting that extra nitrogen is not required as an adult.[83] Similarly, the moth *Zeiraphera canadensis* lived longer when fed on a sugar solution and was more fecund than those fed on water alone. However, the eggs of fed moths had a fertility of only 84% compared with one of 95% in the unfed moths. As overall fecundity was increased, the fed moths laid 88 fertile eggs per female compared with 61 fertile eggs per female for unfed moths.[29] Since the oviposition period was increased but not the oviposition rate, it is possible that the moths ran out of sperm, and, not being in a position to mate for a second time, started to lay infertile eggs. In the case of *P. flammea*, the oviposition period was also increased by adult feeding, but no increase in the daily oviposition rate was seen.[115]

The armyworm *Spodoptera exempta* is unable to realize its potential fecundity without water to achieve hydration and maturation of the oocytes.[85] In addition, *S. exempta* is able to utilize supplementary lipid and protein reserves following carbohydrate uptake and this results in increased fecundity.[86] It might, therefore, be thought that a nitrogen supplement to the diet of adult Lepidoptera would result in increased fecundity. However, adding amino acids to the diet of the butterfly *Euploea core* actually resulted in a decrease in fecundity, whereas sucrose resulted in an increase.[92] Insects that are predators as adults must supplement their nitrogen reserves. The predatory bug *Anthocoris confusus*, although able to lay eggs even when fed at suboptimal rates, are most fecund when fed on aphids at a rate of 6 aphids/day.[68] On the other hand, another aphid predator, the green lacewing *Chrysoperla carnea* was shown to be almost twice as fecund in the absence of prey as when prey was present, although there was no difference in preoviposition period.[196]

a. Adult Longevity and Fecundity

Longevity is a further complicating factor in the size/fecundity relationship for a number of insect species. For example, in the planthopper *Prokelisia marginata*, a significant relationship between eggs produced and adult female weight has been demonstrated.[43] However, this relationship was obtained by dividing total fecundity by total longevity and using the figure obtained as eggs per female per day. This is not strictly correct, as the pre- and postreproductive periods were ignored and both are of great importance in determining insect fecundity.[105,135,197]

Longevity and the factors affecting longevity appear to be the most important factors influencing achieved fecundity in the Lepidoptera. In *P. flammea* there is no relationship between weight and achieved fecundity, but a strong positive correlation between longevity and achieved fecundity exists.[115] In addition, there is no relationship between longevity and weight, but a significant relationship between availability of adult food source and longevity.[115] In *Euxoa messoria* there is a significant relationship between pupal weight and both achieved fecundity and longevity. However, if the moths are fed, the relationship between weight and longevity disappears, although longevity and achieved fecundity remain correlated.[30] Increased fecundity has also been found to be correlated with increased longevity in *H. virescens* and *Sesamia nonagriodes*,[3,143] and with the availability of an adult food source in the pierid butterfly *C. philodice eurytheme*.[191] Larval host plants can also affect longevity and this in turn affects the achieved fecundity of *S. nonagriodes* and *P. flammea* in both the laboratory and the field.[3,116,135] It is interesting to note that virgin moths lived longer than those that had been forced to mate in conditions when suitable oviposition sites were not available.[135] Those mated moths that delayed mating because of being presented with a poor host quality sacrificed life span in favor of producing fewer, but fitter eggs.[127]

b. Adult Longevity and Fertility

The phenomenon of a decline in offspring fitness as the mother ages is not confined to insects; human beings show the same general relationship. In aphids, (which, like humans, give birth to live young) although there is no evidence that the incidence of malformed and dead offspring increases as the mother reaches the end of her lifespan, it has been suggested that small aphids may resorb their embryos to a greater extent than larger aphids.[59] Certainly at high temperatures the frequency of aphid still births is increased.[76]

In Lepidoptera a decline in the fertility of the eggs produced as the mother ages is a common occurrence. Whether this is a result of sperm depletion or the inability of the mother to provision all her eggs is an open question. Fertility of the eggs of the moth *C. cautella* declines with increasing female age, from 80% at day 1 to 20% by day 8.[81] The fertility of eggs of the butterfly *P. aegeria* also decrease with the age of the mother.[224] In the moth *E. postvittana* egg fertility also changes with the age of the mother, but unlike the previous two examples, the relationship is not linear. Egg fertility is greatest halfway through the oviposition period, and is lowest at the beginning and at the end of the lifespan of the moth.[41] In the moth *Choristoneura murinana*, the color of the eggs is indicative of when in the reproductive period of the mother they have been laid; they are green at the beginning of the reproductive period and yellow at the end, and at the same time the fertility decreases from 96 to 65%.[62]

3. Host Plant Quality

Host quality can be defined as those plant attributes, chemical or physical, that contribute either negatively or positively to the fitness of the insect population or individual that feeds upon the plants tissues.[125]

Host quality can affect the fecundity of an herbivorous insect in two ways; it can affect the fecundity (achieved and potential) of the adult arising from the larvae feeding on it, or actually affect the achieved fecundity of the adult insect meeting that individual plant for the first time. The latter case is almost certainly just as common as the former.

The fecundity of the spear-marked black moth, *Rheumaptera hastata*, which feeds on *Pinus nigra*, was affected in two ways by the quality of its host plant. On hosts that had previously been attacked by conspecifics, the sex ratio was skewed in favor of males, and those females that were produced were less fecund than those produced on undamaged host plants.[222] In this case damaged plants were of poorer quality than undamaged plants. The fecundity of the pine processionary moth *Thaumetopoea pityocampa* was also affected by host plant variation in this way.[10]

The grain aphid, *Sitobion avenae*, is more fecund on the ears of cereal plants than on other parts of the plant, and that is where it is most often found.[216] The bird cherry aphid, *R. padi*, is more fecund on young grass and cereal plants than on mature plants, and on mature plants it is most fecund on the stem, and this too, is where it is most often found.[53,129] The cabbage aphid, *Brevicoryne brassicae*, is also strongly affected by host quality, being apparently most fecund on leaves high in nitrogen.[183,208]

In times of nutrient stress, the aphid *Metopolophium dirhodum* reduces embryo growth and resorbs the youngest embryos, so as to safeguard immediate reproduction for when conditions improve.[82] When the quality of the host plant improves, embryo formation begins anew. Host quality greatly affects the fecundity of adult aphids, both in terms of what happens to adults arriving on a novel host plant, and in terms of what they were reared upon. An aphid reared on a poor-quality host plant but feeding on a good-quality host plant will be more fecund than it would have been if it had remained on its original host plant. This has been demonstrated with *R. padi*, transferred from oats to Timothy grass, *Phleum pratense*.[111,122] A further complication occurs depending on the morph of the mother (whether winged or not), the offspring of alate mothers being more fecund than the offspring of apterous mothers. The effect of maternal experience is also seen in Lepidoptera. Larvae of the small ermine moth, *Yponomeuta evonymellus*, survive and develop better when transferred as neonate larvae from poor host plants to good host plants, but do not develop as well as those larvae that have been transferred from good-quality plants to similarly good-quality plants. The reverse is also true, larvae transferred from good-quality plants to poor-quality plants still develop faster than those larvae reared on poor-quality plants for their entire developmental period.[131]

a. Nitrogen

Nitrogen in its various forms is without doubt the most important single foodstuff to an insect, being as it is, the basis for the building blocks of life, protein. Hence, as well as having many important effects on fecundity and fertility *per se*, it also affects reproductive strategies directly and indirectly, by influencing host preferences, life span, etc.[125] The debatable point, particularly when considering herbivorous insects, is the way in which the nitrogen content of a particular food plant is measured, and how what is being measured affects the insect in question. Plant nitrogen content is a major determinant of host quality, although not solely so. Host plant quality affects herbivorous insects in a number of ways, many of which have been reviewed elsewhere.[125]

One aspect of food quality, the level of nutrient nitrogen, is known to be important in several species. For example, the host quality of larval food plant affects the fecundity of the adult gypsy moth, *Lymantria dispar*. Those adults arising from hosts with low nitrogen levels are less fecund than those reared on high nitrogen level hosts.[96] Hosts plants that are high in nitrogen result in more fecund individuals of the armyworm, *S. frugiperda*.[172] Spruce budworm larvae that are reared on old, nitrogen-poor foliage, develop into small adults of reduced fecundity when compared to those arising from larvae reared on young, nitrogen-rich foliage.[21]

When reared on the grass *Holcus lanatus* with artificially enhanced nitrogen levels, the leafhoppers *Dicranotropis hamata*, *Elymana sulphurella*, *Encelis incisus* and *Zyginidia saitellaris* were all more fecund as adults than those reared on plants with low levels of nitrogen.[178] The collembolan *Folsomia candida* was more fecund when fed on a host rich in nitrogen.[205] Adults of the cinnabar moth *Tyria jacobaeae* were larger and produced more eggs in areas where their host plant *Senecio jacobaea* had high nitrogen levels. In addition, larval survival was better on plants high in nitrogen than on those plants which contained low levels of foliar nitrogen.[207]

Nitrogen rich plant parts are important to the plant, and they are often protected by high levels of plant secondary compounds, e.g., tannins.[70] In some instances, plant secondary compounds are also important in determining insect fecundity. The fecundity of the scale insect, *Fiorinia externa*, is increased by terpenoid alcohols as is that of another scale insect, *Nuculaspis tsugae*, both of which feed on species of *Tsuga*.[145] However, the fecundity of *F. externa* is negatively correlated with acyclic terpenoids, whereas that of *N. tsugae* is positively correlated with the same compounds.[146]

b. Seasonal Patterns in Foliar Nitrogen Levels

Seasonal changes in plant nitrogen levels have had a marked effect on the evolution of insect reproductive strategies. This is particularly true in the case of sap-feeding insects, which feed on an extremely dilute solution and have nitrogen acquisition as one of their major objectives. With nitrogen at a premium it is in sap-feeding insects that the importance of host quality to migratory and reproductive strategies may be most easily demonstrated. As herbaceous host plants age, their quality as a food source as measured by aphid performance in terms of growth and fecundity, shows a decline. This is seen in several species; for example, the asparagus aphid *Brachycorynella asparagi* is less fecund on old leaves than it is on young ones,[227] and the flower thrips *Frankliniella occidentalis* also shows a variable fecundity depending on host growth stage.[203]

The soluble nitrogen content of plant tissues varies throughout the year.[52,53,98,220] In the temperate climates of spring, when the leaves are growing rapidly, foliar nitrogen content is high. This falls during the summer months as the leaves mature, and rises again (at least in the phloem) in the autumn, as the leaves begin to senesce and nitrogen reserves are mobilized within the plant in preparation for leaf fall.

As would be expected, this pattern of events has affected the life history strategies of a number of tree-dwelling aphids. For example, the bird cherry-oat aphid *R. padi*, a host alternating aphid, feeds and reproduces on its primary host *Prunus padus* while soluble nitrogen levels are high in the spring.[53] As soluble nitrogen levels fall, winged forms are produced in response to this deterioration in host quality and these winged forms migrate to their secondary graminaceous hosts where host quality is higher.[130] In autumn, as host quality on these grass hosts declines, they return to their primary host whose quality has once again improved.[53,124] The aphid *Uroleucon gravicorne*, although not strictly a host alternating aphid, usually moves from its overwintering host *Solidago* spp., to plants of several annual *Erigeron* species. Like *R. padi,* it responds to changes in soluble nitrogen levels of its host plant and when nitrogen levels are low, reproductive performance is also low. Soluble nitrogen content declines throughout the season on *Erigeron* spp. and growth and developmental rates, as well as the fecundity of *U. gravicorne,* decrease at the same time.[158]

The sycamore aphid *D. platanoidis* also has to face the same deterioration in host quality as that faced by the host alternating *R. padi*. However, instead of migrating to a host plant of a higher host quality status, *D. plantanoidis* reduces its metabolic rate and enters a summer aestivation until the host quality of its food plant once again becomes suitable for growth and reproduction.[50,52] Another tree-dwelling aphid, the birch aphid *Euceraphis punctipennis*, has resolved the same problem by exploiting the fact that birch trees (*Betula* spp.) have a certain number of shoot tips that produce new leaves throughout the spring and summer months. These are high in soluble nitrogen content, and the adult birch aphids track these growing points throughout the season, moving within the tree and even between trees when necessary, in order to deposit their offspring on suitable food material.[226] Aphid species on annual hosts are also greatly affected by the changes in soluble nitrogen concentration which occur as their hosts age. The cabbage aphid *B. brassicae* shows both a steady decline in growth rate as their Brussels sprout host plants age and a corresponding decline in reproductive capacity.[208,209]

Not all sap feeders show this same dependence on young flushing tissue or nitrogen-rich senescent tissue. However, in all cases a nitrogen-rich feeding site is required. Even within the aphid species already discussed, some differences in reproductive strategy are shown. The aphid *R. padi* is frequently found on older cereal plants in the summer months, but is habitually found on the basal parts of the stem where its growth and reproductive rates are highest.[128] This is also the area of the mature cereal plant where nitrogen levels are high.[94]

In the case of other sap suckers such as the delphacid *P. marginata*, which feeds on the same plant species (*Spartina* spp.) the whole year, a response is still seen to soluble nitrogen levels. The insects move to vegetation growing along the sides of streams, which is of a higher quality in the spring and early summer, and return to overwinter on the nutritionally less suitable marshland *Spartina* where greater protection is afforded.[43] Thus, like the birch aphid, these insects track their food plant resources. The psyllids, *Psylla peregrina* and *Psylla subferruginea*, also show this ability to track nitrogen-rich resources. In spring, they are closely associated with the expanding buds of their host *Crataegus monogyna*. After bud burst is completed, the only sources of high nitrogen flush are in the growing shoots and inflorescences, and it is on these points that the psyllids aggregate.[194] Insects that feed throughout the season on one portion of their host plants, such as the membracid *Publilia reticulata* on *Veronia noveboracensis*, must either increase their feeding rate to maintain the same level of activity, go into summer aestivation as in *D. platanoidis*, or restrict their feeding to that portion of the leaf that affords the best nutrition. In the case of *P. reticulata*, feeding is confined to the major cross-veins where sap flow is brisk; this enables the insect to maintain a high reproductive rate.[25] Other membracids with relatively long generation times have hatching times such that nymphal development takes place at those times of the year when soluble nitrogen levels are high. The oak tree hopper *Platycotis vitata* has two generations a year, one in spring and the other in autumn, while adult females diapause in the winter and summer depending on their generation, and the eggs are laid in spring and autumn respectively to take advantage of the flushes in soluble nitrogen in their host trees.[103]

Among chewing insects, a large number of species have adopted flush feeding strategies to achieve high growth rates which lead to increased size and greater fecundity. A problem faced by flush feeding insects is that young plant tissue is often protected by high levels of secondary metabolites, which must be detoxified by the young larvae. The winter moth *Operophtera brumata*, feeding on oak, is a bud feeder.[70] Others begin feeding on flush foliage and then move onto older, more abundant tissues once they have successfully survived the early, highly vulnerable establishment phase. For example, newly hatched larvae of the pine beauty moth *P. flammea* feed on young, newly expanding needles, but then move onto the previous year's growth to complete their development,[218] only returning to the current year's foliage in the late summer if no older needles are available.[219]

C. AGE AT MATING

The age at which an insect mates can have profound influences on a number of life history parameters of the female, for example, the life span of females of the pine beauty moth

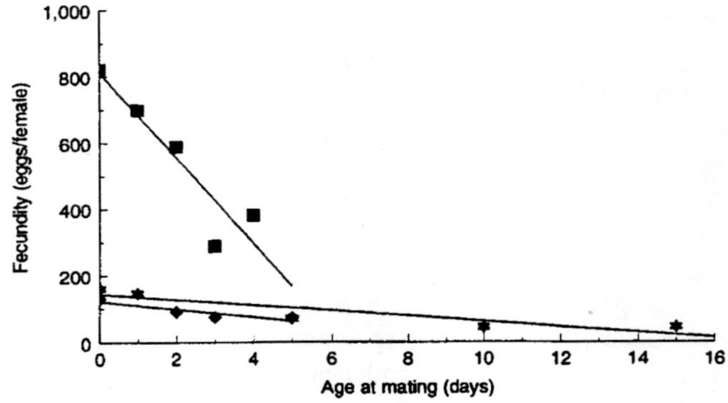

FIGURE 1. The effect of age at mating on the fecundity of the moths *Busseola fusca*, ■, *Panolis flammea*, ♦, and *Pectinophora gossypiella*, ★. (Based on References 135, 141, and 204).

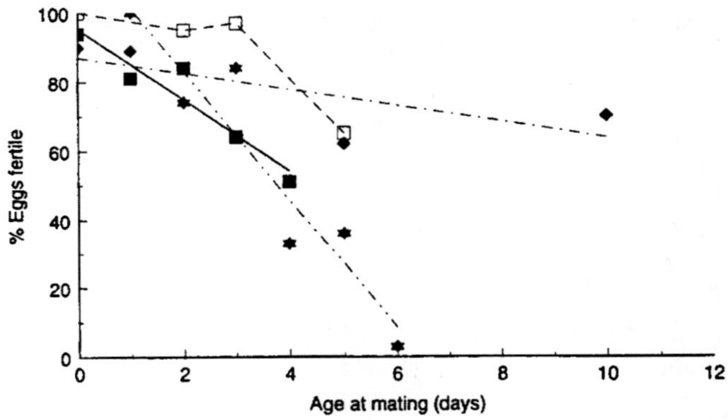

FIGURE 2. The effect of age at mating on the fertility of *Busseola fusca*, ■, *Chilo partellus*, ★, *Panolis flammea*, ♦, and *Pectinophora gossypiella*, □. (Based on References 15, 135, 141, and 204.)

P. flammea is reduced by a delay in mating.[135] Many lepidopteran species display protandry, i.e., the males emerge first. It has been proposed that this ensures that females are mated sooner rather than later and thus minimizes any detrimental effects likely to follow a delay in mating.[74,223] Certainly, in many species of Lepidoptera, a delay in mating results in both a decrease in achieved fecundity (Figure 1) and also in fertility (Figure 2). Interestingly, *P. flammea* exhibits protogyny, i.e., the females emerge before the males, but this is probably a response to the unsettled weather conditions likely to be experienced by this insect in early spring as virgin females that experience a delay in mating realize more of their potential fecundity than those females that mate and then experience a delay in oviposition due to cold weather.[135]

A delay in mating has a fairly straightforward effect on the fertility rate of Lepidoptera (Figure 2), i.e., the greater the delay, the fewer fertile eggs that are laid. However, fecundity, despite showing this same general pattern (Figure 1), can, as the period of delayed mating is extended, show a slight increase; mating imposes a cost in females in term of life span and also causes an increase in prereproductive period as the eggs are fertilized.[127] However, if the number of fertile eggs is calculated, it can be seen that the delay in mating does indeed reduce effective fecundity. It is thus in the interests of a female insect to mate as quickly as possible.

D. COPULATION FREQUENCY

Copulation frequency can have marked effects on achieved fecundity in many insects. Most work has been done on Lepidoptera and accordingly most of the examples presented are from that group. Multiple mating is common in a number of lepidopteran species and this is seen as an investment, in terms of nutrient transfer from the males to the females receiving the spermatophores[24] (Chapter 10). However, in other species there is sperm competition and although multiple mating may take place, the female oviduct may be blocked by the first comer's spermatophore.[108] In the field, as one might expect, fertility is generally high, suggesting that the strategies used by the insects are successful. For example, a 3-year study (1956–1958) of the moth *Rhyacionia buoliana,* showed that egg fertility was 99.64, 94.5 and 100%, respectively.[88]

In the black swallowtail butterfly *Papilio polyxenes asterius*, multiple matings increased both the fecundity and fertility of the females[140] and this was also the case in the moth *E. postvittana.*[41] In other Lepidoptera, multiple matings are required to keep female fecundity at a constant rate; for example, if females of *Euphydryas chalcedona* are only mated once, then egg output steadily declines over the life span of the adult.[185] In other species, the relationship is not simple; for example, in *Adoxophyes orana*, one or two extra copulations increases the fertility of the eggs laid but more than this number of copulations results in a reduction of fertility.[206] In yet other species, multiple mating has no effect on fecundity, e.g., *Danaus plexippus.*[195] In fact, in one study which involved 103 different species of butterfly from around the world, no correlation between multiple mating and high egg counts was found.[64] Multiple mating in the crickets *Gryllodes sigillatus* and *Gryllodes veletis*, offset the costs of reproduction incurred by the females, and also increased fecundity as a result of nutrient transfer.[27]

Copulation frequency on the part of the male may also affect female achieved fecundity. The more times male *H. virescens* have mated, the less sperm they transfer and the more infertile eggs the female will produce.[90] This is also seen in the codling moth *C. pomonella.* The more times a male has mated, the more infertile eggs that the female produces, but the more times a female has mated, the more eggs she will produce.[46] The females, therefore, have to take the risk that by multiple mating they increase their fecundity, but that at the same time their chance of encountering a male that has mated more than once also increases, as does their chance of laying infertile eggs. In some species, the females are able to assess the mating condition of the male and mate more often when they have copulated with males who have repeatedly mated, e.g., *H. charitonius* females are able to assess that the spermatophores are smaller than normal and therefore contain less nitrogen.[22]

1. Sex Ratio

The sex ratio perceived by an insect can have marked effects on its fecundity and fertility. The skewing of sex ratio is basically the idea behind pheromone disruption schemes as it is felt that if the proportion of males is reduced then at least the fertility, and hopefully the fecundity, of the females remaining in the population will be reduced.[11,38,107,225b]

The results are not always straightforward. For example, the pine beauty moth *P. flammea,* which occurs in a 1:1 ratio in nature, is actually most fecund when the females are outnumbered by males 3:1 (Figure 3). On the other hand, the greatest proportion of fertile eggs are produced when the sex ratios are equal (Figure 4). If the appropriate calculation is done, then it can be seen that more fertile eggs are produced overall when the females are present in the same numbers as the males. This apparently strange strategy may be an adaptation to the fact that the females of *P. flammea*, unlike many other Lepidoptera, emerge before the males.[126] In the Sphingid moth *Manduca sexta*, the situation is reversed, here females were most fecund at a sex ratio of 1:1 (Figure 3) but their fertility decreased linearly over the range. Thus, the greatest number of fertile eggs was produced at a sex ratio of five males to one female (Figure 3). In another Lepidopteran species, *Spodoptera litura,* (unfortunately not tested over sex ratios where males were in excess) fecundity declined in much the same way as the previous two species,[166] but fertility increased with increasing female bias (Figure 3). In another lepidopteran species, *Plodia interpunctella*, the number of unmated females was recorded in

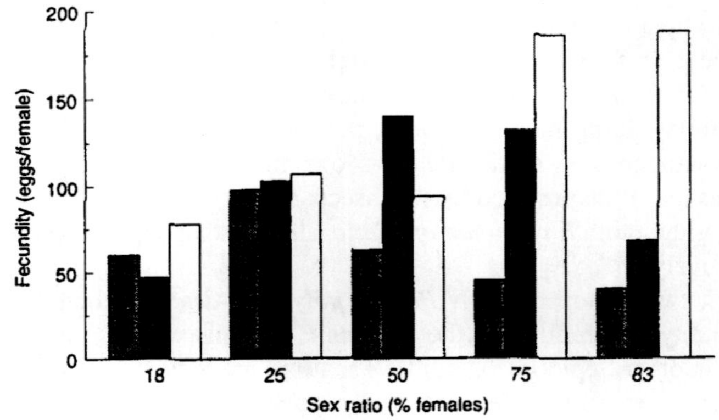

FIGURE 3. The effect of sex ratio on the fecundity of *Heliothis virescens,* □, *Manduca sexta,* ■ and *Panolis flammea,* ▩. (Based on References 84, 123, and 192.).

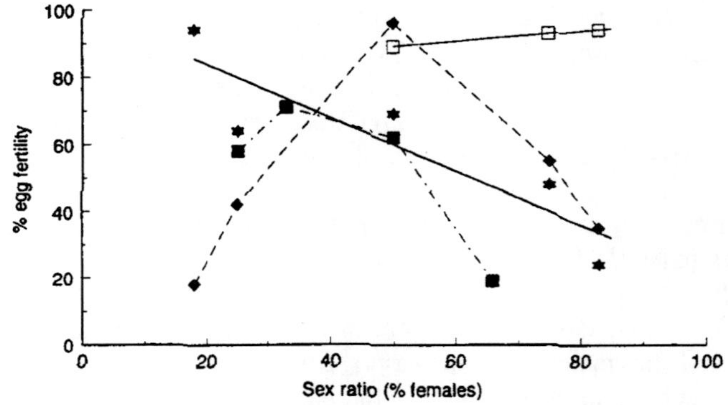

FIGURE 4. The effect of sex ratio on the fertility of *Epiphyas postvittana,* ■, *Manduca sexta,* ★, *Panolis flammea,* ◆, and *Spodoptera* litura, □. (Based on References 41, 123, 166, and 192).

respect to sex ratio. Over the range of six males to each female to an even sex ratio, all females were successfully mated. However, once the ratio was biased in favor of females, the proportion of unmated females rose to 20%. This was not a linear response.[26]

Sex ratio thus has marked effects on female fecundity and fertility of eggs deposited by adult Lepidoptera. This is likely to apply to other insect groups.

E. ABIOTIC FACTORS

1. Temperature

Temperature is perhaps the most important abiotic factor influencing living organisms. It is particularly important in affecting insect life history parameters such as developmental period, life span, pre-reproductive delay, fecundity, and fertility, as well as flight and behavior.

Basically speaking, all insects show the same response to temperature: there is a lower threshold below which reproduction does not occur and an upper threshold beyond which reproduction ceases. Fecundity increases between these two thresholds up to an optimal point, usually linearly, with the turnover point of the curve being related to the normal climatic distribution of the insect (Figure 5). For example, aphids from temperate regions have an optimal temperature for reproduction at about 20°C whereas aphids from subtropical regions have an optimal temperature for reproduction and development at 25°C. This is also seen within other groups, e.g., the moth *S. nonagriodes* has an optimum temperature for reproduction at 21°C while *P. flammea* has an optimum temperature for reproduction at 18°C (Figure 5). The response shown between the lower threshold and the optimal point can be extremely linear; for example, in *P. flammea,* the correlation between temperature and egg hatch over this range is 99.99%.[136]

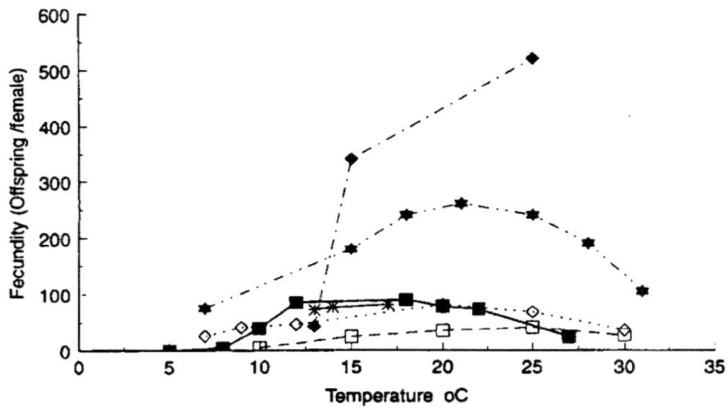

FIGURE 5. The effect of temperature on the fecundity of insects. (*Acyrthosiphon pisum*, ◇, *Diuraphis noxia*, *, *Ostrinia nubilalis*, ◆, *Panolis flammea*, ■, *Rhopalosiphum padi*, □, *Sesamia nonagriodes*, ★. (Based on References 1, 3, 80a, 104, 111, and 116.)

Fertility can also be affected by high temperatures; for example, the number of malformed and still born nymphs born to aphids increases as the upper temperature threshold is approached.[76]

2. Photoperiod

Photoperiod is one of the most reliable indicators available for insects to distinguish forthcoming changes in season. It is thus not surprising to find that photoperiod has marked effects on insects. In aphids, photoperiod affects the morph produced[59,104] so that the morphs leading on to the diapause stage are initiated; it acts as signal for diapause preparation in the moth *Cotesia rubecula*[163] and also for several butterfly and beetle species[133].

Photoperiod also has marked effects on the reproductive behavior of insects, in particular, aphids. In the case of aphids, this is more a result of the fact that as day length decreases, aphids produce sexual and sexual-producing morphs which are generally less fecund than the spring and summer forms.[55] Temperature modifies the effect of photoperiod and the response to temperature and photoperiod is not the same for all aphid species. For example, the aphid *Myzus persicae* is more fecund under long day lengths at 24°C than at 18°C. *Brachycaudus helichrysi* is more fecund under short day conditions at 18°C than at 24°C, but the situation is reversed under long day conditions.[227a] However, the reproductive activity of the aphid *Hyperomyzus lactucae* at 22°C is unaffected by changes in photoperiod.[142]

The reproductive behavior of Lepidoptera is also affected by photoperiod. For example, the moth *Plutella maculipennis* will only copulate in the dark but oviposition behavior is unaffected by light conditions.[151] However, fecundity in this species is affected by the photoperiod experienced by the adult. For example under short day conditions, i.e., 12 h light, the mean fecundity is only 37 eggs per female, whereas under long day conditions (16 h light), fecundity is twice that.[87] The photoperiod experienced by moths during their larval development can also affect their fecundity. For example both *C. pomonella* and *Grapholitha funebrana* are more fecund when exposed to long day lengths during their larval periods than if they are exposed to short day lengths.[47]

Photoperiod is thus an important factor in determining the fecundity of an insect species, but as photoperiod is seasonally and geographically determined, the effects are inextricably linked with particular stages of the life cycle that are seasonally adapted to a particular photoperiod.

3. Humidity

Humidity has been shown to have marked effects on insect fecundity and fertility, although the results are not always straightforward. For example, apterous (nonwinged) adults of the

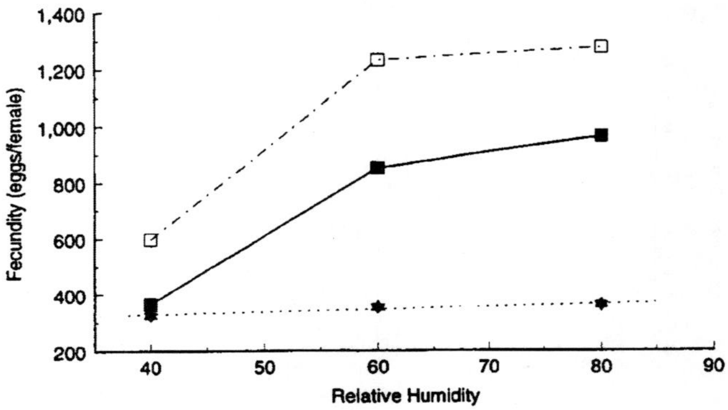

FIGURE 6. The effect of relative humidity on the fecundity of *Heliothis zea* at 21.1°C, ■, 26.7°C, □, and 32.6°C, ★. (Based on Ellington, J.J. and El-Sokkari, A., *S.W. Entomol.*, 11, 177, 1986.)

aphid *R. padi* were shown in choice chamber experiments to prefer more humid conditions (>70% RH).[111] However, in experiments in which fecundity was measured, the aphids were shown to be more fecund at 35% RH.[117a] The experimental host plants on which the aphids were reared all received adequate water, and it is possible that this somehow altered host plant quality so as to benefit the aphids feeding on plants at low relative humidities.

Adults of the moth *H. zea* are more fecund at higher relative humidities (Figure 6) than at lower relative humidities and the percentage hatch of its eggs is also greater at higher relative humidities than at lower.[65] However, this may not be a result of reduced fertility at the lower relative humidities but rather an expression of the ability of the eggs to survive desiccation.

F. FLIGHT AND DISPERSAL

As discussed earlier, migration, whether it be by active flight, ballooning or by walking must impose a cost on reproduction, if only because of the energy used for this purpose rather than for offspring production. Foraging and oviposition behavior, which are components of flight and dispersal, are also affected by the egg load of the mother. In general, females with a high egg load search more intensively, accept hosts of a lower quality, spend less time searching the host that they land on, and, in gregarious species, lay larger egg clutches.[153]

It is interesting to note that in many insects the decision to invest in a migratory form is usually caused by a deterioration in host quality. For example, the light brown apple moth *E. postvittana* produces smaller adults when the food available to the larvae is suboptimal. These smaller insects have a significantly lower wing loading than large individuals and are more adapted to dispersal and presumably the ability to locate additional suitable or better quality hosts for their offspring.[39,40] This is seen in a number of other lepidopteran species. The velvetbean caterpillar *Anticarsia gemmatalis*, if deprived of food for any period during its larval development, produces adult moths with a low wing loading and consequently greater powers of dispersal. The fall armyworm *S. frugiperda*, the corn earworm *Helicoverpa zea*, and the armyworm *Spodoptera latifascia*, also show this response to larval starvation.[5]

One of the more obvious responses to changes in host quality evinced by those insect species which exhibit alary polymorphism, is the switch from apterous nonwinged forms to winged forms. As well as being influenced by changes in photoperiod and temperature, wing formation in these insects is also influenced by changes in food quality and by rearing density. Crowding during development in nongregarious species is usually an indicator that host quality is likely to decline in the near future and the production of a winged dispersive form in these circumstances is adaptive. This change to the production of winged forms is common in many species of Homoptera. For example, many planthoppers, including *P. marginata*, produce winged forms in response to nymphal crowding, as do many species of aphids.[42,55] It would be reasonable to expect that insects capable of dispersal respond to a deterioration in host quality by producing either a specifically adapted morph, or utilizing a set of behavioral patterns to escape adverse conditions. This then would maximize reproductive potential while operating under the constraints of crowding or poor food. To illustrate this hypothesis, examples of a few, but well-documented examples from the literature will be used.

In those aphid species where the possession of wings is the norm, migratory flight is taken when host quality deteriorates locally. For example, the sycamore aphid *D. platanoidis* responds to crowded conditions during development by leaving its host plant with little or no reproduction taking place, preferring to reproduce on an unexploited resource.[51] In those aphids that show alary polymorphism, the winged morphs also colonize habitats of a higher host quality than that on which they were reared. On arrival at the new host plant they produce apterous offspring which are able to exploit the new habitat effectively.[221] These apterae are also able to detect and anticipate deterioration in their habitat through cues such as crowding and lowered foliage nitrogen levels, they, in turn produce alatae and are able to fly to other hosts.[57]

Host location is, however, a risky process, and although the quality of their present host plant may be deteriorating, it will not always become totally unsuitable for reproduction and feeding. Thus, it is not to the advantage of the species as a whole for all of the alatae to disperse from the current host plant. A flexible response should be an advantage. This is certainly the case within some aphids. Winged *R. padi* and *S. avenae* are produced with a variable number of ovarioles,[139] that is, the reproductive investment of the offspring of an apterous mother on a host plant of deteriorating quality is not a constant. The reproductive investment of aphids is linked with their migratory strategies. Alate aphids with a small number of ovarioles take off more readily, do so at a steeper angle, and delay wing muscle autolysis for longer than those aphids with a greater number of ovarioles. In addition, those aphids with a lower number of ovarioles possess more chemosensory organs and are thus better able to locate host plants. Thus, those individuals with a low reproductive investment are better suited to long distance dispersal than those with a greater reproductive investment.[210]

In addition to these differences, aphids with smaller reproductive investments are able to withstand starvation for longer periods of time than those with a large reproductive investment.[210] Moreover, aphids with a large number of ovarioles (i.e., greater reproductive investment) are more likely to die before reaching the adult stage.[213] The number of individuals within each ovariole class produced by the mother is adjusted so as to maximize the fitness of their offspring. Thus, on poor quality hosts where survival of apterous forms is likely to be low, a higher proportion of those that are born have a low number of ovarioles when compared to those born on high quality hosts.[211] Alate aphids landing on high-quality hosts produce proportionately more offspring with a large number of ovarioles than do the wingless apterae. In addition, when aphids autolyze their wing muscles, their reproductive output becomes similar to that of apterous morphs.[55]

When apterous aphids come under the influence of nutritional stress, they are less able to avoid the effects than winged forms. They are able to disperse to a certain extent by walking, but this is likely to be costly in terms of time, and the aphid is likely to undergo a period of starvation. The black bean aphid *A. fabae* can survive starvation as an adult for up to 3 days. This was shown to reduce life time fecundity and longevity. However, to compensate for this potential loss of fitness, the adult aphids increased their rate of reproduction above normal, when favourable host conditions became available.[134]

After being produced in response to host adversity, active winged insects migrate in search of new habitats. Not all the plants selected by migrants are of the same quality, and the postalighting response of a migrating insect is influenced by features of the host it has landed on. Characteristically, a complex suite of responses is found in herbivorous insects.[125] For example, it has been shown that prealighting behavior often depends on different cues than oviposition behavior. The butterflies *Eurema brigitta* and *Eurema herla*, which are monophagous on *Cassia mimosoides*, land on plants with leaves of similar shapes, size, and color. They then use chemical and textural cues to confirm their choice of host, leaving for a further period of search if the appropriate oviposition stimulants are not received.[144] Similar or more complex patterns occur in other plant-feeding insects. In aphids, plants are probed and sampled; the decision to remain is a result of the balance between food plant stimuli, resources remaining for migration, and those available for reproduction.[20] This is itself strongly influenced by the availability of wing muscles that can be autolyzed after settling, as is common in many species. In the aphid *R. padi*, alates that land on poor quality hosts are relatively less likely to autolyze their wing muscles, thus giving themselves the option to produce a few offspring and still make a further migratory flight.[138] It is noticeable that the emigrants of this species, which are produced in early summer when habitat quality is generally higher, are less variable in terms of ovariole number and do not show this adaptation.[138]

The seasonal adaptations of aphids to their expected environment are best demonstrated by looking at two studies involving the bird cherry aphid *R. padi*. The dispersal and reproductive strategies of the three winged morphs can be compared. These are (1) the emigrants (arising from the primary host and migrating to the secondary host in spring), (2) the alate exules (which migrate between secondary hosts during the summer), and (3) the gynoparae (arising from the secondary host and migrating to the primary host in autumn).[54,138] Several dispersal and reproductive characteristics have been examined. The emigrants, which migrate into a high-quality habitat that is nitrogen-rich with rapidly growing graminaceous hosts, have a high reproductive investment with most of their embryos in an advanced stage of development and wing muscles that tend to autolyse sooner, rather than later. The alate exules on the other hand, which are produced in times of adversity, i.e., crowded nymphal development or low host plant quality) and are likely to be surrounded by equally poor-quality hosts, have a low reproductive investment and wing muscles that are not programmed to autolyse quickly. They are thus able to delay wing muscle autolysis if the host on which they land is of a poorer quality than that from which they emigrated. The environment to which they migrate is unpredictable and they are adapted to this unpredictability. The gynoparae, on the other hand, which are migrating to a less common host, (of good, but temporally limited quality due to the imminence of leaf fall), have a low fecundity but produce their offspring very quickly once they have settled. Their fat reserves are high which is appropriate, as they may take some time to locate a suitable host. Each morph appears to be well adapted to fit the expected environment.

Although aphids are so versatile, they are not the only insects that can respond to host quality in such a flexible manner. The pine beauty moth *P. flammea* is also able to assess host quality by monitoring the monoterpene composition of the plant foliage,[117,120] and it will avoid poor quality hosts if possible. If the choice of host plant is restricted, then the female moths will eventually lay eggs on the low-quality hosts, but only after a prolonged prereproductive delay.[135] Among butterflies, if the reproductive cost of dispersal is low, then many species will move between host patches so as to maximize offspring fitness. This has been shown in the cabbage butterfly *Pieris rapae*, which, after initiating oviposition on a suitable host plant, will eventually leave for another host patch when its supply of chorionated eggs is exhausted.[34] Habitat quality, that is, the relative proportion of suitable host plants, also affects dispersal. The checkerspot butterfly *E. editha*, for example, will leave patches with few or none of its preferred host, *Pedicularis semibarbata*, but will remain in those patches where *P. semibarbata* is abundant.[199] Thus, herbivorous insects respond to detrimental changes in the quality of their host plants in a number of different ways, all of which are likely to maximize their fitness and that of their offspring. After dispersal has occurred, the insect is faced with another problem: the exploitation of the host to which it has migrated.

III. OVIPOSITION AND LARVIPOSITION

Upon arrival at its new host, a reproducing female insect must deposit its young, be they immobile eggs or mobile larvae. A number of decisions have to be taken at this stage and these are affected by the plant, not just in terms of its suitability as a host but also its attractiveness to the female insect. Although it is in the interest of the insect to produce as many offspring as possible in as short a time as possible, its responses are mediated by the quality of the host plant. The relationship between oviposition preference in insect herbivores and their offspring performance is recorded to range from good to poor, although most evidence points to a good correlation.[86a,200,201] The reasons for why this relationship is not as good as human observers would expect is probably linked to how preference and performance are related genetically, and has been fully reviewed elsewhere.[200] The concept of perceived host quality has been discussed in some detail,[125] but it is graphically illustrated by the oviposition behavior of the two chrysomelid beetles, *Phratora vitellinae* and *Galerucella lineola,* which both feed on *Salix* spp. *G. lineola* lays its eggs on species of *Salix* which are low in salicylates and on which the larvae are able to grow and develop better than if they were feeding on salicylate-rich plants. *P. vitellinae,* on the other hand, lays its eggs on willow species high in salicylates on which its larvae are at a disadvantage in terms of growth and development. However, the larvae of *P. vitellinae* have a better defense against predators as they are able to sequester salicylates from their host plant and produce a secretion which is high in these distasteful compounds, when attacked. Larvae of *G. lineola* and *P. vitellinae* feeding on hosts with low concentrations of salicylates are, however, virtually defenseless.[44] It is also felt that although the correlation between oviposition preference and offspring performance is not perfect, the majority of insects requiring specialist hosts can depend on plant chemistry in order to select the most suitable host for the development and survival of their offspring.[31,184]

A. DETERRENTS AND ATTRACTANTS

It is probable that one of the major ways in which an insect assesses the quality of a particular plant is by the volatile compounds released by the tissues of that plant. These compounds can act as either deterrents or attractants, or also as oviposition stimulants. It would be expected that those compounds acting as deterrents would be correlated with poor herbivore performance and that those acting as attractants would be positively correlated with herbivore performance. However, evidence is mounting to indicate that deterrents are not always correlated with negative postingestion effects as would logically be expected.[16] Instead, it has been suggested that avoidance responses to these chemicals have evolved under selection pressures that are not related to herbivore performance in respect to the deterrent.[17] Insects with a wide host range appear to be markedly affected by some of these compounds, e.g., iridoid glycosides, which often have strong deterrent effects.[182]

The situation with those compounds that act as oviposition stimulants and/or attractants is somewhat more clear-cut. It is possible that certain volatile chemicals, although not directly acting as nutrition sources, are correlated with the presence or absence of true nutrients, such as protein and amino acids. For example, the spruce budworm *C. fumiferana* is stimulated to oviposit on host plants by the presence of the monoterpenes alpha and beta pinene,[189] as is the pine beauty moth *P. flammea* which lays more eggs on pine plants with high beta to alpha pinene ratios.[117,120] The pyralid moth *Dioryctria amatella* lays more eggs on pine trees containing high levels of monoterpenes.[69] The cinnabar moth, *T. jacobaeae* on the other hand, lays its eggs on plants high in organic nitrogen and sugars, once it has orientated on a host plant using the secondary compounds as primary stimulants.[207] The butterfly *P. rapae* lays its eggs on tall plants high in nitrogen, although the primary stimulants are various glucosinolates,[144] and the monarch butterfly is primarily stimulated to oviposit by the presence of cardenolide glucosides.[167]

B. OFFSPRING FITNESS

Offspring fitness can be measured in a number of ways, but generally speaking, high larval growth rates, along with rapid larval development and high larval survival (leading to a large fecund adult) are indicative of high fitness and a good quality host plant.

1. Maternal Choice

There is much debate as to whether egg distribution on a single host plant indicates sites where offspring fitness is enhanced. A recent review suggests that ovipositing insects are relatively poor at discerning suitable host plants for their offspring,[35] and this is supported to some extent by other authors.[200] However, the real area of debate is whether plant chemistry is the major determinant of host plant selection and offspring fitness. The suggestion that deterrents may have little to do with overall offspring fitness[16] is indeed a tenable hypothesis, but chemical attractants do appear to be more positively correlated with offspring fitness.[117,118]

It has been suggested that rather than host plant chemistry (equalling quality) being the criterion for host plant selection, other abiotic factors may be the determining measure.[35] Moore, Myers, and Eng,[156] working with the western tent caterpillar *Malacosoma californicum pluviale*, suggested that microclimate was more important in determining host selection than host quality. They questioned the idea that egg distribution reflected host quality variation within the host tree, as they found that eggs were laid mostly on the sunny side of the tree. From this they suggested that it was not host chemistry that determined oviposition site but temperature. However, leaves on the sunny side of the tree had higher nitrogen levels and the male pupae produced from those leaves were heavier than those reared on leaves from the shady side of the tree. It is thus possible that the female moths were somehow responding to the nitrogen content of the leaves rather than to the microclimate.

Some insects show a relatively straightforward relation between host quality and host plant selection; for example, adults of the leaf-folding sawfly *Phyllocolpa* spp., lay their eggs on those *Salix lasiolepis* bushes on which their offspring will grow and develop fastest.[77] In other insects, the relationship between the stimuli attracting the insect to oviposit and the suitability of the host for offspring growth and survival is not so apparent. An excellent example of this type of relationship between oviposition preference and host suitability is shown by the monarch butterfly *D. plexippus*. Adult females are stimulated to oviposit by cardenolides in their milkweed host plants. However, the larvae are adversely affected by very high levels of cardenolides in individual *Asclepias* spp. Females lay their eggs on plants with intermediate cardenolide content, which is also most suitable for the survival rate of their larvae.[229] Many insects appear to be capable of making quite subtle choices of host in relation to plant chemistry. The chrysomelid beetle *Paroposis atomaria*, for example, has larvae that perform best in terms of growth and survival on *Eucalyptus* hosts that are high in nitrogen irrespective of the content of essential oils and the adult beetles choose oviposition sites accordingly.[161]

The pine beauty moth *P. flammea* is greatly affected by the quality of the pine trees it encounters. Like many species of Lepidoptera, it is able to detect differences among hosts (the cues used involve the monoterpene profiles of the plants, more eggs being laid on plants with a high beta to alpha pinene ratio than on plants with a low ratio[117,120]) and to select those individual plants on which their offspring have a greater fitness in terms of growth and fecundity when they become adults.[135] Adults of both *P. flammea* and *Crocidosema plebejana* have been shown to have marked preferences for host plants that will maximize their capacity for increase (Figure 7) and this clearly implies that females select their host plants in a way that maximizes larval growth and survival. There is no evidence available yet concerning oviposition preferences relative to phenology in the former case as the eggs are laid when the plants are still dormant. Adults of the cinnabar moth, *Tyria jacobaeae*, which oviposit on growing plants of tansy ragwort *S. jacobaea*, lay more eggs on individual plants with high nitrogen contents,[207] and the noctuid moth *Helicoverpa armigera* uses cues such as flowering status of the various host plants to determine its oviposition choices. Flower structures are more suitable for larval growth and survival than vegetative structures due to their high nitrogen content and low levels of alkaloids and other secondary compounds.[71]

However, sometimes the ovipositing insect does indeed appear to be unable to distinguish the host plant of higher quality. For example, larvae of the pine beauty moth *P. flammea* grow, develop, and have higher survival rates on pine trees that are water stressed.[217] The adult

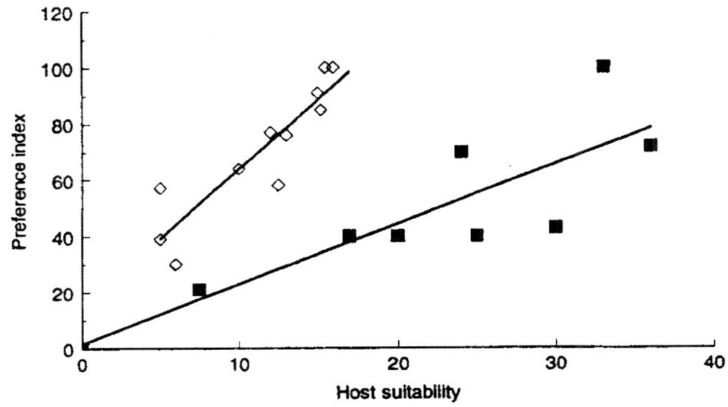

Figure 7. Relationship between preference index of adults of the moth *Panolis flammea*, ■, and *Crocidosema plebejana*, □, and suitability of the tree for growth of their larvae. (Based on References 86a and 117.)

moths, however, although able to distinguish between pine trees of differing suitability because of genetic differences or because they have previously suffered insect attack,[117,137] are unable to distinguish between plants of the same genetic material which are in different states of water stress.[217] This may, however, be because the host plant stimuli on which *P. flammea* relies upon to determine the quality of its host (monoterpenes) is unaffected by water stress. On the other hand, some insects are more adept when faced with this situation. There is, for example, a strong relationship between oviposition preference and larval performance in the shoot-galling sawfly, *Euura lasiolepis*.[37] In this case, plants that are under water stress are not suitable hosts for the larvae of *E. lasiolepis*, and the adult females are able to distinguish between stressed and unstressed plants.[179] This may be because *E. lasiolepis* lays its eggs inside the host plant and can use its ovipositor to detect stimuli relating to host plant quality, whereas *P. flammea* lays its eggs externally on the needles and has a less intimate association with the host plant.

Plants may appear suitable for offspring in one aspect, such as high nitrogen levels, but they may have secondary disadvantages. For example, host quality can be reduced by the feeding activity of other insects. The weevil *Cyrtobagous salvinae* and the moth *Samea multiplicalis*, both feed on the aquatic plant *Salvina molesta*, and both prefer to lay eggs on buds high in nitrogen. However, *C. salvinae* avoids plants that have been damaged, particularly if they have been damaged by *S. multiplicalis*. The weevils also select their host plants by assessing the number of buds available and avoiding plants with few buds so as to maximize the potential food supply of their offspring.[75,198] Other insects show similar sensitivities. The bruchid beetle *Callosobruchus maculatus*, which feeds on cowpea, lays its eggs in the field before harvest. Younger, smaller pods are preferred as oviposition sites, and these are indeed more suitable for the larvae.[149] In addition, the female is able to determine whether another beetle has already laid its eggs on a pod and to assess the number of eggs already present. The adult females avoid laying eggs on pods with a high egg load and thus ensure that larval competition is reduced. Similarly, the moth *Cactoblastis cactorum*, lays its eggs on those *Opuntia* plants that its larvae grow well on. Unlike some insects, it lays more eggs on previously attacked plants than unattacked plants. In this case, the signs of a successful previous attack may be an indicator of a high quality host plant.[161]

The perception of host plant quality can be changed by the previous experience of an insect. In a choice test, the leaf-mining fly *Liriomyza trifolii* preferred tomato plants which had a high foliar nitrogen content on which their larvae survived and developed better than foliage which was low in nitrogen. However, flies with no experience of high quality hosts (those high in nitrogen), showed no preference at first. After exposure to a high quality plant, however, a strong preference for high quality plants was evinced. It was postulated that in nature, *L. trifolii* uses plants that maintain or increase in acceptability, and disperses relatively quickly from areas containing plants of a low acceptability (and quality). As conditions worsen and host deprivation increases, then nutritionally subthreshold plants once again become acceptable as host plants.[152]

It would appear from the evidence presented above that, on the whole, mother does indeed know best and oviposition errors may be attributable to factors other than host quality — to mothers making the best of a bad job or a result of factors unappreciated by the observer. Whatever the reason or reasons, further study in this area is required.

2. Host Plant Phenology

Plant phenology also affects insect oviposition strategies. The tephritid flies *Tephritis bardanae* and *Cerajocera tussilaginis* time their oviposition so that the establishment and rapid growth of their early instar larvae coincides with the two main periods of physiological activity of their host — the development of the flower heads of the burdock *Arctium minus*. *T. bardanae* lays its eggs in the flower heads at flowering, and *C. tussilaginis* at achene maturation. At both these times, nutrient availability is at a maximum and the young larvae position themselves in those structures where the greatest nutrient flux occurs.[193] Another tephritid, *Eurosta solidaginis*, samples the ramets of its host, *Solidago altissimma,* and lays its eggs on those hosts at the phenological stage most suitable for the growth of its larvae.[4]

Adult leaf miners must be extremely sensitive to leaf quality since their offspring are confined to one leaf for the whole of their development, and a poor selection will mean very high mortality. In addition to the usual details of host selection, they may have to deal with the fact that plants often show a defensive response to attack by leaf miners by shedding leaves earlier than normal.[180,181] For these reasons, as well as nutritional ones, the age of the leaf which they lay their eggs on or in, is very important in oviposition choice. *Lithocolletis quercus*, a miner of oak leaves, selects its leaves in a way that appears to minimize the chance of early death for its larvae.[7] Other insects are constrained in their choice of host plant due to lack of suitable leaves. Those that feed on deciduous trees in the autumn are in a particularly difficult position. For example, the aphid *Periphyllus californiensis* only colonizes those maple trees whose leaves are orange-yellow. Trees with red leaves are avoided. Orange-yellow leaves remain on the tree longer, and this ensures that the aphids are able to complete their development before leaf fall.[78] On the other hand, the bird cherry aphid *R. padi*, which colonizes the bird cherry tree *P. padus* in autumn, shows no preference for different ages of leaves.[112] However, it times the production of the morphs that colonize trees in the autumn in response to the temperature it experiences in July. This cue is, in fact, an excellent predictor of leaf fall.[212] In addition, although these autumn morphs do not feed as adults,[113] they are able to distinguish the relative suitability of their tree hosts for their offspring's offspring. If the host is unsuitable they will leave, if able to do so, or if unable (due to wing muscle autolysis), they will produce offspring that lay fewer eggs on that particular tree.[118]

The tropical satyrid butterfly *Mycalesis perseus* lays its eggs on plants that are high in nitrogen, but only chooses the youngest leaves on that plant as oviposition sites. It inhabits savannah grassland, which is prone to premature drying, and this behavior ensures that eggs are laid only on those grasses that are likely to live longest. This gives its offspring the maximum possible time in which to develop.[155] A similar scenario is seen in the oviposition strategy of the butterfly *E. chalcedona*. In laboratory trials, the females prefer to lay their eggs on the shrub *Scrophularia californica*. This is nutritionally superior in terms of larval growth rates to an alternative host plant, the shrub *Diplacus auranticus*. However, *D. auranticus* is more drought-tolerant than *S. californica,* and in the wild *E. chalcedona* will lay a sizeable proportion of its eggs on the former shrub. Although *D. auranticus* is nutritionally inferior, its drought-tolerance means that it is more persistent in dry years and means that in those years larval survival will be assured. Thus host use reflects a tradeoff between nutritional quality and resource availability (persistence).[225]

Once the insect has made the decision to produce its offspring on a particular host plant, then a number of reproductive options are still open to it, and these are greatly dependent on the quality of that host plant.

3. Clutch Size

Clutch size, i.e., the number of eggs or offspring produced in a reproductive event, has attracted a great deal of attention from evolutionary ecologists. The general conclusion among vertebrate zoologists is that large clutches are characteristic of short-lived, fast developing animals exhibiting little parental care and high offspring mortality. Small clutches, on the other hand, are thought to be characteristic of long-lived, slow maturing animals with a high investment in parental care and high offspring survival.[202] Clutch size in birds, for example, is often closely related to prey/food availability; in years following good food conditions, clutch size tends to be higher.[173] If clutch size is high and the food conditions experienced by the adults are poor, then there will be an increased mortality among the hatchlings since the more eggs that are laid by the bird, the smaller each hatchling will be and its chances of surviving are decreased.[174] The evolution of insect clutch size has undoubtedly been shaped by similar factors, and this assumption is clearly apparent in the literature.

Clutch size in insects represents a trade off between offspring fitness and adult dispersal. A detailed review of the costs and benefits of varying clutch size in insects is presented by Godfray.[80] It is worth considering, however, some of the basic assumptions behind current models.

Clutch size is related to a number of factors: the frequency of oviposition or larviposition, the size of the mother, the number of eggs/offspring produced, and the quality and amount of the host. Several authors have addressed the question of clutch size using models,[147,169,170] and their conclusions are in broad agreement with those reached by the vertebrate zoologists. A major difference is that some authors believe that one of the most important factors determining clutch size in insects is not offspring success, but the need to achieve maximum egg or larval deposition.[170] This does not contradict the theory that host quality is a major factor in determining clutch size as it has been shown that host quality affects all the factors listed above.[127]

a. Adult Life Span

The rate of offspring production is not a constant over the life of the insect, and the patterns of clutch size variation are to a certain extent determined by this fact. In nature, female insects do not generally achieve their maximum possible life span,[121] and the offspring produced at the beginning of the reproductive period are therefore of greater value than those produced later, (and thus, in some species, those represented by development of eggs in the upper reproductive tract). If this is true, then most female insects would be expected to have a large initial burst of reproduction, and then to have a reproductive rate that would fall away gradually with increasing age. In fact, female insects usually show a pattern of reproduction which starts at a low level as the female first begins to reproduce, builds to a peak (the time taken depending on the probable life span of the insect), and then drops off to a low level again. Despite this pattern, however, the result is that approximately half the reproductive capacity of a female insect is achieved in the first third of its reproductive life. This pattern is seen in long-lived Lepidoptera such as *D. plexippus*,[228] in relatively short-lived Lepidoptera such as *P. flammea*,[127] and also in short-lived aphids and bugs such as *A. fabae* and *Oncopeltus fasciatus*.[59,187,197] Those insects with a high probability of early death lay most of their eggs either in one clutch or on the first night of oviposition. An example of this strategy is seen in *Chilo partellus*, which lays most of its eggs on its first night of oviposition, with clutch size becoming smaller each succeeding night.[15] Some insects with relatively immobile females are constrained to lay their eggs in one large reproductive burst. For example, the gypsy moth

L. dispar[154] and the bird cherry ermine moth *Y. evonymellus*[119] both lay their eggs clutches in one large reproductive event. Such insects adjust their reproductive effort in other ways and must also show specific adaptations to avoid predation on the large egg masses.[190] Female insects that lay their eggs in such large clusters tend to have a high realized fecundity.[33] This may compensate for the risks entailed in laying all their eggs on the same host rather than spreading the risk by moving from one host to another, and the increased probability of death for the mother before oviposition is completed.

Cluster-laying (i.e., the deposition of offspring in aggregations) in butterflies is associated with aposematic coloration of eggs and larvae. A large number of cluster-laying butterflies have been shown to be relatively unpalatable as both larvae and adults (e.g., *Pieris brassicae, Euphydryas, Battus* spp. and *Eumeas* spp.).[32] However, this is not the whole story as some aposematic butterflies lay eggs singly and this has been suggested to be a response to competition for limited larval food resources.[14]

b. Host Quality

Does host quality affect clutch size? If offspring produced earlier in the reproductive life of an insect are of greater value, then on a good quality host it would be expected that peak offspring production would be reached sooner rather than later, as it is to the advantage of the female to deposit its offspring on a good quality host before she dies. On a poor quality host, the peak should be achieved later, as the insect has to trade off the risks of waiting for an improvement in host quality vs. an investment in dispersal to seek out a better quality host. Some insects are not as flexible in their strategies as others. For example, in the flies *Dacus jarvisi* and *Dacus tryoni*, clutch size is related to food availability. *Dacus jarvisi* normally feeds on large fruit and lays large egg clutches, whereas *D. tryoni* normally feeds on small fruit and consequently lays only a few eggs on each. When presented with the fruit of the opposite size class, they are unable to change their strategies.[73] The pine beauty moth *P. flammea,* on the other hand, when laying eggs on a good host plant, has a short prereproductive delay, lays a greater proportion of its eggs in the first three days of reproduction, and lives longer than when ovipositing on a poor quality host.[135] The pipevine butterfly *Battus philenor* is also able to adjust its clutch size in response to differences in host quality, with more eggs per clutch being deposited when host quality is good.[175,176]

Thus, as predicted, herbivorous insects tend to lay large clutches early in their reproductive life when host quality is high, which will maximize both female reproductive success and offspring fitness. On poor quality hosts, clutches are small and are produced at a slower rate than on good quality hosts. In general, clutch size is adjusted so as to maximize reproductive effort. This can be done by adjusting the size of the offspring produced.[127]

4. Offspring Size

One of the ways in which an insect can manipulate its reproductive investment is by having different sizes of offspring. In some insects, for example, large females produce large eggs, e.g., *Parapediasia teterella,*[148] large females being produced as a result of good larval nutrition. Egg size variability within a species is generally a response to host or habitat quality. Theoretically, the more investment there is in reproductive effort per egg or larva (e.g., by having parental care or large offspring), the greater the likelihood of survival of those offspring.[28] Models show that the more eggs produced per clutch, the smaller those eggs will be.[6] Theory states that the larger the offspring are at birth, the greater are the advantages to be gained. Large egg size in Lepidoptera and other insects confers an advantage on the offspring, in that they are able to combat poorer conditions.[224] Thus, on a poor quality host, or in situations where the eggs are laid at a distance from the host, insects hatching from large eggs or being deposited as large live young will have greater reserves and be able to establish more successfully than smaller insects.[101] In two grasshopper species, *Chorthippus brunneus* and *Chorthippus parallelus*, insects arising from big eggs do indeed grow at a faster rate than those arising from small eggs.[186] On a good quality host, large size is not as critical as rapid growth and development is afforded by the host plant. Thus, theoretically, insects depositing offspring on a poor host should deposit fewer, but larger offspring than those depositing offspring on a good quality host. The pine beauty moth *P. flammea* lays fewer, larger eggs on Alaskan lodgepole pine (a poor host), than it does on south coastal lodgepole pine (a good host).[127] The feeding habitats of the insect can also affect the size of eggs laid. For example, in the Japanese Skipper butterflies (Lepidoptera: Hesperiidae) where some species are polyphagous and others less so, the more polyphagous species (which are likely to encounter a wider variety of hosts of differing quality), lay larger eggs and are less fecund than the species with the more restricted diets.[162] Although there are many exceptions, there is a general tendency for larger insect species to produce relatively large eggs,[45] and for generalist feeders to produce larger eggs than specialists.[215] The phenomenon of producing many offspring but of a small size on a good host is also seen within as well as between species. Aphids, in general, produce fewer but larger offspring when host quality is poor.[58]

Egg size in insects generally shows a decrease in size as the mother ages, e.g., in Lepidoptera such as *H. zea* and *Spodoptera ornithogalli*[2] and in the pentatomid bug, *Graphosoma lineatum*.[110] In the bug *Dysdercus fasciatus*, both clutch size and egg size decrease as the mother ages.[102] However, in the moth *P. teterella*, although clutch size declines over the life span of the mother, egg size remains constant.[148] In this species, fertility is inversely related to egg size in that larger eggs are less viable, however, despite this, large females are fitter than small females due to the fact that although they produce larger eggs overall, they produce more viable eggs than small females.[148]

The pine beauty moth *P. flammea* provides a very good illustration of the effects of both maternal age and host quality on egg size. The size of the eggs laid decreases as the mother ages.[127] When confined to hosts of differing quality, the female moths respond by delaying the onset of oviposition on some hosts, and when they do begin to oviposit, their eggs are larger on the poor quality hosts than on the better quality hosts. In addition, the greater the clutch size, the smaller the eggs produced.[127] This phenomenon is also found in other insect groups. Egg size of the beech leaf-mining weevil, *Rhynchaenus fagi* is smaller when adult nutrition is poor, but is much greater when food quality improves.[9] Modeling work on two grasshopper species, *Myrmeleotettix maculatus* and *C. brunneus*, also shows that egg size should decline with maternal age as reproductive investment becomes increasingly risky.[12]

Egg size is frequently affected by the size of the mother as well as by her age. For example, in the butterfly *Euploea core corrinne*, egg weight decreases with maternal age and is also correlated with maternal weight.[92] The maternal weight is in turn an indicator of host quality of the larvae.[92,93] However, another study on the same species showed no correlation between larval diet and the weight of eggs produced by the resultant females.[92] Egg weight in the butterfly *P. aegeria* is also positively correlated with female size and decreases as the mother ages.[100] Further, in the butterflies *D. plexippus*, *Lasiommata petropolitana*, *Lasiommata maera*, *Lasiommata megera*, *Lopinga achine*, and *P. rapae crucivora*, egg weights decrease as the mother ages.[100,106,195] This is seen in a large number of insects in other groups. For example, in the beetle *C. maculatus*, egg size decreases with maternal age and large eggs are more viable than small ones.[214]

If host quality is ignored by the observer, then apparently anomalous results can be found. For example, the size of eggs of the butterfly *P. rapae* were found in one study to be positively correlated with maternal age and inversely correlated with adult size, which appears to run contrary to the general rule.[99] Large individuals were more fecund but laid smaller eggs. However, the small butterflies that then arose were generally present as adults in poor quality habitats and thus produced large eggs that produced larvae able to withstand the poor conditions.[99] Another apparent contradiction is shown by *L. dispar*. Small eggs were produced in forests where there were large numbers of defoliated trees and hosts were of poor quality.[109] This was explained by the fact that small eggs gave rise to less mobile larvae which tended to remain on the hosts on which they were deposited. Since suitably foliated host trees are in short supply under these conditions, it is to the advantage of the larvae to remain where they were laid, rather than use energy dispersing to another tree which is just as likely to be of poor host quality as the one they had left. Large eggs are produced when conditions are good and hosts are in abundance, as is the case when a small insect population is in the process of building up. Under these circumstances, it may be to the advantage of the larvae to disperse to new hosts. Large larvae are more mobile and thus disperse successfully.[109]

In the Lepidoptera, the relationship between offspring size and fitness is not immediately clear. In the butterfly *P. aegeria*, no fitness function was found in relation to egg size, that is, large larvae did not survive better than small larvae. However, egg size did decrease with maternal age,[224] possibly allowing more eggs to be laid. The checkerspot butterfly *E. editha*, when fed as adults, were found in one study to keep their egg weights constant over their entire life span, in contrast to unfed females which showed the expected decline in egg weight.[157] However, in other studies Lepidoptera were also shown to have a decline in egg weight with maternal age even when fed.[127] Clearly, egg weight can alter in response to selection pressure over evolutionary time, but the variability with relation to the host plant within a species or individual is also surprisingly large and apparently adaptive. In this context, it is relevant to note that egg size in Lepidoptera may be related to the number of larval instars and the size of the adult. Dyar[63] showed that in many lepidopteran species there is an approximate doubling of volume between each instar. Thus, by having an egg double the size of a related species, the number of larval instars can be reduced by one and an adult of the same size still produced.

IV. CONCLUSIONS

Despite the large amounts of literature pertaining to the subject, it is apparent that many of the factors affecting fecundity, fertility, and offspring deposition in insect are still relatively little understood. Although the mechanics of oviposition and many of the biological constraints, e.g., sex ratio, larval nutrition, etc., affecting fecundity and fertility are well known, the effect that the various interactions between host quality and abiotic factors have on them requires further research.

ACKNOWLEDGMENT

It is a pleasure to thank Nigel Straw for his perceptive and helpful comments on earlier drafts of this chapter.

REFERENCES

1. **Aalbersberg, Y.K., Du Toit, F., van der Westinizen, M.C., and Hewitt, P.H.**, Development rate, fecundity and lifespan of apterae of the Russian wheat aphid, *Diuraphis noxia* (Mordvilko) (Hemiptera: Aphididae), under controlled conditions. *Bull. Entomol. Res.*, 77, 629, 1987.
2. **Adler, P.H., Willey, M.B., and Bowen, M.R.**, Temporal oviposition patterns of *Heliothis zea* and *Spodoptera ornithogalli*. *Entomol. Exp. Appl.*, 58, 159, 1991.
3. **Al Salti, M.N.**, Influence de la temperature pendant la vie imaginale sur les potentialites reproductrices de l'espece *Sesamia nonagriodes* Lefebvre (Lep., Noctuidae). *Acta Oecol. Oecol. Appl.*, 5, 103, 1984.
4. **Anderson, S.S., McCrea, K.D., Abrahamson, W.G., and Hartzel, L.M.**, Host genotype choice by the ball gallmaker *Eurosta solidagensis* (Diptera: Tephritidae). *Ecology*, 70, 1048, 1989.
5. **Angelo, M.J. and Slansky, F.**, Body building insects: trade-offs in resource allocation with particular reference to migration. *Fla. Entomol.*, 67, 22, 1984.
6. **Atkinson, D.A. and Begon, M.**, Reproductive variation and adult size in two co-occurring grasshopper species. *Ecol. Entomol.*, 12, 119, 1987.

7. **Auerbach, M. and Simberloff, D.**, Oviposition site preference and larval mortality in a leaf-mining moth. *Ecol. Entomol.*, 14, 131, 1989.
8. **Auger, M.A., Geri, C., Jay-Allemand, C., and Bastien, C.**, Comestibilité de differénts clones de pin sylvestre pour *Diprion pini* L. (Hym., Diprionidae). 1. Incidence de la consommation des aiguilles de differénts clones de pin sylvestre sur le developpement de *Diprion pini* L. *J. Appl. Entomol.*, 110, 489, 1990.
9. **Bale, J.S.**, Bud burst and success of the beech weevil, *Rhynchaenus fagi*: feeding and oviposition. *Ecol. Entomol.*, 9, 139, 1984.
10. **Battisti, A.**, Host-plant relationships and population dynamics of the pine processionary caterpillar *Thaumetopoea pityocampa* (Dennis and Schiffermuller). *J. Appl. Entomol.*, 105, 393, 1988.
11. **Beevor, P.S., Mumford, J.D., Shah, S., Day, R.K., and Hall, D.R.**, Observations on pheromone baited mass trapping for control of cocoa pod borer, *Conopomorpha cramerella*, in Sabah, East Malaysia. *Crop Prot.*, 12, 134, 1993.
12. **Begon, M. and Parker, G.A.**, Should egg size and clutch size decrease with age? *Oikos*, 47, 293, 1988.
12a. **Bellinger, R.G. and Pienkowski, R.L.**, Interspecific variation in ovariole number in Melanopline grasshoppers (Orthoptera: Acrididae). *Ann. Entomol. Soc. Am.*, 78, 127, 1985.
13. **Bennetova, B. and Fraenkel, G.**, What determines the number of ovarioles in a fly ovary? *J. Insect. Physiol.*, 27, 403, 1981.
14. **Benson, W.W.**, Resource partitioning in passion vine butterflies. *Evolution*, 32, 493, 1978.
15. **Berger, A.**, Egg weight, batch size and fecundity of the spotted stalkborer, *Chilo partellus* in relation to weight of females and time of oviposition. *Entomol. Exp. Appl.*, 50, 199, 1989.
16. **Bernays, E.A. and Chapman, R.F.**, Evolution of plant deterrence to insects. In *Perspectives in Chemoreception and Behavior*, (Eds., R.F. Chapman, E.A. Bernays, and J.G. Stoffolano), Springer-Verlag, New York, 1987, 159.
17. **Bernays, E.A. and Cornelius, M.**, Relationship between deterrence and toxicity of plant secondary compounds for the alfalfa weevil *Hypera brunneipennis*. *Entomol. Exp. Appl.*, 64, 289, 1993.
18. **Bevan, D. and Brown, R.M.**, *Pine Looper Moth*, Forestry Commission Forest Record 119, HMSO, London, 1978.
19. **Bintcliffe, E.J.B. and Wratten, S.D.**, Antibiotic resistance in potato cultivars to the aphid *Myzus persicae*. *Ann. Appl. Biol.*, 100, 382, 1982.
20. **Blackman, R.L.**, Specificity in aphid/plant genetic interactions with particular reference to the role of the alate colonizer. In *Aphid-Plant Genotype Interactions*, (Eds., R.K. Campbell and R.D. Eikenbary), Elsevier, Amsterdam, 1990, 251.
21. **Blais, J.R.**, Effects of the destruction of the current year's foliage of Balsam fir on the fecundity and habits of flight of the spruce budworm. *Can. Entomol.*, 85, 446, 1953.
22. **Boggs, C.L.**, Selection pressures affecting male nutrient investment at mating in heliconiine butterflies. *Evolution*, 35, 931, 1981.
23. **Boggs, C.L.**, Ecology of nectar and pollen feeding in Lepidoptera. In *Nutritional Ecology of Insects, Mites, Spiders, and Related Invertebrates* (Eds., F. Slansky and J.G. Rodriguez), John Wiley & Sons, New York, 1987, 369.
24. **Boggs, C.L. and Watt, W.B.**, Population structure of Pierid butterflies. IV. Genetic and physiological investment in offspring by male *Colias*. *Oecologia*, 50, 320, 1981.
25. **Bristow C. M.**, Differential benefits from ant attendance to two species of Homoptera on New York iron weed. *J. Anim. Ecol.*, 53, 715, 1984.
26. **Brower, J.H.**, *Plodia interpunctella*: effect of sex ratio on reproductivity. *Ann. Entomol. Soc. Am.*, 68, 847, 1975.
27. **Burpee, D.M. and Sakaluk, S.K.**, Repeated matings offset costs of reproduction in female crickets. *Evol. Ecol.*, 7, 240, 1993.
28. **Calow, P.**, Economics of ontogeny — adaptational aspects. In *Evolutionary Ecology*, (ed. B. Shorrocks), Blackwell Scientific Publications, Oxford, 1984, 81.
29. **Carroll, A.L. and Quiring, D.T.**, Sucrose ingestion by *Zeiraphera canadensis* Mut. and Free. (Lepidoptera, Tortricidae) increases fecundity and lifetime fecundity, but not oviposition rate. *Can. Entomol.*, 124, 335, 1992.
30. **Cheng, H.H.**, Oviposition and longevity of the dark-sided cutworm, *Euxoa messoria* (Lepidoptera: Noctuidae), in the laboratory. *Can. Entomol.*, 104, 919, 1972.
31. **Chew, F.S.**, Coevolution of pierid butterflies and their cruciferous food plants. II. The distribution of eggs on potential foodplants. *Evolution*, 31, 568, 1977.
32. **Chew, F.S and Robbins, R.K.**, Egg laying in butterflies. In *The Biology of Butterflies*, (Eds., R.I. Vane-Wright and P.R. Ackery), Symposium of the Royal Entomological Society of London, Number 11, Academic Press, London, 1984, 65.
33. **Courtney, S.P.**, The evolution of egg clustering by butterflies and other insects. *Am. Nat.*, 123, 276, 1984.
34. **Courtney, S.P.**, Why insects move between host patches — some comments on risk-spreading. *Oikos*, 47, 112, 1986.

35. **Courtney, S.P. and Kibota, T.K.**, Mother doesn't know best: selection of hosts by ovipositing insects. In *Insect-Plant Interactions, Volume II* (Ed., E.A. Bernays), CRC Press, Boca Raton, 1990, 161.
36. **Craig, T.P., Itami, J.K., and Price, P.W.**, A strong relationship between oviposition preference and larval performance in a shoot-galling sawfly. *Ecology*, 70, 1691, 1989.
37. **Craig, T.P., Price, P.W., and Itami, J.K.**, Resource regulation by a stem-galling sawfly on the arroyo willow. *Ecology*, 67, 419, 1986.
38. **Critchley, B.R., Chamberlain, D.J., Campion, D.G., Attique, M.R., Ali, M., and Ghaffar, A.**, Integrated use of pink bollworm pheromone formulations and selected conventional insecticides for the control of the cotton pest complex in Pakistan. *Bull. Entomol. Res.*, 81, 371, 1991.
39. **Danthanarayana, W.**, Factors determining variation in fecundity of the light brown apple moth, *Epiphyas postvittana* (Walker) (Tortricidae). *Aust. J. Zool.*, 23, 439, 1975.
40. **Danthanarayana, W.**, Environmentally cued size variation in light-brown apple moth, *Epiphyas postvittana* (Walk.) (Tortricidae), and its adaptive value in dispersal. *Oecologia*, 26, 121, 1976.
41. **Danthanarayana, W. and Gu, H.**, Multiple mating and its effect on the reproductive success of female *Epiphyas postvittana* (Lepidoptera: Tortricidae). *Ecol. Entomol.*, 16, 169, 1991.
42. **Denno, R.F. and Grissell, E.E.**, The adaptiveness of wing-dimorphism in the salt marsh inhabiting planthopper *Prokelisia marginata* (Homoptera: Delphacidae). *Ecology*, 60, 221, 1979.
43. **Denno, R.F. and McCloud, E.S.**, Predicting fecundity from body size in the planthopper, *Prokelisia marginata* (Homoptera: Delphacidae). *Environ. Entomol.*, 14, 846, 1985.
44. **Denno, R.F., Larsson, S., and Olmstead, K.L.**, Role of enemy-free space and plant quality in host-plant selection by willow beetles. *Ecology*, 7, 124, 1990.
45. **Derr, J.A., Alden, B., and Dingle, H.**, Insect life histories in relation to migration, body size and host plant assay: a comparative study of *Dysdercus*. *J. Anim. Ecol.*, 50, 181 1981.
46. **Deseo, K.V.**, Study of factors influencing the fecundity and fertility of codling moth, (*Laspeyresia pomonella* L. Lepid.; Tortr.). *Acta Phytopathol. Acad. Sci. Hung.*, 6, 247, 1971.
47. **Deseo, K.V. and Saringer, G.**, Photoperiodic effect on fecundity of *Laspeyresia pomonella*, *Grapholitha funebrana* and *G. molesta*: the sensitivity period. *Entomol. Exp. Appl.*, *18*, 187, 1975.
48. **Dewar, A.M.**, Assessment of methods for testing varietal resistance to aphids in cereals. *Ann. Appl. Biol.*, 87, 183, 1977.
49. **Dixon, A.F.G.**, Reproductive activity of the sycamore aphid, *Drepanosiphum platanoides* (Schr.) (Hemiptera, Aphididae). *J. Anim. Ecol.*, 32, 33, 1963.
50. **Dixon, A.F.G.**, The effect of population density and nutritive status of the host on the summer reproductive activity of the sycamore aphid, *Drepanosiphum platanoides* (Schr.). *J. Anim. Ecol.*, 35, 105, 1966.
51. **Dixon, A.F.G.**, Population dynamics of the sycamore aphid *Drepanosiphum platanoides* (Schr.) (Hemiptera: Aphididae): migratory and trivial flight activity. *J. Anim. Ecol.*, 38, 585, 1969.
52. **Dixon, A.F.G.**, Quality and availability of food for a sycamore aphid population. In *Animal Populations in Relation to their Food Resources*, (Ed., A. Watson), Blackwell Scientific Publications, Oxford, 1970, 277.
53. **Dixon, A.F.G.**, The life cycle and host preferences of the bird cherry-oat aphid, *Rhopalosiphum padi* (L.) and their bearing on the theories of host alternation in aphids. *Ann. Appl. Biol.*, 68, 135, 1971.
54. **Dixon, A.F.G.**, Reproductive strategies of the alate morphs of the bird cherry-oat aphid *Rhopalosiphum padi* (L.). *J. Anim. Ecol.*, 45, 1976.
55. **Dixon, A.F.G.**, *Aphid Ecology*, Blackie, London, 1985, Chapter 6.
56. **Dixon, A.F.G. and Dharma, T.R.**, Number of ovarioles and fecundity in the black bean aphid, *Aphis fabae*. *Entomol. Exp. Appl.*, *28*, 1, 1980.
57. **Dixon, A.F.G. and Glen, D.M.**, Morph determination in the bird cherry-oat aphid, *Rhopalosiphum padi* (L.). *Ann. Appl. Biol.*, 68, 11, 1971.
58. **Dixon, A.F.G. and Wellings, P.W.**, Seasonality and reproduction in aphids. *Int. J. Inv. Rep.*, 5, 83, 1982.
59. **Dixon, A.F.G. and Wratten, S.D.**, Laboratory studies on aggregation size and fecundity in the black bean aphid, *Aphis fabae* (Scop.). *Bull. Entomol. Res.*, 61, 97, 1971.
60. **Dunlap-Pianka, H.L.**, Ovarian dynamics in Heliconius butterflies: correlations among daily oviposition rates, egg weights, and qualitative aspects of oogenesis. *J. Insect Physiol.*, 25, 741, 1979.
61. **Dunlap-Pianka, H.L., Boggs, C.L., and Gilbert, L.E.**, Ovarian dynamics in Heliconiine butterflies: programmed senescence versus eternal youth. *Science*, 197, 487, 1977.
62. **Du Merle, P. and Brunet, S.**, From green to yellow or yellowish white: egg-colour changes in relation to oviposition rank in the fir budworm *Choristoneura murinana* (Hb.) (Lep., Tortricidae). *J. Appl. Entomol.*, 111, 342, 1991.
63. **Dyar, H.G.**, The number of moults of lepidopterous larvae. *Psyche*, 5, 420, 1890.
64. **Ehrlich, A.H. and Ehrlich, P.R.**, Reproductive strategies in the butterflies: mating frequency, plugging and egg number. *J. Kans. Entomol. Soc.*, 51, 666, 1978.
65. **Ellington, J.J. and El-Sokkari, A.**, A measure of the fecundity, ovipositional behavior and mortality of the bollworm, *Heliothis zea* (Boddie) in the laboratory (Lepidoptera, Noctuidae). *S.W. Entomol.*, 11, 177, 1986.

66. **Elliott, W.M.**, A method of predicting short term population trends of the green peach aphid, *Myzus persicae* (Homoptera: Aphididae), on potatoes. *Can. Entomol.*, 105, 11, 1973.
67. **El-Sherif, S., Gomaa, A.A., and Hemeida, I.A.**, Effect of adult diet and mating on egg laying capacity and longevity of potato tuber moth, *Phthorimaea operculella* Zeller. *Z. Angew. Entomol.*, 87, 170, 1979.
68. **Evans, H.F.**, The population dynamics of *Anthocoris confusus* in a laboratory cage system. *J. Anim. Ecol.*, 45, 773, 1976.
69. **Fatzinger, C.W. and Merkel, E.P.**, Oviposition and feeding preferences of the southern pine coneworm (Lepidoptera, Pyralidae) for different host-plant materials and observations on monoterpenes as an oviposition stimulant. *J. Chem. Ecol.*, 11, 689, 1985.
70. **Feeny, P.P.**, Seasonal changes in oak leaf tannins and nutrient as a cause of spring feeding by winter moth caterpillars. *Ecology*, 51, 565, 1970.
71. **Firempong, S. and Zalucki, M.P.**, Host plant selection by *Helicoverpa armigera* (Hubner) Lepidoptera: Noctuidae); the role of certain plant attributes. *Aust. J. Zool.*, 37, 675, 1990.
72. **Fitt, G.P.**, Variation in ovariole number and egg size of species of *Dacus* (Diptera: Tephritidae and their relation to host specialization. *Ecol. Entomol.*, 15, 255, 1990.
73. **Fitt, G.P.**, Comparative fecundity, clutch size, ovariole number and egg size of *Dacus tryoni* and *D. jarvisi*, and their relationship to body size. *Entomol. Exp. Appl.*, 55, 11, 1990.
74. **Ford, E.B.**, *Butterflies*. Collins, London, 1945.
75. **Forno, I.W. and Bourne, A.S.**, Oviposition by the weevil *Cyrtobagous salviniae* Calder and Sands when its host plant, *Salvinia molesta* is damaged. *J. Appl. Entomol.*, 106, 85, 1988.
76. **Frazer, B.D.**, Life tables and intrinsic rates of increase of apterous black bean aphids and pea aphids, on broad bean (Homoptera: Aphididae). *Can. Entomol.*, 104, 1717, 1972.
77. **Fritz, R.S. and Nobel, J.**, Host plant variation in mortality of the leaf-folding sawfly on the arroyo willow. *Ecol. Entomol.*, 15, 25, 1990.
78. **Furuta, K.**, Host preferences and population dynamics in an autumnal population of the maple aphid, *Periphyllus californiensis* Shinji (Homoptera: Aphididae). *Z. Angew. Entomol.*, 102, 93, 1986.
79. **Gilbert, L.E.**, The biology of butterfly communities. In *The Biology of Butterflies*, (Eds. R.I. Vane-Wright and P.R. Ackery), Symposium of the Royal Entomological Society Number 11, Academic Press, 1984, 41.
80. **Godfray, H.C.J.**, The evolution of clutch size in invertebrates. *Oxford Surveys in Evolutionary Biology*, 4, 117, 1987.
80a. **Gohari, H. and Hawlitzky, N.**, Acivité reproductrice de lay pyrale du mais, *Ostrinia nubilalis*, Hbn. (Lep., Pyralidae) à basses températures constantes. *Agronomie*, 6, 911, 1986.
81. **Gordon, D.M. and Stewart, R.K.**, Demographic characteristics of the stored products moth *Cadra cautella*. *J. Anim. Ecol.*, 57, 627, 1988.
82. **Gruber, K. and Dixon, A.F.G.**, The effect of nutrient stress on development and reproduction in an aphid. *Entomol. Exp. Appl.*, 47, 23, 1988.
83. **Gu, H. and Danthanarayana, W.**, The role of availability of food and water to the adult *Epiphyas postvittana*, the light brown apple moth, in its reproductive performance. *Entomol. Exp. Appl.*, 54, 101, 1990.
84. **Guerra, A.A., Wolfenbarger, A., and Garcia, R.D.**, Factors affecting reproduction of the tobacco budworm in the laboratory. *J. Econ. Entomol.*, 65, 1341, 1974.
85. **Gunn, A. and Gatehouse, A.G.**, Effects of the availability of food and water on reproduction in the African armyworm, *Spodoptera exempta*. *Phys. Entomol.*, 10, 53, 1985.
86. **Gunn, A. and Gatehouse, A.G.**, The effect of adult feeding on lipid and protein reserves in African armyworm, *Spodoptera exempta*, moths before and during reproduction. *Phys. Entomol.*, 11, 423, 1986.
86a. **Hamilton, J.G. and Zalucki, M.P.**, Interactions between a specialist herbivore, *Crocidosema plebejana*, and its host plants *Malva parviflora* and cotton, *Gossypium hirsutum* — oviposition preference. *Entomol. Exp. Appl.*, 66, 207, 1993.
87. **Harcourt, D.G. and Cass, L.M.**, Photoperiodism and fecundity in *Plutella maculipennis* (Curt.). *Nature*, 210, 217, 1966.
88. **Harris, P.**, Natural mortality of the pine shoot moth *Rhyacionia buoliana* (Schiff.) in England. *Can. J. Zool.*, 38, 755, 1960.
89. **Heliovaara, K. and Vaisanen, R.**, Changes in population dynamics of pine insects induced by air pollution. In *Population Dynamics of Forest Insects*, (Eds., A.D. Watt, S.R. Leather, M.D. Hunter, and N.A.C. Kidd), Intercept Press, Andover, 1990, 209.
90. **Henneberry, T.J. and Clayton, T.E.**, Tobacco budworm moths (Lepidoptera: Noctuidae): effect of time of emergence, male age, and frequency of mating on sperm transfer and egg viability. *J. Econ. Entomol.*, 78, 379, 1985.
91. **Heron, R.J.**, The reproductive capacity of the larch sawfly and some factors of concern in its measurement. *Can. Entomol.*, 98, 561, 1966.
92. **Hill, C.J.**, The effect of adult diet on the biology of butterflies. 2. The common crow butterfly, *Euploea core corinna*. *Oecologia*, 81, 258, 1989.

93. **Hill, C.J. and Pierce, N.E.**, The effect of adult diet on the biology of butterflies. 1. The common imperial blue, *Jalmenus evagoras*. *Oecologia*, 81, 249, 1989.
94. **Honek, A.**, Host plant energy allocation to and within ears, and abundance of cereal aphids. *J. Appl. Entomol.*, 110, 68, 1990.
95. **Horton, D.R.**, Performance of a willow-feeding beetle, *Chrysomela knabi* Brown, as affected by host species and dietary moisture. *Can. Entomol.*, 121, 777, 1989.
96. **Hough, J.A. and Pimentel, D.**, Influence of host foliage on development, survival and fecundity of the gypsy moth. *Environ. Entomol.*, 7, 97, 1978.
97. **Imms, A.D.**, *A General Textbook of Entomology*, Ninth ed., Chapman & Hall, London, 1957.
98. **James, W.O.**, *An Introduction to Plant Physiology*, Oxford University Press, Oxford, 1973, Chapter 7.
99. **Jones, R.E., Hart, J.R., and Bull, G.D.**, Temperature, size and egg production in the cabbage butterfly, *Pieris rapae* L. *Aust. J. Zool.*, 30, 223, 1982.
100. **Karlsson, B.**, Variation in egg weight, oviposition rate and reproductive reserves with female age in a natural population of the speckled wood butterfly, *Pararge aegeria*. *Ecol. Entomol.*, 12, 473, 1987.
101. **Karlsson, B. and Wiklund, C.**, Egg weight variation in relation to egg mortality and starvation endurance of newly hatched larvae in some satyrid butterflies. *Ecol. Entomol.*, 10, 205, 1985.
102. **Kasule, F.K.**, Egg size increases with maternal age in the cotton stainer bugs *Dysdercus fasciatus* and *D. cardinalis* (Hemiptera: Pyrrhocoridae). *Ecol. Entomol.*, 16, 345, 1991.
103. **Keese, M.C. and Wood, T.K.**, Host-plant mediated geographic variation in the life history of *Platycotis vittata* (Homoptera: Membracidae). *Ecol. Entomol.*, 16, 63, 1991.
104. **Kenten, J.**, The effect of photoperiod and temperature on reproduction in *Acyrthosiphon pisum* (Harris) and on the forms produced. *Bull. Entomol. Res.*, 46, 599, 1955.
105. **Kidd, N.A.C. and Tozer, D.J.**, On the significance of post-reproductive life in aphids. *Ecol. Entomol.*, 10, 357, 1985.
106. **Kimura, K. and Tsubaki, Y.**, Egg weight variation associated with female age in *Pieris rapae crucivora* Boisduval (Lepidoptera; Pieridae). *Appl. Entomol. Zool.*, 20, 500, 1985.
107. **Koyama, J., Teruya, T., and Tanaka, K.**, Eradication of the Oriental fruit fly (Diptera: Tephritidae) from the Okinawa Islands by male annihilation technique. *J. Econ. Entomol.*, 77, 468, 1984.
108. **Labine, P.A.**, The population biology of the butterfly, *Euphydryas editha*. IV. Sperm precedence — a preliminary report. *Evolution*, 20, 580, 1966.
109. **Lance, D.R.**, Host-seeking behavior of the gypsy moth: the influence of polyphagy and highly apparent host plants In *Herbivorous Insects — Host Seeking Behaviour and Mechanisms* (Ed., S. Ahmad), Academic Press, New York, 1983, 201.
110. **Larsson, F.K.**, Female longevity and body size as predictors of fecundity and egg length in *Graphosoma lineatum* L. *Dtsch. Entomol. Z.*, *36*, 329, 1989.
111. **Leather, S.R.**, *Aspects of the Ecology of the Bird Cherry-Oat Aphid, Rhopalosiphum padi (L.)*. Unpublished Ph.D. Thesis, University of East Anglia, Norwich, U.K., 1980.
112. **Leather, S.R.**, Reproduction and survival: a field study of gynoparae of the bird cherry-oat aphid, *Rhopalosiphum padi* (Homoptera; Aphididae) on its primary host *Prunus padus*. *Ann. Entomol. Fenn.*, 47, 131, 1981.
113. **Leather, S.R.**, Do gynoparae and males need to feed? An attempt to allocate resources in the bird cherry oat aphid, *Rhopalosiphum padi*. *Entomol. Exp. Appl.*, *31*, 386, 1982.
114. **Leather, S.R.**, Evidence of ovulation after adult moult in the bird cherry-oat aphid, *Rhopalosiphum padi*. *Entomol. Exp. Appl.*, *33*, 348, 1983.
115. **Leather, S.R.**, The effect of adult feeding on the fecundity, weight loss and survival of the pine beauty moth, *Panolis flammea* (D & S). *Oecologia*, 65, 70, 1984.
116. **Leather, S.R.**, Factors affecting pupal survival and eclosion in the pine beauty moth, *Panolis flammea* (D & S). *Oecologia*, 63, 75, 1984.
117. **Leather, S.R.**, Oviposition preferences in relation to larval growth rates and survival in the pine beauty moth, *Panolis flammea*. *Ecol. Entomol.*, 10, 213, 1985.
117a. **Leather, S.R.**, Atmospheric humidity and aphid reproduction. *Z. Angew. Entomol.*, 100, 510, 1985.
118. **Leather, S.R.**, Host monitoring by aphid migrants: do gynoparae maximise offspring fitness? *Oecologia*, 68, 367, 1986.
119. **Leather, S.R.**, Insects on bird cherry. I. The bird cherry ermine moth, *Yponomeuta evonymellus* (L.) (Lepidoptera: Yponomeutidae). *Entomol. Gaz.*, 37, 209, 1986.
120. **Leather, S.R.**, Pine monoterpenes stimulate oviposition in the pine beauty moth, *Panolis flammea*. *Entomol. Exp. Appl.*, *43*, 295, 1987.
121. **Leather, S.R.**, Size, reproductive potential and fecundity in insects. Things aren't as simple as they seem. *Oikos*, 51, 386, 1988.
122. **Leather, S.R.**, Do alate aphids produce fitter offspring? The influence of maternal rearing history and morph on life-history parameters of *Rhopalosiphum padi* (L.). *Funct. Ecol.*, 3, 237, 1989.
123. **Leather, S.R.**, Sex-ratio and reproductive success in the pine beauty moth, *Panolis flammea* (Den. and Schiff.) (Lep., Noctuidae). *J. Appl. Entomol.*, 109, 200, 1990.

124. **Leather, S.R.**, The role of host quality, natural enemies, competition and weather in the regulation of autumn and winter populations of the bird cherry aphid. In *Population Dynamics of Forest Insects*, (Eds., A.D.Watt, S.R. Leather, M.D. Hunter, and N.A.C. Kidd), Intercept Press, Andover, 1990, 35.
125. **Leather, S.R.**, Life history traits of insect herbivores in relation to host quality. In *Insect-Plant Interactions, Volume V*, (Ed., E.A. Bernays), CRC Press, Boca Raton, 1994, 176.
126. **Leather, S.R. and Barbour, D.A.**, The effect of temperature on the emergence of pine beauty moth, *Panolis flammea* Schiff. (Lep., Noctuidae). *Z. Angew. Entomol.*, 96, 445–448, 1983.
127. **Leather, S.R. and Burnand, A.C.**, Factors affecting life-history parameters of the pine beauty moth, *Panolis flammea* (D & S): the hidden costs of reproduction. *Funct. Ecol.*, 1, 331, 1987.
128. **Leather, S.R. and Dixon, A.F.G.**, The effect of cereal growth stage and feeding site on the reproductive activity of the bird-cherry aphid, *Rhopalosiphum padi*. *Ann. Appl. Biol.*, 97, 135, 1981.
129. **Leather, S.R. and Dixon, A.F.G.**, Growth, survival and reproduction of the bird-cherry aphid, *Rhopalosiphum padi*, on its primary host. *Ann. Appl. Biol.*, 99, 115, 1981.
130. **Leather, S.R. and Dixon, A.F.G.**, Secondary host preferences and reproductive activity of the bird cherry-oat aphid, *Rhopalosiphum padi*. *Ann. Appl. Biol.*, 101, 219, 1982.
131. **Leather, S.R. and Mackenzie, G.A.**, Factors affecting the population development of the bird cherry ermine moth, *Yponomeuta evonymella* (L.). *The Entomologist*, 113, 81, 1994.
132. **Leather, S.R. and Wellings, P.W.**, Ovariole number and fecundity in aphids. *Entomol. Exp. Appl.*, 30, 128, 1981.
133. **Leather, S.R., Walters, K.F.A., and Bale, J.S.**, *The Ecology of Insect Overwintering*, Cambridge University Press, Cambridge, U.K., 1993.
134. **Leather, S.R., Ward, S.A., and Dixon, A.F.G.**, The effect of nutrient stress on life history parameters of the black bean aphid, *Aphis fabae* Scop. *Oecologia*, 57, 156, 1983.
135. **Leather, S.R., Watt, A.D., and Barbour, D.A.**, The effect of host plant and delayed mating on the fecundity and lifespan of the pine beauty moth, *Panolis flammea* (Dennis and Schiffermuller) (Lepidoptera: Noctuidae): their influence on population dynamics and relevance to pest management. *Bull. Entomol. Res.*, 75, 641, 1985.
136. **Leather, S.R., Watt, A.D., and Entwistle, P.F.**, The effect of temperature on oviposition and egg hatch in the pine beauty moth, *Panolis flammea* (D & S): simulation and prediction of field and laboratory populations. *For. Ecol. Manage.*, in press.
137. **Leather, S.R., Watt, A.D., and Forrest, G.I.**, Insect-induced changes in young lodgepole pine (*Pinus contorta*): the effect of previous defoliation on oviposition, growth and survival of the pine beauty moth, *Panolis flammea*. *Ecol. Entomol.*, 12, 275, 1987.
138. **Leather, S.R., Wellings, P.W., and Dixon, A.F.G.**, Habitat quality and the reproductive strategies of the migratory morphs of the bird cherry-oat aphid, *Rhopalosiphum padi* (L.), colonizing secondary host plants. *Oecologia*, 59, 302, 1983.
139. **Leather, S.R., Wellings, P.W., and Walters, K.F.A.**, Variation in ovariole number within the Aphidoidea. *J. Nat. Hist.*, 22, 381, 1988.
140. **Lederhouse, R.C.**, The effect of female mating frequency on egg fertility in the black swallowtail, *Papilio polyxenes asterius* (Papilionidae). *J. Lepid. Soc.*, 35, 266, 1981.
141. **Lingren, P.D., Warner, W.B., and Henneberry, T.J.**, Influence of delayed mating on egg production, egg viability, mating and longevity of female pink bollworm, (Lepidoptera: Gelechiidae). *Environ. Entomol.*, 17, 86, 1988.
142. **Liu, S.S. and Hughes, R.D.**, The influence of temperature and photoperiod on the development, survival and reproduction of the sowthistle aphid, *Hyperomyzus lactucae*. *Entomol. Exp. Appl.*, 43, 31, 1987.
143. **Lukefahr, M.J. and Martin, D.F.**, The effect of various larval and adult diets on the fecundity and longevity of the bollworm, tobacco budworm and the cotton leafworm. *J. Econ. Entomol.*, 57, 233, 1964.
144. **Mackay, D.A. and Jones, R.E.**, Leaf shape and the host-finding behaviour of two ovipositing monophagous butterfly species. *Ecol. Entomol.*, 14, 423, 1989.
145. **McClure, M.S.**, Reproduction and adaptation of exotic hemlock scales (Homoptera: Diapsidae) on their new and native hosts. *Environ. Entomol.*, 12, 1811, 1983.
146. **McClure, M.S. and Hare, J.D.**, Foliar terpenoids in Tsuga species and the fecundity of scale insects. *Oecologia*, 63, 185, 1984.
147. **Mangel, M.**, Oviposition site selection and clutch size in insects. *J. Math. Biol.*, 25, 1, 1987.
148. **Marshal, L.D.**, Intraspecific variation in reproductive effort by female *Parspediasia teterella* (Lepidoptera: Pyralidae) and its relation to body size. *Can. J. Zool.*, 68, 44, 1990.
149. **Messina, F.J.**, Influence of cowpea pod maturity on the oviposition choices and larval survival of a bruchid beetle *Callosobruchus maculatus*. *Entomol. Exp. Appl.*, 35, 241, 1984.
150. **Miller, W.E.**, Reproductive enhancement by adult feeding effects of honeydew in imbibed water on spruce budworm. *J. Lepid. Soc.*, 43, 167, 1989.
151. **Moller, J.**, Investigations on a laboratory culture of the diamond-back moth, *Plutella maculipennis* (Curt.) (Lep., Tineidae). II. Influence of larval density on life history parameters of larvae, pupae and adults. *J. Appl. Entomol.*, 105, 425, 1988.

152. **Minkenberg, O.P.J.M. and Fredrix, M.J.J.**, Preference and performance of an herbivorous fly, *Liriomyza trifolii* (Diptera: Agromyzidae), on tomato plants differing in leaf nitrogen. *Ann. Entomol. Soc. Am.*, 82, 350, 1989.
153. **Minkenberg, O.P.J.M., Tatar, M., and Rosenheim, J.A.**, Egg load as a major source of variability in insect foraging and oviposition behaviour. *Oikos*, 65, 134, 1992.
154. **Montgomery, M.E. and Wallner, W.E.**, The gypsy moth — a westward migrant. In *Dynamics of Forest Insect Populations: Patterns, Causes, Implications*, (Ed., A.A. Berryman), Plenum Press, New York, 1988, 353.
155. **Moore, G.J.**, Host plant discrimination in tropical satyrine butterflies. *Oecologia*, 70, 592, 1986.
156. **Moore, L.V., Myers, J.H., and Eng, R.**, Western tent caterpillars prefer the sunny side of the tree, but why? *Oikos*, 51, 321, 1988.
157. **Moore, R.A. and Singer, M.C.**, Effects of maternal age and adult diet on egg weight in the butterfly *Euphydryas editha*. *Ecol. Entomol.*, 12, 401, 1987.
158. **Moran, N.A.**, Seasonal shifts in host usage in *Uroleucon gravicorne* (Homoptera: Aphididae) and implications for the evolution of host alternation in aphids. *Ecol. Entomol.*, 8, 371, 1983.
159. **Morrow, P.A. and Fox, L.R.**, Effects of variation in Eucalyptus essential oil yield on insect growth and grazing damage. *Oecologia*, 45, 209, 1980.
160. **Murphy, D.D., Launer, A.E., and Ehrlich, P.R.**, The role of adult feeding in egg production and population dynamics of the checkerspot butterfly *Euphydryas editha*. *Oecologia*, 56, 257, 1983.
161. **Myers, J.H., Monro, J., and Murray, N.**, Egg clumping, host plant selection and population regulation in *Cactoblastis cactorum* (Lepidoptera). *Oecologia*, 51, 1981.
162. **Nakasuji, F.**, Egg size of skippers (Lepidoptera: Hesperiidae) in relation to their host specificity and to leaf toughness of host plants. *Ecol. Res.*, 2, 175, 1987.
163. **Nealis, V.G.**, Diapause and the seasonal ecology of the introduced parasite *Cotesia (Apanteles) rubecula* (Hymenoptera: Braconidae). *Can. Entomol.*, 117, 333, 1985.
164. **Newton, C. and Dixon, A.F.G.**, Methods of hatching the eggs and rearing the fundatrices of the English grain aphid, *Sitobion avenae*. *Entomol. Exp. Appl.*, *45*, 277, 1987.
165. **Noguchi, H.**, Mating frequency, fecundity, and egg hatchability of the smaller tea tortrix moth, *Adoxyphyes sp.* (Lepidoptera: Tortricidae). *Jpn. J. Appl. Entomol. Zool.*, 25, 259, 1981.
166. **Otake, A. and Sakuratami, Y.**, Egg production of the adult tobacco cutworm, *Spodoptera litura* (Lepidoptera: Noctuidae) under different sex ratios and population processes of the larvae. *Appl. Entomol. Zool.*, 7, 190, 1972.
167. **Oyeyele, S.O. and Zalucki, M.P.**, Cardiac glycosides and oviposition by *Danaus plexippus* on *Asclepias fruiticosa* in south-east Queensland (Australia), with particular notes on the effect of plant nitrogen content. *Ecol. Entomol.*, 15, 177, 1990.
168. **Packer, M.J. and Corbet, P.S.**, Size variation and reproductive success of female *Aedes punctor* (Diptera: Culicidae). *Ecol. Entomol.*, 14, 297, 1989.
169. **Parker, G.A. and Begon, M.**, Optimal egg size and clutch size: effects of environment and maternal phenotype. *Am. Nat.*, 128, 573, 1986.
170. **Parker, G.A. and Courtney, S.P.**, Models of clutch size in insect oviposition. *Theor. Popul. Biol.*, 26, 27, 1984.
171. **Peckarovsky, B.L. and Cowan, C.A.**, Consequences of larval intraspecific competition to stonefly growth and fecundity. *Oecologia*, 88, 277, 1991.
172. **Pencoe, N.L. and Martin, P.B.**, Fall armyworm (Lepidoptera: Noctuidae) larval development and adult fecundity on five grass hosts. *Environ. Entomol.*, 11, 720, 1982.
173. **Petty, S.J.**, Fur and feather — tawny owls in 1988. *Entopath News*, 92, 7, 1989.
174. **Pianka, E.R.**, *Evolutionary Ecology*, 2nd ed., Harper & Row, New York, 1978, Chapter 5.
175. **Pilson, D. and Rausher, M.D.**, Clutch size adjustment by a swallowtail butterfly. *Nature*, 333, 361, 1988.
176. **Pilson, D. and Rausher, M.D.**, In response to Tabor. *Oikos*, 55, 136, 1989.
177. **Pitnick, S.**, Male size influences mate fecundity and remating interval in *Drosophila melanogaster*. *Anim. Behav.*, 41, 735, 1991.
178. **Prestidge, R.A.**, Instar duration, adult consumption, oviposition and nitrogen utilization efficiencies of leafhoppers feeding on different quality food (Auchenorrhyncha: Homoptera). *Ecol. Entomol.*, 7, 91, 1982.
179. **Preszler, R.W. and Price, P.W.**, Host quality and sawfly populations: a new approach to life table analysis. *Ecology*, 69, 2012, 1988.
180. **Pritchard, I.M. and James, R.**, Leaf mines: their effect on leaf longevity. *Oecologia*, 64, 132, 1984.
181. **Pritchard, I.M. and James, R.**, Leaf fall as a source of leafminer mortality. *Oecologia*, 64, 140, 1984.
182. **Puttick, G.M. and Bowers, M.D.**, Effect of qualitative and quantitative variation in allelochemicals on a generalist insect: iridoid glycosides and the southern armyworm. *J. Chem. Ecol.*, 14, 335, 1988.
183. **Raworth, D.A., McFarlane, S., Gilbert, N., and Frazer, B.D.**, Population dynamics of the cabbage aphid, *Brevicoryne brassicae* (Homoptera: Aphididae) at Vancouver, British Columbia. III. Development, fecundity and morph determination vs aphid density and plant quality. *Can. Entomol.*, 116, 879, 1984.

184. **Renwick, J.A.A.**, Chemical ecology of oviposition in phytophagous insects. *Experientia*, 45, 223, 1989.
185. **Rutowski, R.L. and Gilchrist, G.W.**, Courtship, copulation and oviposition in the chalcedon checkerspot, *Euphydryas chalcedona* (Lepidoptera: Nymphalidae). *J. Nat. Hist.*, 21, 1109, 1987.
186. **Sibly, R. and Monk, K.A.**, A theory of grasshopper life cycles. *Oikos*, 48, 186, 1987.
187. **Slansky, F.**, Food consumption and reproduction as affected by tethered flight in female milkweed bugs (*Oncopeltus fasciatus*). *Entomol. Exp. Appl.*, 28, 1980.
188. **Springer, P. and Boggs, C.L.**, Resource allocation to oocytes: heritable variation with altitude in *Colias philodice eriphyle* (Lepidoptera). *Am. Nat.*, 127, 252, 1986.
189. **Stadler, E.**, Host plant stimuli affecting oviposition behavior of the eastern spruce budworm. *Entomol. Exp. Appl.*, *17*, 176, 1974.
190. **Stamp, N.E.**, Egg deposition patterns in butterflies: why do some species cluster their eggs rather than deposit them singly? *Am. Nat.*, 115, 367, 1980.
191. **Stern, V.M. and Smith, R.F.**, Factors affecting egg production and oviposition in populations of *Colias philodice eurytheme* Boisduval (Lepidoptera: Pieridae). *Hilgardia*, 29, 411, 1960.
192. **Stewart, J.G. and Philogene, B.J.R.**, Reproductive potential of laboratory-reared *Manduca sexta* (Lepidoptera: Sphingidae) as affected by sex ratio. *Can. Entomol.*, 115, 295, 1983.
193. **Straw, N.A.**, The timing of oviposition and larval growth by two tephritid fly species in relation to host-plant development. *Ecol. Entomol.*, 14, 443, 1989.
194. **Sutton, R.D.**, The effect of host plant flowering on the distribution and growth of hawthorn psyllids (Homoptera: Psylloidea). *J. Anim. Ecol.*, 53, 37, 1984.
195. **Svard, L. and Wiklund, C.**, Fecundity, egg weight and longevity in relation to multiple matings in females of the monarch butterfly. *Behav. Ecol. Sociobiol.*, *23*, 39, 1988.
196. **Tauber, C.A. and Tauber, M.J.**, Phenotypic plasticity in *Chrysoperla* — genetic variation in the sensory mechanism and in correlated reproductive traits. *Evolution*, 46, 1754, 1992.
197. **Taylor, L.R.**, Longevity, fecundity and size; control of reproductive potential in a polymorphic migrant, *Aphis fabae* Scop. *J. Anim. Ecol.*, 44, 135, 1975.
198. **Taylor, M.F.J. and Forno, I.W.**, Oviposition preferences of the Salvinia moth *Samea multiplicalis* Guenee (Lep., Pyralidae) in relation to host plant quality and damage. *J. Appl. Entomol.*, 104, 73, 1987.
199. **Thomas, C.D.**, Butterfly larvae reduce host plant survival in vicinity of alternative host plants. *Oecologia*, 70, 113, 1986.
200. **Thompson, J.N.**, Evolutionary ecology of the relationship between oviposition preference and performance of offspring in phytophagous insects. *Entomol. Exp. Appl.*, *47*, 3, 1988.
201. **Thompson, J.N.**, Coevolution and the evolutionary genetics of interactions among plants and insects and pathogens. In *Pests, Pathogens and Plant Communities*, (Eds., J.J. Burdon and S.R. Leather), Blackwell Scientific Publications, Oxford, 1990, 249.
202. **Tinkle, D.W., Wilbur, H.M., and Tilley, S.G.**, Evolutionary strategies in lizard reproduction. *Evolution*, 24, 55, 1969.
203. **Trichilo, P.J. and Leigh, T.F.**, Influence of resource quality on the reproductive fitness of flower thrips (Thysanoptera: Thripidae). *Ann. Entomol. Soc. Am.*, 81, 64, 1988.
204. **Unnithan, G.C. and Paye, S.O.**, Factors involved in mating, longevity, fecundity and egg fertility in the maize stem-borer, *Busseola fusca*, (Fuller) (Lep., Noctuidae). *J. Appl. Entomol.*, 109, 295, 1990.
205. **van Amelsvoort, P.A.M. and Usher, M.B.**, A method for assessing the palatability of senesced leaf litter using *Folsomia candida* (Collembola: Isotomidae). *Pedobiologia*, 33, 193, 1989.
206. **van der Kraan, C. and van der Straten, M.**, Effects of mating rate and delayed mating on the fecundity of *Adoxophyes orana*. *Entomol. Exp. Appl.*, 48, 15, 1988.
207. **van der Meijden, E., van Zoolen, A.M., and Soldaat, L.L.**, Oviposition by the cinnabar moth, *Tyria jacobaeae*, in relation to nitrogen, sugars and alkaloids of ragwort, *Senecio jacobaea*. *Oikos*, 54, 337, 1989.
208. **van Emden, H.F. and Bashford, M.A.**, A comparison of the reproduction of *Brevicoryne brassicae* and *Myzus persicae* in relation to soluble nitrogen concentration and leaf age (leaf position) in the Brussels sprout plant. *Entomol. Exp. Appl.*, *12*, 351, 1969.
209. **van Emden, H.F. and Bashford, M.A.**, The performance of *Brevicoryne brassicae* and *Myzus persicae* in relation to plant age and leaf amino acids. *Entomol. Exp. Appl.*, *14*, 349, 1971.
210. **Walters, K.F.A. and Dixon, A.F.G.**, Migratory urge and reproductive investment in aphids: variation within clones. *Oecologia*, 58, 70, 1983.
211. **Walters, K.F.A., Brough, C., and Dixon, A.F.G.**, Habitat quality and reproductive investment in aphids. *Ecol. Entomol.*, 13, 337, 1988.
212. **Ward, S.A., Leather, S.R., and Dixon, A.F.G.**, Temperature prediction and the timing of sex in aphids. *Oecologia*, 62, 230, 1983.
213. **Ward, S.A., Wellings, P.W., and Dixon, A.F.G.**, The effect of reproductive investment on pre-reproductive mortality in aphids. *J. Anim. Ecol.*, 52, 305, 1983.
214. **Wassermann, S.S. and Asami, T.**, The effect of maternal age upon fitness of progeny in the southern cowpea weevil, *Callosobruchus maculatus*. *Oikos*, 45, 191, 1985.

215. **Wassermann, S.S. and Mitter, C.**, The relationship of body size to breadth of diet in some Lepidoptera. *Ecol. Entomol.*, 3, 115, 1978.
216. **Watt, A.D.**, The effect of cereal growth stages on the reproductive activity of *Sitobion avenae* and *Metopolophium dirhodum*. *Ann. Appl. Biol.*, 91, 147, 1979.
217. **Watt, A.D.**, The performance of the pine beauty moth on water-stressed lodgepole pine plants: a laboratory experiment. *Oecologia*, 70, 578, 1986.
218. **Watt, A.D.**, The effect of shoot growth stage of *Pinus contorta* and *Pinus sylvestris* on the growth and survival of *Panolis flammea* larvae. *Oecologia*, 72, 429, 1987.
219. **Watt, A.D. and Leather, S.R.**, The pine beauty in Scottish lodgepole pine plantations. In *Dynamics of Forest Insect Populations*, (Ed. A. A. Berryman), Plenum Publishing Corporation, New York, 1988, 243.
220. **Webb, J.W. and Moran, V.C.**, The influence of the host plant on the population dynamics of *Acizzia russellae* (Homoptera: Psyllidae). *Ecol. Entomol.*, 3, 313, 1978.
221. **Wellings, P.W., Leather, S.R., and Dixon, A.F.G.**, Seasonal variation in reproductive potential: a programmed feature of aphid life cycles. *J. Anim. Ecol.*, 49, 975, 1980.
222. **Werner, R.A.**, Influence of host foliage on development, survival, fecundity and oviposition of the spear-marked black moth, *Rheumaptera hastata* (Lepidoptera: Geometridae). *Can. Entomol.*, 111, 317, 1979.
223. **Wiklund, C. and Fagerström, T.**, Why do males emerge before females? A hypothesis to explain the incidence of protandry in butterflies. *Oecologia*, 31, 153, 1977.
224. **Wiklund, C. and Persson, A.**, Fecundity, and the relation of egg weight variation to offspring fitness in the speckled wood butterfly *Pararge aegeria*, or why don't female butterflies lay more eggs? *Oikos*, 40, 53, 1983.
224a. **Willers, J.L., Schneider, J.C., and Ramaswamy, S.B.**, Fecundity, longevity and caloric patterns in female *Heliothis virescens*: changes with age due to flight and supplemental carbohydrate. *J. Insect. Physiol.*, 33, 803-808, 1987.
225. **Williams, K.S.**, The coevolution of *Euphydryas chalcedona* butterflies and their larval host plants. III. Oviposition behavior and host plant quality. *Oecologia*, 56, 336, 1983.
225a. **Willson, H.R. and Trammel, K.**, Sex pheromone trapping for control of codling moth, Oriental fruit moth, lesser appleworm and three tortricid leafrollers in a New York apple orchard. *J. Econ. Entomol.*, 73, 291, 1980.
226. **Wratten, S.D.**, Aggregation in the birch aphid, *Euceraphis punctipennis* (Zett.) in relation to food quality. *J. Anim. Ecol.*, 43, 191, 1974.
227. **Wright, L.C. and Cone, W.W.**, Population statistics for the asparagus aphid, *Brachycorynella asparagi* (Homoptera: Aphididae) on different ages of asparagus foliage. *Environ. Entomol.*, 17, 699, 1988.
227a. **Wyatt, I.J. and Brown, S.**, The influence of light intensity, daylength and temperature on increase rates of four glasshouse aphids. *J. Appl. Ecol.*, 14, 391, 1977.
228. **Zalucki, M.P.**, The effects of age and weather on egg laying in *Danaus plexippus* L. (Lepidoptera: Danaidae). *Res. Popul. Ecol.*, 23, 318, 1981.
229. **Zalucki, M.P., Brower, L.P., and Malcolm, S.B.**, Oviposition by *Danaus plexippus* in relation to cardenolide content of three *Asclepias* species in the southeastern USA. *Ecol. Entomol.*, 15, 231, 1990.

Chapter 8

PROTANDRY AND MATE ACQUISITION

Christer Wiklund

CONTENTS

I. INTRODUCTION

Mate acquisition can be a problem for females as well as males. However, as a rule sexually active males greatly outnumber females, and hence, the male-biased operational sex ratio makes mate finding a more severe problem for the average male compared to the average female.

A variety of mate location systems have been observed in insects and include behaviors where males spend there entire life in search of females to territoriality and lek polygyny.[1] Territoriality and lekking are mate location systems that usually entail the opportunity for interference competition between males for females, and hence, can lead to selection for large male body size as in many mammals where males as a rule are larger than females.[2-11] Mate location systems where males search incessantly for females can be regarded as a situation similar to scramble competition, where large size does not seem to confer any obvious advantage.[12-13] Although the size of males may have important consequences for mate acquisition, males also compete for females in the time dimension. In short, the fact that the

0-8493-6695-X/95/$0.00+$.50

operational sex ratio is almost invariably male biased means that receptive females are in shortage and hence males should time their appearance so that they maximize the number of receptive females they encounter during their lifetime.

This was realized by Darwin[14] who wrote,

> It is certain that amongst almost all animals there is a struggle between the males for the possession of the female. This fact is so notorious that it would be superfluous to give instances. Hence, the females have the opportunity of selecting one out of several males,... Thus the males of our migratory birds generally arrive at their places of breeding before the females, so that many males are ready to contend for each female. ... The majority of the male salmon in our rivers, on coming up from the sea, are ready to breed before the females. So it appears to be with frogs and toads. Throughout the great class of insects the males almost always are the first to emerge from the pupal state, so that they generally abound for a time before any females can be seen. The cause of this difference between the males and females in their periods of arrival is sufficiently obvious. Those males which first migrated into any country, or which in the spring were first ready to breed, or were the most eager, would leave the largest number of offspring; and these would tend to inherit similar instincts and constitutions. It must be borne in mind that it would have been impossible to change very materially the time of sexual maturity in the females, without at the same time interfering with the period of the production of the young — a period which must be determined by the seasons of the year.

In addition, Darwin[14] quotes Wallace saying, "that in proportion as the individual moth (*Bombyx mori*) is finer, so is the time required for its metamorphosis longer; and for this reason the female, which is the larger and heavier insect, from having to carry her numerous eggs, will be preceded by the male, which is smaller and has less to mature". Darwin goes on to say

> Now as most insects are short-lived, and as they are exposed to many dangers, it would manifestly be advantageous to the female to be impregnated as soon as possible. This end would be gained by the males being first matured in large numbers ready for the advent of the females; and this again would naturally follow, as Mr. A.R. Wallace has remarked, through natural selection; for the smaller males would be first matured, and thus would procreate a large number of offspring which would inherit the reduced size of their male parents, whilst the larger males from being matured later would leave fewer offspring.

This quotation may be taken as a case in point supporting Alan Grafen's[15] assertion that "Darwin (1871) discovered almost everything important now known about sexual selection and did so without measurement."

Darwin's general statement that males usually emerge before females "throughout the great class of insects" might make further verification superfluous, but for the sake of record I still feel that it can be useful to point out that this pattern, known as protandry, seems to be virtually ubiquitous in most insect taxa, e.g., mayflies,[16] katydids,[17-18] solitary bees and wasps,[19-21] butterflies,[22-26] and mosquitoes.[27]

II. PROTANDRY THEORY

A. GENERAL

Although Darwin[14] advanced a perfectly sound explanation for protandry, his appreciation of the phenomenon seems not to have been generally realized, and from the late 19th up to the middle of the 20th century a variety of hypotheses were advanced to account for the relatively early emergence of males, including (1) prevention of inbreeding,[22] (2) enhancement of the selective process by which unfit males are removed during the prereproductive period,[19] (3) reduction of prereproductive death of females through facilitation of their rapid fertilization after eclosion,[28-29] and (4) improvement of male reproductive success because individuals emerging early would have access to a larger number of receptive females than males emerging late, a "rediscovery" of Darwin's idea.[23]

Wiklund and Fagerström[26] reviewed these ideas and pointed out that prevention of inbreeding would be equally well served if females emerged before males, and so does not seem to be capable of explaining the ubiquity of the relatively earlier emergence of males. The idea that males emerge early so as to put themselves on the real "cutting edge" of selection facilitating the weeding out of the relatively unfit is a notorious "for the good of the species" explanation and as such utterly un-Darwinistic. Moreover, the idea that males emerge before females so as to minimize prereproductive death of females is also group-selectionist, because it explicitly states that early emerging males are risking their own prereproductive death so as to minimize that of the females, the female sex being recognized as the more important for the propagation of the species.

Although Darwin's appreciation of protandry has obvious merit, there still are aspects of the early emergence/appearance of males that warrant further analysis; for example, the statement that "those males which annually first migrated into any country, or which in the spring were first ready to breed, or were the most eager, would leave the largest number of offspring". On face value, this statement implies that early males do better than late males, and hence, natural selection should lead to a situation where all males appear before the first female. Hence, Darwin's theory, in a strict sense, predicts that the whole male population should emerge before the onset of the female emergence period. This leaves the field open to alternative predictions of how the male population should emerge in relation to that of the female population. The prediction of how male and female emergence curves should be timed relative to one another is greatly facilitated by mathematical modeling, and from the late 1970s up to the present, a number of models have been advanced to explain the adaptive significance of protandry.[26,30-36] These models all ask the question, given that we know the shape of the emergence curve for the female population, how should the male population emerge in time. Basically, these models fall into two categories that can be called "fixed distribution models" and "flexible distribution models", respectively. The fixed distribution models assume that the distribution of emergence times has a fixed shape,[26,30-31,34-36] while the flexible distribution models assume that there are absolutely no constraints on the shape of the male emergence curve.[32-33,37] The fixed distribution models assume that, although the mean time of male emergence is under selective control, the shape of the male and female emergence curves are not. Instead, the shape of the emergence curves is the result of genetic factors interacting with environmental factors like temperature.[26,30,34-36] By contrast, the flexible distribution models assume that, in addition to the location in time of the emergence curve, the shape is also determined by selection.[32-33,36] The fixed distribution models conclude that the fitness distribution around the male emergence peak is somewhat asymmetrical, which means that the mean emergence time that is favored is not the emergence time that results in the highest fitness (although it is reasonably close). The flexible distribution models result in a predicted emergence curve for the male population where all males are expected to have equal mating success, and hence, represents an Evolutionarily Stable Strategy (ESS).

Apart from the different optimization criteria, the two kinds of models are based on the same set of assumptions: viz., that (1) male and female emergence curves are under genetic control; (2) all females become fertilized (but see Zonneveld and Metz[34]); (3) Females mate only once; (4) females always mate on the day of eclosion; (4) males are capable of multiple mating; and (5) males are endowed with a given mortality rate which is independent of age. Moreover, both models also assume that all males active on the same day have equal opportunity of finding and mating with a virgin female, and hence, both kinds of models

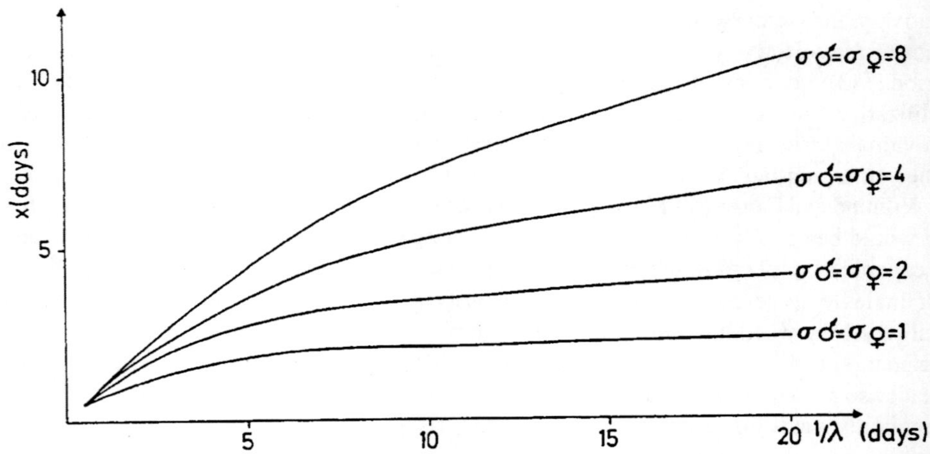

FIGURE 1. Protandry calculated according to an optimality model assuming that the mean time of emergence for the male population is under selective control, whereas the shape of the emergence curves of male and females are not. Protandry (x) as a function of death rate of males (1/λ) when the duration of male and female emergence periods are equal (here given by σ♂ and σ♀, the standard deviations of the population eclosion curves; hence 95% of the population emerged within a time span equaling 4σ). (From Wiklund, C. and Fagerström, T. 1977. *Oecologia (Berlin)*, 31:153–158. With permission.)

assume use of the optimization criterion that males are selected for maximizing the number of matings during their lifetime, an end which is proportional to the number of females they encounter and inversely proportional to the number of competing males.

Before proceeding to the quantitative predictions of the different models, it is necessary to point out that the domain of protandry in insects is limited to systems where species have discrete non-overlapping generations. Indeed, protandry cannot occur in environments where reproductively active males and females are present throughout the year.

B. MALE BENEFIT OF PROTANDRY

The fixed distribution model presented by Wiklund and Fagerström,[26] which assumed the variance around the mean time of eclosion to be given, yielded the result that males emerging at the peak of the male emergence curve, should maximize their expected number of matings if the male population emerged some time before the female population (Figure 1). Moreover, protandry should increase with the life expectancy of males, and with the duration of the emergence period. In species where the life expectancy is only about 3 days, and virtually all females and males emerge during a 5-day period, males should, on average, maximize the number of expected matings by having their emergence curve peaking 1 day before that of females. On the other hand, in species where the life expectancy of males is some 20 days, and the duration of the emergence period of the male and the female population approaches a month, the difference in time of mean emergence between the sexes should be approximately 10 days (Figure 1).

The flexible distribution models by Iwasa et al.[32] and Parker and Courtney,[33] assuming that the distribution of male emergence times can respond without any constraints to sexual selection (so leading to an "ideal free" distribution), arrive at very similar predictions, hence, giving considerable robustness to the general hypothesis that protandry can be viewed as a mate acquisition strategy by males. The main difference between the quantitative predictions of the two kinds of models is that the fixed distribution models predict that the very early and very late males have lower fitnesses, whereas the flexible distribution models predict a slightly smaller amount of protandry, and that the male eclosion curve should exhibit a rather abrupt truncation point some time after the peak of male emergence because the loss of male mating

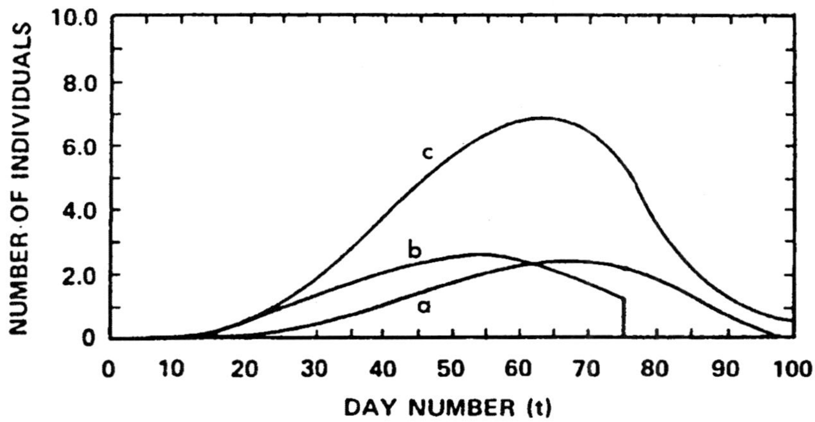

FIGURE 2. Protandry calculated according to a game theory model assuming that both the location and the shape of the male emergence curve are under selective control. Curve a is the female emergence curve, curve b is the male emergence curve, and curve c is the male presence curve, calculated for the case where the total number of males and females are 100, and the difference in preemergence and postemergence mortality is small. (From Iwasa, Y., Odendaal, F.J., Murphy, D.D., Ehrlich, P.R., and Launer, A.E. 1983. *Theor. Popul. Biol.* 23:363–379. With permission.)

opportunity is more severe for males emerging several days after the male emergence peak than for males emerging as many days before (Figure 2).

C. FEMALE BENEFIT OF PROTANDRY

The models discussed up to now have all been male chauvinistic, asking the question of how males should time their emergence to maximize their mating opportunities when the emergence curve of the female population is given. However, even if the approximate time of year when females should emerge may reasonably be given by the timing of the appearance of suitable host plants etc., cf. Darwin's[14] assertion that "it would have been impossible to change very materially the time of sexual maturity in the females, without at the same time interfering with the period of the production of the young — a period which must be determined by the seasons of the year", it seems reasonable to assume that also females can be selected to time their period of emergence relative to that of males. So females may be selected to emerge so as to minimize death before reproduction. This is best achieved by females emerging at the time when there is a maximum of living males in the population. Since even very short-lived insects do have *some* life expectancy, the peak of the male presence curve must always occur some time after that of the male eclosion curve (cf. Figure 2), and hence, it is easy to understand that females benefit from emerging later than males. In other words, protandry may be viewed also as a female mating strategy by which the time they spend unmated is minimized.[30]

D. CONFLICTS BETWEEN MALES AND FEMALES

Although both males and females benefit from protandry, it is conceivable that there could be considerable conflict between the sexes as to how profound the difference in emergence between the sexes should be. However, when comparing the optimal degree of protandry between the sexes, Wiklund and Fagerström[30] found that there seemed to be virtually no conflict between the sexes for cases where the eclosion periods were similar for the two sexes (Figure 3). Using a slightly different approach, were ESS models for optimal time of emergence where calculated for males and females separately, Zonneveld and Metz,[35] also arrived at the same general conclusion, i.e., that although male and female ESS's were not identical, there is only a small conflict of interest.

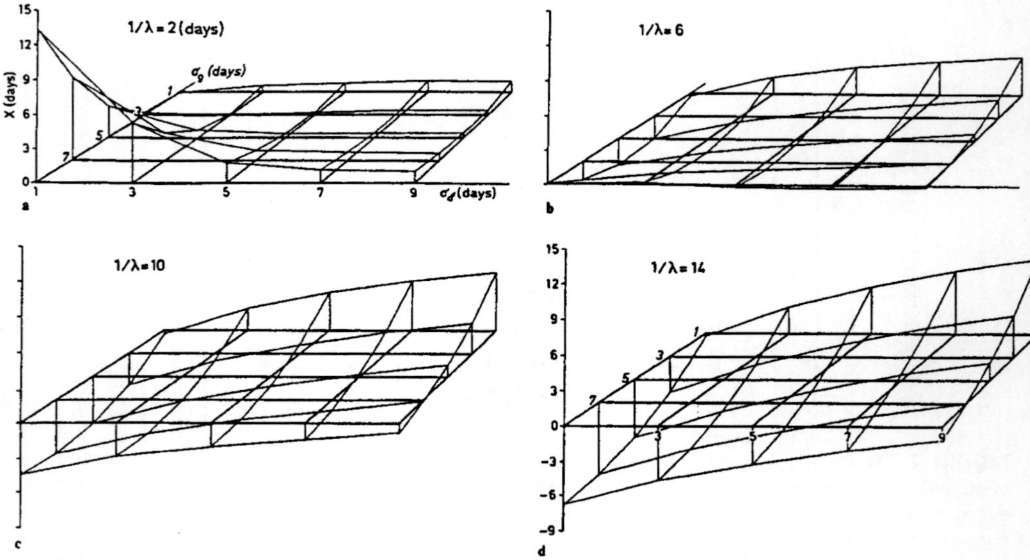

FIGURE 3. The degree of conflict between the sexes in the decision when to emerge (x) as a function of the standard deviation of male (σ♂) and female (σ♀) eclosion curves for different values of the life expectancy of males (1/λ). Positive x-values signify that the degree of protandry that is optimal for the males is less than that being optimal for the females; negative x-values signify the reverse. (From Fagerström, T. and Wiklund, C. 1982. *Oecologia (Berlin)* 52:164–166. With permission.)

III. TESTS OF PROTANDRY THEORY

A. COMPARISON BETWEEN THEORETICAL MODELS

There are several ways in which protandry theory can be tested, but unfortunately the exact quantitative predictions are difficult to test in the field because the development time of insects is so dependent on temperature. Natural selection can certainly act on how insects respond to temperature, and so there will be individual variation in growth rate even under constant temperature; however, since insects are so strongly ectothermic, variation in growth rate between individuals will be superseded by the action of temperature. For instance, the actual observation of protandry in nature bears witness to the fact that the development time of males is shorter than that of females (Figure 4). However, the exact difference in time between the peak of male and female eclosion is likely to be strongly influenced by temperature. If a drastic cold spells occurs just after the peak of male emergence, the difference in mean time of emergence is likely to be quite accentuated. However, if the weather is cold up to the point of the male emergence peak, but warm thereafter, the difference in time of emergence between the sexes is likely to be considerably smaller.

The flexible distribution models of Iwasa et al.[32] and Parker and Courtney,[33] and the optimization model of Wiklund and Fagerström[26] and Fagerström and Wiklund,[30] were compared by Bulmer,[31] who found difficulty in accepting that natural selection could result in a emergence curve that is abruptly truncated after its maximum, as predicted by the flexible distribution models. Bulmer says, "It is difficult for selection to alter the *shape* of the distribution of a polygenic metric character because recombination tends to restore approximate normality each generation (Bulmer[37]). It is particularly difficult to envisage a mechanism which could produce a distribution with an abrupt truncation point as required in the ideal free distribution". Moreover, even if the prediction of a specific truncation of the male emergence curve is somewhat relaxed, Bulmer[31] argues that "it seems unlikely that either the shape or the

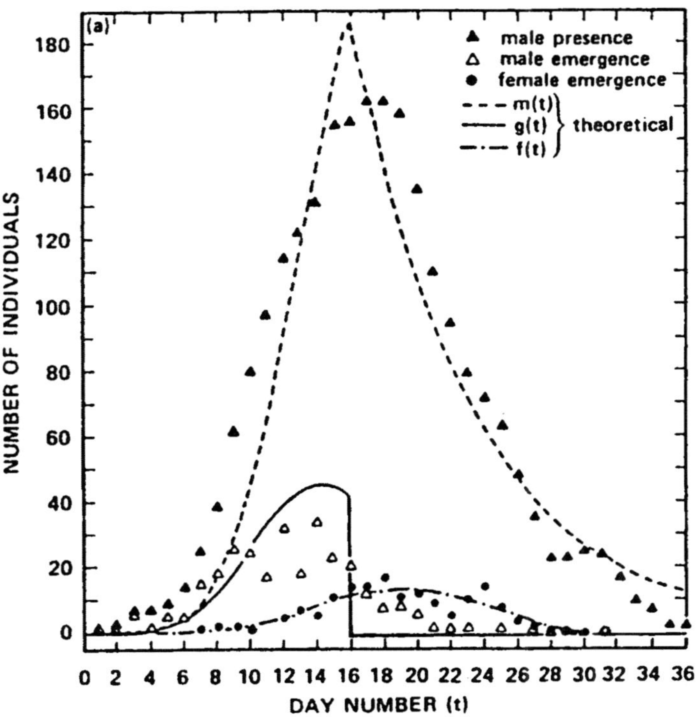

FIGURE 4. Number of males present, emerging females, and emerging males each day for *Euphydryas editha* data (points) and theoretical values (lines) when the preemergence mortality is 0.04 (From Iwasa, Y., Odendaal, F.J., Murphy, D.D., Ehrlich, P.R., and Launer, A.E. 1983. *Theor. Popul. Biol.* 23:363–379. With permission.)

variance of the distribution of male emergence times will be able to respond to selection in the way predicted by (flexible distribution model) theory" [(flexible distribution model) added for clarification by the present author].

Moreover, examination of emergence times for males and females from the field have consistently failed to give evidence of any truncation point in the eclosion of the male population (Figures 4 and 5; Iwasa et al.;[32] Parker and Courtney;[33] Baughman et al.[36]). Hence, the exact prediction of the flexible distribution models about the shape of the male distribution curve have not been corroborated by observation in the field. Moreover, the empirical data on emergence of male and female populations show that they are strongly overlapping, hence, lending little support to the strict interpretation of Darwin's statement predicting that the whole male population should emerge before the onset of female emergence. Additionally, Baughman[38] was able to test the mating success of early, intermediately, and late emerging males of *Euphydryas editha* by brushing male genitalia with fluorescent dust of three different colors, and later collecting females to see if the dust color from the early emerging group of males was more commonly found on the outer part of female genitalia. However, the result was that male mating success appeared to be independent of time of emergence, showing again that Darwin's strict prediction that early emerging males have higher mating success does not hold true. However, Baughman's results do not allow any conclusion as to whether the flexible distribution models of Iwasa et al.[32] and Parker and Courtney,[33] or the fixed distribution models of Wiklund and Fagerström[26] and Fagerström and Wiklund,[30] hold true, as the first predicts (or rather is based on the assumption) that all males have equal mating success, the latter assumes that males emerging at the peak of the male emergence curve should have maximal mating success.

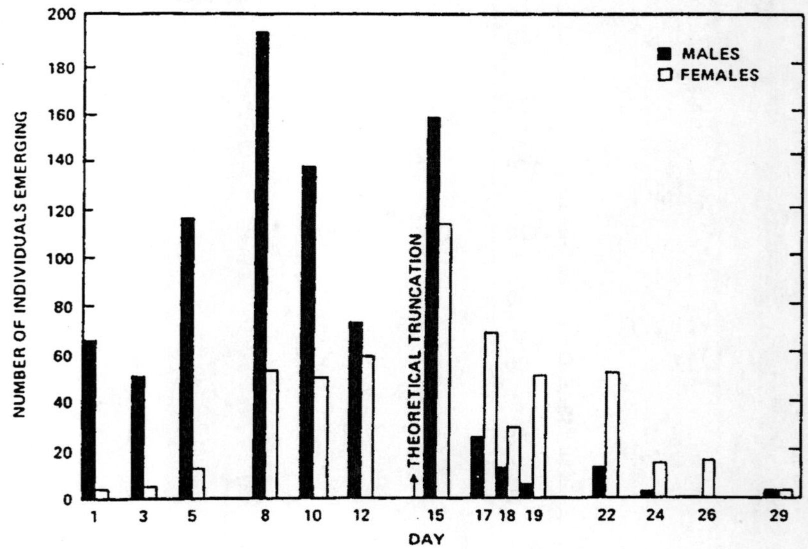

FIGURE 5. Graphical representation of male and female emergence patterns for *Euphydryas editha* in 198 Baughman, J.F., Murphy, D.D., and Ehrlich, P.R. 1988. *Theor. Popul. Biol.* 33:102–113. With permissio

B. TESTS OF ASSUMPTIONS OF PROTANDRY THEORY

Apart from testing quantitative predictions, another way to test the general hypothe protandry is a mate acquisition strategy is to test the relevance of its assumptions. As out previously, males cannot emerge before females in environments when reprodu active males and females are found throughout the year. Hence, insofar as the earlie gence of males relative to females is associated with a cost of some kind, males sho be expected to have shorter development time compared to females in populations discrete generations (i.e., where there can be no protandry). A case in point is served speckled wood butterfly *Pararge aegeria*, which has a Palearctic distribution. In s Sweden and England, this species can spend the winter either as half grown larvae o pupal stage, both of which have a bivoltine life cycle (Goddard;[39] Wiklund, unpub However, although *P. aegeria* consequently exhibits four flight periods annually, the d generations are largely nonoverlapping. In Spain, adults of the speckled wood cannot b from November to February, but are multivoltine through the rest of the year with d generations overlapping quite strongly (Garcia-Barros, unpublished). On the island deira, *P. aegeria* flies throughout the year with little, if any, seasonality. In accordan assumptions of protandry theory, populations from southern Sweden and England strong protandry, whereas populations from Spain and Madeira do not exhibit sig protandry at all (Figure 6; Nylin et al.[40]). This is quite remarkable as there is no vari the degree of sexual size dimorphism between the four populations — males weigh some 80% of female weight in all four populations (Table 1; Nylin et al.[40]).

Another general prediction of protandry theory, again assuming that shorter devel time and higher growth rate carries a cost, is that male development time should be than that of females only when individuals become sexually active immediately fo development. Hence, in butterflies that overwinter as adults and do not mate un hibernation, males should not be expected to emerge before females in the summer. H protandry should be expected in the spring with males emerging from hibernation e the season relative to females. This kind of life cycle is exhibited by the brimstone b which emerges from the pupa in midsummer but where individuals do not mate un hibernation. As predicted, the development time from egg to adult is not shorter fo

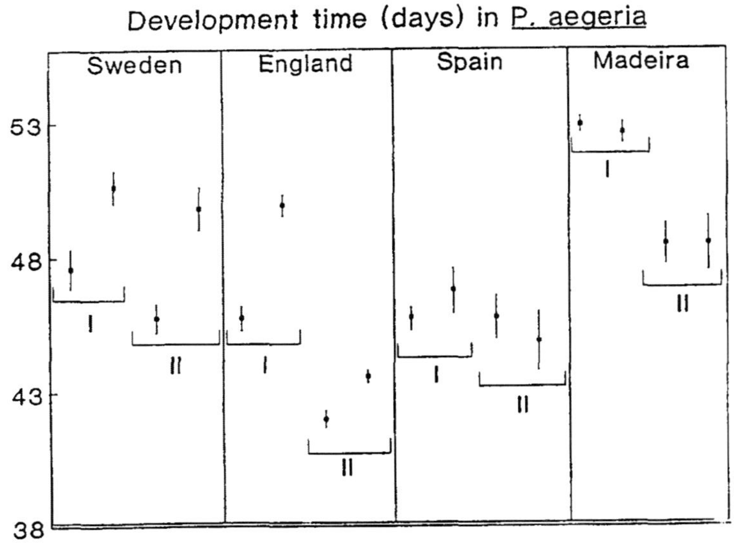

FIGURE 6. Total development time (mean +/– S.E.) for the two sexes in four potentially multivoltine populations of *Pararge aegeria* as measured in two replicates. Males are shown to the left in each pair (cohort) and the differences between the sexes show the degree of protandry. Populations from south Sweden and England represent more strongly seasonal environments that Spain or Madeira, and here protandry was significant. (From Nylin, S., Wiklund, C., Wickman, P.-O., and Garcia-Barros, E. *Ecology,* 74: 1414–1427. With permission.)

TABLE 1
Sexual Differences in Weight (means + S. E.) at Different Developmental Stages in Cohorts of Five Populations of *Pararge aegeria*

Cohort	Sex	N	Pupal weight (mg)	Female/Male pupal weight ratio
Central Sweden	Males	15	140.7 + 2.9	
	Females	20	173.0 + 1.9	1.23
South Sweden I	Males	24	137.9 + 3.2	
	Females	30	165.7 + 3.2	1.20
South Sweden II	Males	13	153.5 + 3.7	
	Females	10	172.0 + 3.8	1.12
England I	Males	34	148.9 + 1.5	
	Females	24	185.5 + 2.8	1.25
England II	Males	10	137.3 + 1.3	
	Females	14	163.6 + 2.0	1.19
Spain I	Males	20	134.4 + 2.4	
	Females	25	169.1 + 3.2	1.25
Spain II	Males	12	142.6 + 3.3	
	Females	7	170.1 + 5.2	1.19
Madeira	Males	24	132.7 + 2.0	
	Females	20	153.1 + 2.1	1.15

Note: Sexual dimorphism in weight was highly significant ($p < 0.01$ - $p < 0.001$; ANOVA) in all experiments. Roman numerals indicate replicate number.

From Nylin, S., Wiklund, C., Wickman, P.-O., and Garcia-Barros, E. *Ecology,* 74: 1414–1427. With permission.

TABLE 2
Pupal Weights (mean + S. E) of Diapausing and Nondiapausing *Leptidea sinapis* at 23°C

		Pupal weight (mg)	N
Diapause	Females	58.6 + 0.9	26
	Males	56.1 + 1.2	21
	Difference	2.5	
Non-diapause	Females	55.5 + 0.9	22
	Males	51.7 + 1.2	21
	Difference	3.8	

From Wicklund, C. and Solbreck, C. 1982. *Evolution* 36:56–62. With permission.

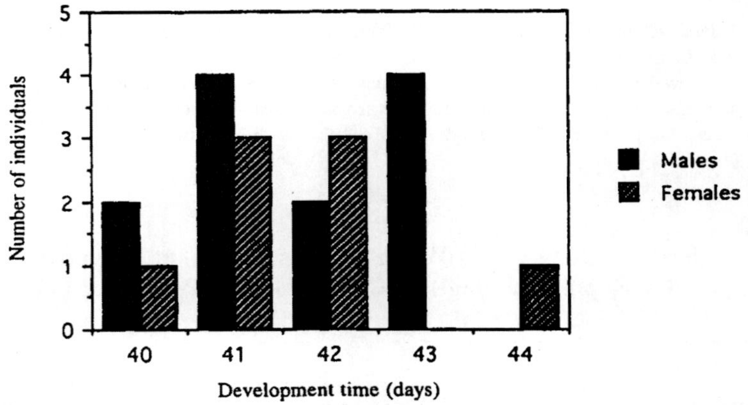

FIGURE 7. Development time of *Gonepteryx rhamni* from egg to adult when reared at 20°C and a 20 h daylength. The mean development time was 41.7 +/– 0.3 days for the 12 males and 41.6 +/- 0.4 days for the 8 females. Note that this butterfly emerges during summer and does not mate until after hibernation, and so protandry is not relevant at the time when the adults emerge (Wiklund and Lindfors, in preparation.)

compared to females, but in the spring males can be seen on the wing much earlier than females (Table 2; Figures 7 and 8; Wiklund and Lindfors, unpublished).

Yet another way to test whether protandry is a mate acquisition strategy or not is to compare alternative hypothesis for the shorter development time of males relative to females. This can be done in bivoltine species in which protandry is effectuated differently in the two generations. The hypothesis that protandry is a mate acquisition strategy is based on the idea that the difference in time of emergence between the sexes is selected for, per se. However, in most insects males are smaller than females, and so the shorter development time of males might be explained by the simple fact that it takes less time to acquire a smaller soma. These two hypotheses yield different predictions for the degree of protandry in the two generations for a butterfly that overwinters in the pupal stage like the wood white *Leptidea sinapis*. In Sweden, the first generation of this butterfly emerges in the middle of May, and the first eggs are laid in late May/early June. These early eggs give rise to larvae that develop directly and produce pupae from which a second generation of adults is formed that can be seen on the wing in late July/early August (late eggs laid by first generation females yield larvae that develop into overwintering pupae). The eggs laid by the second generation females give rise to larvae that

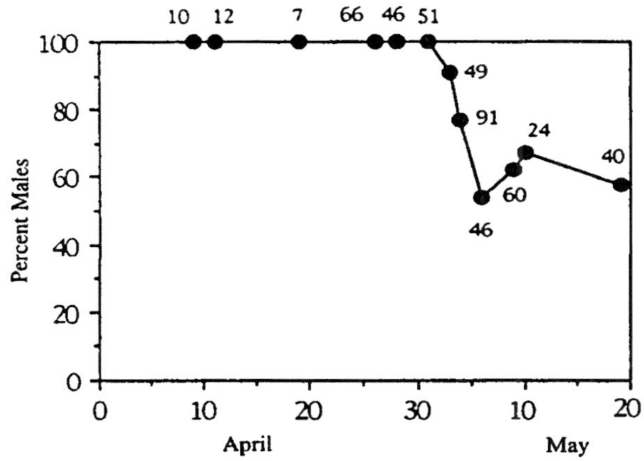

FIGURE 8. Number of male and female *Gonepteryx rhamni* observed during 1 hour counts by car along roads in central Sweden during April (days 1–30) and 1–10 of May (days 31–40) in 1992. The first males were observed on April 9, whereas the first females were observed on May 3, i.e., protandry measured as the difference in time of appearance amounted to more than 3 weeks. (Wiklund and Lindfors, in preparation.)

produce overwintering pupae from which the adult butterflies emerge the following spring. The wood white can only survive the winter in the pupal stage, and interestingly, development is synchronized during the winter. This is evidenced both by the observation that the pupae spend the winter in an undeveloped stage, being semitransparent when held against a strong light, and because there is no difference in time of emergence between the pupae formed in June/July by the offspring from the first generation of adults and those formed in August by the offspring from the second generation of adults. This means that the early emergence of males in the directly developing summer generation is brought about both by male larvae and pupae, having shorter development time compared to females. However, in the overwintering generations, the earlier emergence of males is brought about only because of the difference in postdiapause development time between male and female pupae. Hence, if males emerge before females only because they are smaller, the difference in time of appearance between the sexes should be considerably smaller in the overwintering spring generation than in the directly developing summer generation. However, if the difference between the sexes in time of emergence is selected for, per se, protandry should be expected to be equal in the two generations, given equal costs, regardless of the fact that it is brought about differently. Indeed, when reared in the laboratory under a constant temperature of 23°C, the males emerge some 2 to 2.5 days earlier than females both in the diapausing and the nondiapausing generations (Figure 9). Hence, the hypothesis that protandry in the wood white is an incidental side effect of the sexual size dimorphism was not supported, whereas the hypothesis that the difference betwen the sexes in time of emergence is selected for, per se, was supported. The observation that protandry is similar in directly developing and diapausing generations (and that it is brought about by a relative increase in the sex difference in postdiapause development time) has also been made in the black swallowtail butterfly *Papilio polyxenes* (Lederhouse et al.[41]) and for a number of Swedish pierid and satyrid butterflies (Wiklund and Forsberg[42]).

C. MALE VS. FEMALE COOPERATION/CONFLICTS OVER PROTANDRY

As previously mentioned, theoretical models predict that males and females both have similar interests relating to how many days earlier males should emerge prior to females. The case of the bivoltine wood white butterfly can also be used to analyze this aspect of protandry because the results suggest that the two sexes take turns in realizing protandry. When overwintering pupae of *L. sinapis* are kept at 4°C diapause is broken after approximately 2.5 months, after which the duration of the time pupae are kept at low temperature does not seem

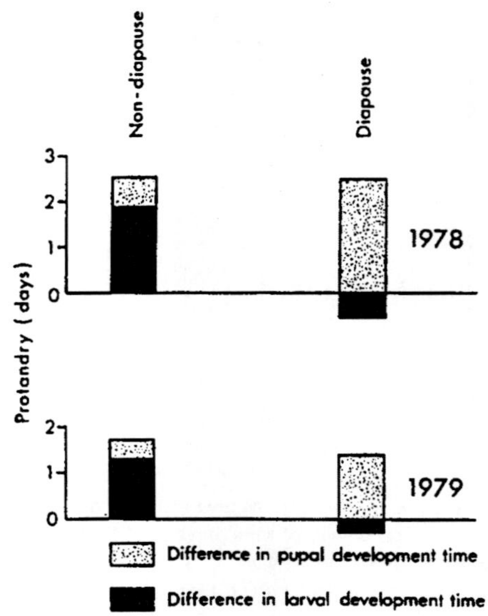

FIGURE 9. Sex differences in egg-larval and pupal development times for *Leptidea sinapis* under diapause and nondiapause development, respectively. The degree of protandry is similar in the two generations although it is brought about only by the sex difference in postdiapause development time in the diapausing generation, whereas it is the accumulated sex difference in development time from egg, larval, and pupal development in the non-diapausing generation. (From Wiklund, C., and Solbreck, C. 1982. *Evolution* 36:56–62. With permission.)

to affect the postdiapause development time of male pupae (Figure 10). However, postdiapause development time is strongly dependent on the time spent at low temperature, and the statement that female pupae take some 2 to 2.5 days longer to produce adults after hibernation holds true only when pupae are kept at 4°C between 2.5 to 4 months. When female pupae are kept at low temperature for longer than that, postdiapause development time becomes shorter with increasing time spent in 4°C (Figure 10). Hence, when females are kept at low temperature for 8 months, the difference in postdiapause development time between the sexes is only about 0.5 days, i.e., equal to the sex difference in pupal development time in the directly developing generation (cf. Figure 9). The observation that females, but not males, vary their postdiapause development time indicates that it is the female sex that effectuates protandry in the diapausing generation by not developing a maximum speed for intermediate times spent at low temperature. Given that the same time difference in the appearance of the sexes is optimal in both generations of the wood white butterfly, it appears that the degree of protandry brought about by females corresponds to the optimal one of 2 to 2.5 days when the pupae hibernate for an intermediately long period of time. Wiklund and Solbreck[43] have argued that up to 4 months spent at 4°C corresponds to the average winter duration in Sweden, whereas longer times at low temperatures suggest that winter is longer than usual. This implies that spring will be later than usual, and given that it is beneficial for females to emerge so early in the season that their offspring will have time to develop directly, females may have to make a trade off between emerging early and effectuating optimal protandry. According to this scenario, females opt for optimal protandry for average duration of the winter period, whereas they effecuate suboptimal protandry after unusually long winters to avoid the penalty of emerging too late in the season.

According to the idea that it is important for females not to emerge too late in the season, the second generation of females will be even more pressed for time because their offspring must complete development and pupate before the autumn temperatures get too cold for development. Hence, it is unlikely that the female sex will effectuate protandry in the directly

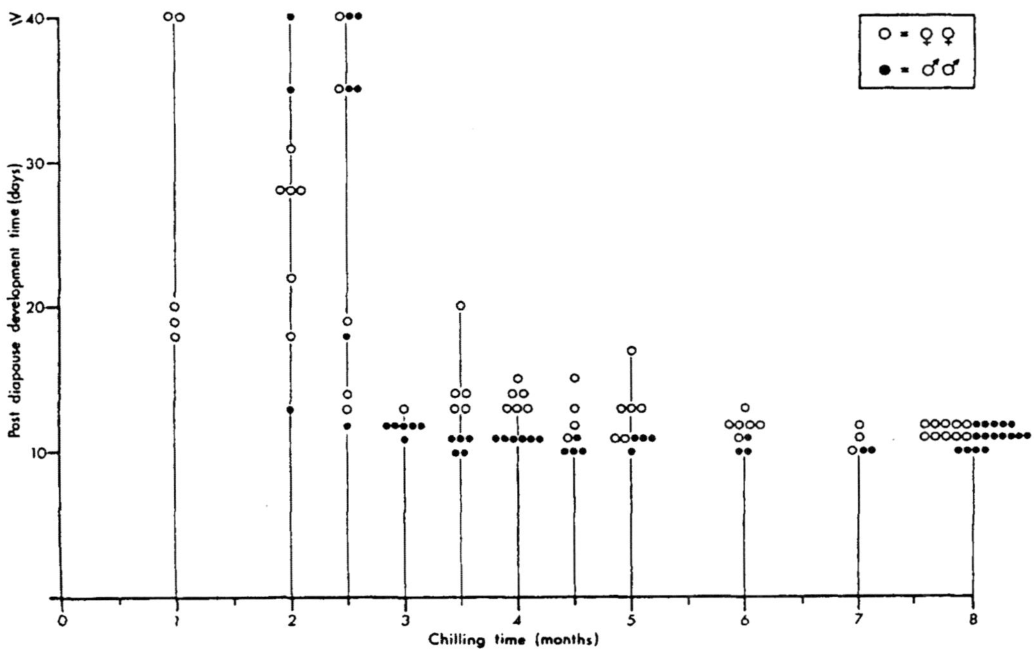

FIGURE 10. Postdiapause development times for male and female pupae of *Leptidea sinapis* in relation to the length of chilling (+ 4°C) period. (From Wiklund, C. and Solbreck, C. 1982. *Evolution* 36:56–62. With permission.)

developing generation. Hence, the males will have to shorten their development time if they are to emerge before the females and, as pointed out by Singer,[44] given that males and females grow at a similar rate, protandry can only be achieved by the male sex becoming smaller. Although few issues in evolutionary biology have proved less easily tractable than that of optimal size (cf. Sibly and Calow[45]), insofar as optimal male size is concerned, it ought to be achieved by the individuals from the first generation of wood whites that enter the diapause developmental pathway and have some 2 months longer to complete development before the winter compared to the offspring from the second generation. Indeed, when comparing development times of males and females from the two generations of wood whites, it turns out that the difference in egg-larval development times between the sexes was only some 0.5 days for the diapausing individuals, whereas it was some 1.5 for the directly developing larvae (cf. Figure 9). When comparing the difference in weight between male and female pupae, the difference was considerably larger in the directly developing generation (albeit not significantly so), which suggests the tendency that it is the male sex that shortens its development time relative to the female sex, and in doing so becomes smaller (Table 2). Hence, it may be that males in the directly developing generation have to make a trade off between optimal size and achieving optimal protandry. Since the degree of protandry achieved in the directly developing generation was very similar to that in the diapausing generation, this can be interpreted as the males favoring the achievement of optimal protandry to that of optimal size.

This section has been based on studies of one butterfly, but the pattern that both sexes achieve protandry is also exhibited by the green-veined white butterfly *Pieris napi*, in which females also seem to achieve to effectuate protandry in the overwintering generations, with males effectuating protandry in the directly developing generation (according to the same argumentation as for the wood white; cf. Forsberg and Wiklund[46]). However, there is one interesting difference between protandry in the two generations of the two butterflies, insofar as the time difference in emergence of the sexes is smaller in the directly developing generation of *P. napi*, whereas the difference in size between the sexes is just as profound as

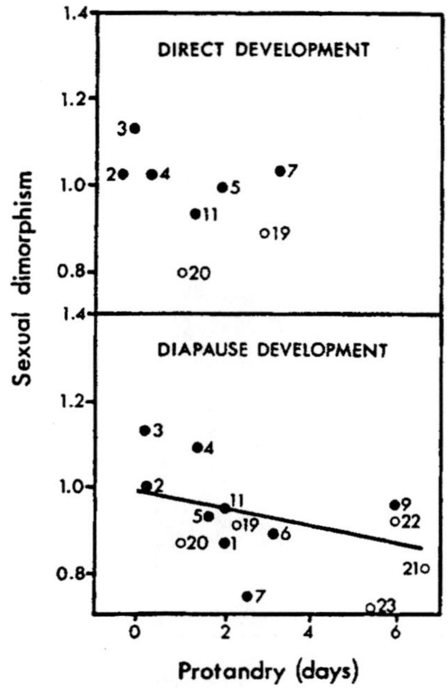

FIGURE 11. The relationship between male/female size dimorphism and protandry. For directly developing generations: y = 0.02x + 0.03; r = –0.327; n = 8; p <0.3: For diapause developing generations: y = –0.02x + 0.99; r = –0.511; n = 14; p <0.05, one-tailed test. Values for pierids are indicated by (●) and values for satyrids are indicated by (○): The numbers refer to the following species (2) *Pieris brassicae*; (3) *Pieris rapae*, (4) *Pieris napi*, (5) *Pontia daplidice*, (6) *Anthocharis cardamines*, (7) *Colias hyale*, (9) *Colias palaeno*, (11) *Leptidea sinapis*, (19) *Pararge aegeria*, (20) *Lasiommata megera*, (21) *Lasiommata maera*, (22) *Lasiommata petropolitana*, (23) *Lopinga achine*. (From Wiklund, C. and Forsberg, J. 1991. *Oikos* 60:373–381. With permission.)

in the diapausing generation. This may suggest that males of the green-veined white, when having to make a trade off between achieving optimal size and protandry, favor achieving optimal size to achieving optimal protandry. This, in turn, may be understandable in terms of the different mating systems and the influence of relative male size on reproductive success in different matings systems, as will be discussed below.

IV. PROTANDRY AND SEXUAL SIZE DIMORPHISM

In two families of butterflies, Wiklund and Forsberg[42] have shown that relative male size increases with the degree of female polygamy. The causal mechanism underlying this pattern is suggested to be sperm competition, where large males are reproductively favored because the size of the spermatophore that males transfer to females at mating is proportional to male size, and the time following mating, during which females are unreceptive to male courtship is, in turn, proportional to the size of the spermatophore the female received at mating (Rutowski;[47] Oberhauser;[48] Svärd and Wiklund;[49] Kaitala and Wiklund, submitted). As pointed out by Singer,[44] "If males are to emerge before females as a result of selection for protandry, they must have shorter development times than females, and hence will be smaller, if they grow at the same rate as larvae". Hence, a negative relationship between male/female size dimorphism and protandry should be expected, especially among directly developing generations where development is uninterrupted and males can only emerge before females by shortening their development time, in the process becoming smaller than females. However, a comparative study of six pierid and two satyrid butterflies failed to give any evidence of a negative relationship between male/female size dimorphism and protandry (Figure 11; Wiklund and Forsberg[42]).

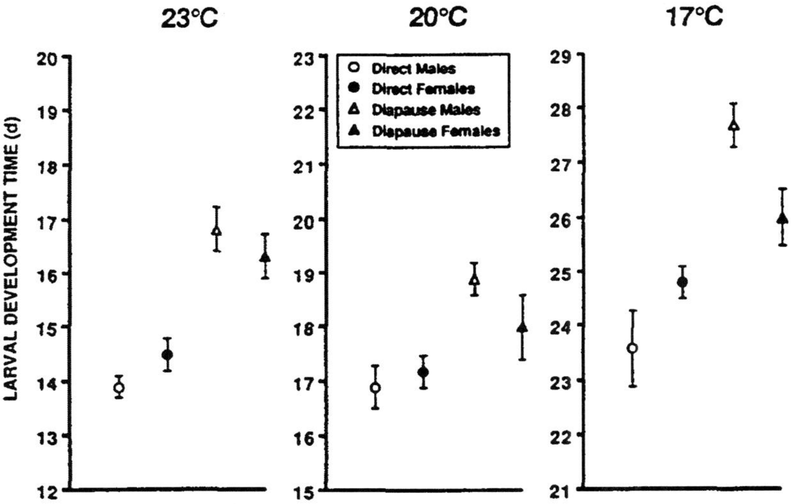

FIGURE 12. Larval development times of males and females of *Pieris napi* under direct and diapause development, respectively, at three different temperatures 17°C and a 22 h daylength, at 20°C and a 22 h daylength, and pooled results for larvae reared at 23°C and 18:00, 18:15, 18:30, 18:45 and 19:00 h daylengths. Although directly developing males complete their development several days before their brothers, the mean weight of directly developing pupae is always higher than that of diapausing pupae (although usually not significantly heavier). (From Wiklund, C., Nylin, S., and Forsberg, J. 1991. *Oikos* 60:241–250. With permission.)

The absence of such a negative relationship can only be understood if Singer's[44] assumption of male and female larvae growing at the same rate is violated. This explanation is corroborated by a study on the growth rate and development times of male and female larvae of the green-veined white butterfly P. *napi*, under direct and diapause development. In this species, where males are actually larger than females (i.e., they have larger mass and longer wings compared to females at eclosion), the development time of males is indeed longer than that of females in individuals that develop according to a univoltine life cycle and diapause in the pupal stage before emerging as sexually active adults the following spring. However, when individuals develop directly, male development is considerably shorter than that of females, although the male/female size dimorphism remains unaltered. In fact, sibling males that develop directly become larger than brothers that enter diapause development while still completing their development a couple of days earlier (Figure 12). Wiklund et al.[50] showed this by rearing larvae of *P. napi* at the critical photoperiod, at which a proportion of the larvae entered direct development and part entered diapause development, which means that the results shown in Figure 12 relate to differences between individuals that were experiencing virtually identical temperature (and photoperiodic) conditions being reared in the same climatic chamber. These results show that larvae of *P. napi* exhibit considerable sex-related plasticity in growth rate, a phenomenon that has been demonstrated also in *P. aegeria* (Nylin et al.[51]). This capacity for varying growth rate is surprising, in view of the belief that insect larvae as a rule are heavily preyed upon and hence might be expected to develop at some maximum rate. However, the demonstration that butterfly larvae do have this ability can only be understood in terms of high growth rates carrying some kind of cost, as has been indeed shown in birds and mammals (Clutton-Brock et al.[8]). Wiklund et al.[50] envisage a cost in terms of high growth rates being associated with higher vulnerability to disease and higher susceptability to death from starvation, but these hypotheses are as yet untested.

A negative relationship between male/female size dimorphism should also be expected among diapausing generations, but here the causal relationship could easily be the reverse of that relevant for directly developing generations. For example, among species that overwinter in the pupal stage, protandry results only because of the sex difference in postdiapause

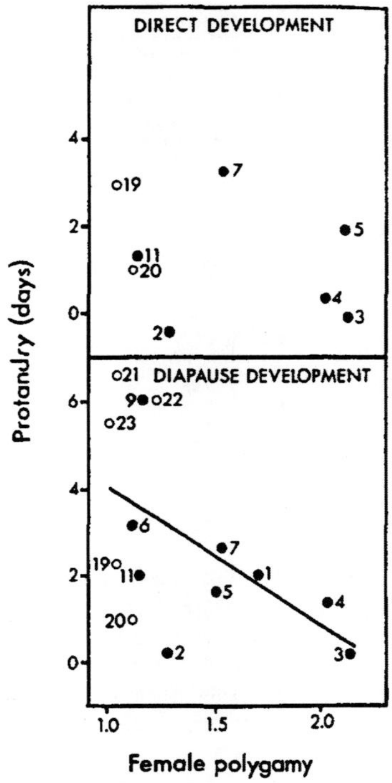

FIGURE 13. The relationship between protandry and female polygamy: For directly developing generations: y = –0.69x + 2.3; r = –0.248; n = 8; p <0.3. For diapause generations: y = –3.14x + 7.14; r = –0.521; n = 14; p <0.05. Values for pierids are indicated by (●) and values for satyrids are indicated by (○), the numbers refer to species as in Figure 11. (From Wiklund, C. and Forsberg, J. 1991. *Oikos* 60:373–381. With permission.)

development time, and it stands to reason that the morphogenesis of an adult butterfly from the undifferentiated state of an overwintering pupa takes less time for a small soma compared to a larger one. Indeed, a comparative study of nine pierid and five satyrid butterflies did demonstrate a negative relationship betwen male/female size dimorphism and protandry in diapausing generations (Figure 11). In view of the phenotypic plasticity in growth rate of larvae in the directly developing generations, the demonstrated negative relationship between male/female size dimorphism in the diapausing generations suggest that post-diapause development rate of pupae is less phenotypically plastic.

V. PROTANDRY IN RELATION TO MONANDRY/POLYANDRY

All of the first generation of protandry models use the important assumption that females mate only once. When females are monandrous and always mate soon after eclosion, this strongly compresses the time when males can find females to mate with, and so favors early male emergence because the penalty of emerging late (when no receptive females can be found) is more severe than emerging early (when males face the risk of prereproductive death before the appearance of the first females). In a general sense, selection for protandry should be relaxed under mating systems where females mate more than once. Hence, a negative relationship might be expected between protandry and the degree of female polygamy.

A comparative analysis of protandry in relation to female polygamy failed to give any significant relationship for directly developing generations of six pierid and two satyrid butterflies, whereas a weak negative relationship was found under diapause development in nine pierid and five satyrid butterfly species (Figure 13; Wiklund and Forsberg[42]). Although the benefit conferred upon males by emerging early decreases with increasing degree of females polygamy, selection for protandry may still be important insofar as mating with virgin females is more beneficial than mating with nonvirgin females. The observation that the number of eggs laid per day decreases with age in many insects suggests this (cf. Labine;[52] Karlsson and Wiklund;[53,54] Svärd and Wiklund;[55,56] Zonneveld[35]), and may hence explain the relative lack of a strong negative relationship between protandry and polyandry. Indeed, a recent theoretical model shows that the influence of female polygamy on protandry seems to be relatively small, as a result of the fact that the reproductive value of females decreases with female age (Zonneveld[35]). Moreover, Zonneveld's model shows that the influence of female polygamy also decreases with decreasing life expectancy of females, which is but another way in which the importance of mating with young females is accentuated on the part of males.

It is obvious that protandry, sexual size dimorphism, and female polygamy are interrelated in several ways, conferring that the direction of causality may be difficult to discern. However, in some cases the direction of causality seems clear. For instance, it is conceivable that the mating system maintained by females affects the optimal time of male eclosion in relation to females, whereas it seems less likely that the degree of protandry should affect female mating frequency.

However, the relationship between development time (and protandry) and size may be more variable between species. For obvious reasons, development time should have consequences for the size acquired, and if growth rate were constant, a given development time would result in a given predictable size. Hence, in species where there is a conflict between both optimal development time and size, "trade offs" have to be made. In this situation, it is conceivable that the mating system maintained may have consequences for the order of priority. For example, it is possible that males belonging to strictly monandrous species place priority on emerging early (i.e., favoring protandry at suboptimal size), whereas males belonging to polyandrous species may put priority on achieving large size and emerging at a suboptimal time. Data on the direction of "trade offs" made in the directly developing generations of the monandrous *L. sinapis* and the polyandrous *P. napi* are consistent with this view (Wiklund and Solbreck;[43] Forsberg and Wiklund;[46] Wiklund and Forsberg[42]).

VI. PROTANDRY IN RELATION TO SEX RATIO

In temperate areas, directly developing generations may be severely pressed for time, and as a consequence many alleged bivoltine species are likely to be only partially so. The establishment of what proportion of individuals that do develop directly requires some sort of rearing, because the relatively common observation that the second flight generation holds more individuals than the first does not allow the conclusion that all of the offspring from the first generation developed directly. Indeed, in Sweden, the second flight generation of *P. napi* seems to be larger than the first, but outdoor rearing of larvae showed that all cohorts of eggs laid by females in the first generation produced some individuals that entered the univoltine developmental pathway and produced diapausing pupae (Wiklund et al.[50]). The fact that many second generations that develop directly are hard pressed for time is of relevance here because it may interfere with the ability of second generation males to achieve protandry, and, as a result, may influence male propensity to develop directly.

The flexible distribution models for protandry of Iwasa et al.[32] show that not only should males start to emerge before females, but also cease to emerge before females (Bulmer;[31] Iwasa et al.;[32] Parker and Courtney[33]). More specifically, even if the female eclosion curve is smooth with one peak, the male eclosion curve should be abruptly truncated before female emergence has ceased (cf. Figure 2). In temperate insects having partial second generations, we predict that males should enter diapause development at an earlier date than should females, as a result of late emerging males being penalized in terms of fewer mating opportunities, and so postpone emergence until next season. This prediction follows logically from

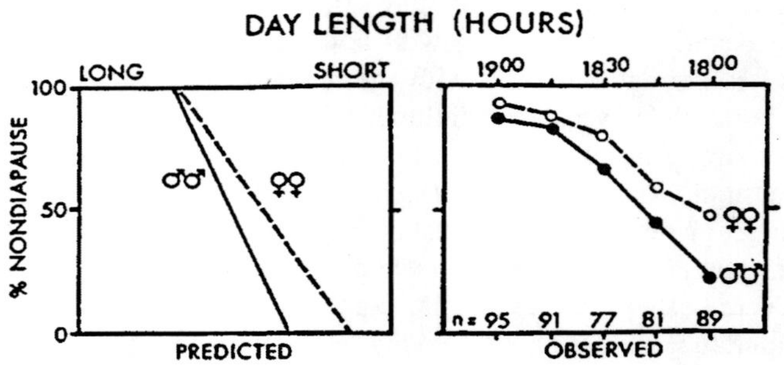

FIGURE 14. The difference between males and females in the propensity to develop directly in response to photoperiod, (left) as predicted by theory, and (right) as shown by *Pieris napi* reared at 23°C and the daylengths 18:00, 18:15, 18:30, 18:45 and 19:00 h. The number of individuals reared are shown at the bottom of the figure. (From Wiklund, C., Wickman, P.-O., and Nylin, S. 1992. *Evolution* 46:519–528. With permission.)

theories on optimal timing of eclosion of males relative to females, when applied to partially bivoltine insect populations.

The decision to diapause or to develop directly is usually mediated by response to environmental stimuli of which daylength is the most important, but also temperature can play a role (Danilevskii;[57] Tauber et al.;[58] Danks[59]). Hence, we predict that the mechanism by which males enter diapause development at an earlier date than females will be that of the male propensity to enter diapause development being shifted towards longer day lengths compared to that of females (Figure 14). More specifically, we predict that the difference in diapause propensity between the sexes should increase when approaching the point where all individuals enter diapause development, because directly developing males will be most severely penalized when few or no females develop directly. This means that a higher percentage of females should develop directly at any given photoperiod in the interval of day lengths that produce mixed broods, i.e., broods in which some of the individuals develop directly and others enter diapause development. As a result, partial second generations should be female biased, on the assumption that the primary sex ratio is unity. As a corollary, first generations of populations exhibiting partial bivoltinism should be male biased because some males of the first generation are from the previous year.

These predictions have now been supported in a number of ways. Firstly, Wiklund et al.[60] showed that a higher proportion of females of *P. napi* developed directly when reared in climatic chambers under constant photoperiod and temperature conditions that produced mixed broods. The higher propensity for females to develop directly was also demonstrated by outdoor rearings of cohorts of larvae of *P. napi* and *Pieris rapae* that were initiated throughout the season (Figure 15). Moreover, the prediction that partial second generations should be female biased was supported by laboratory rearings at constant temperature of *P. napi* (Pieridae), *P. aegeria* (Satyridae), and *Polygonia c-album* (Nymphalidae) under critical day length conditions, producing female-biased sex ratio under direct, and male-biased sex ratio under diapause development (Table 3).

VII. TIME AND LIFE HISTORY

This whole chapter has really been about time. In Stephen Vogel's[61] book *Life's Devices,* Woody Allen is quoted as having said that "Time is nature's way of keeping everything from happening all at once", but as may have been demonstrated by this chapter, time may play other roles in the lives of insects. In short-lived animals that die before they show any evidence of senescence (as evidenced by the good fit of models that assume daily survival, probability is independent of age), it is important that all of the important events of life happen, if not "all at once", still within a very short time period.

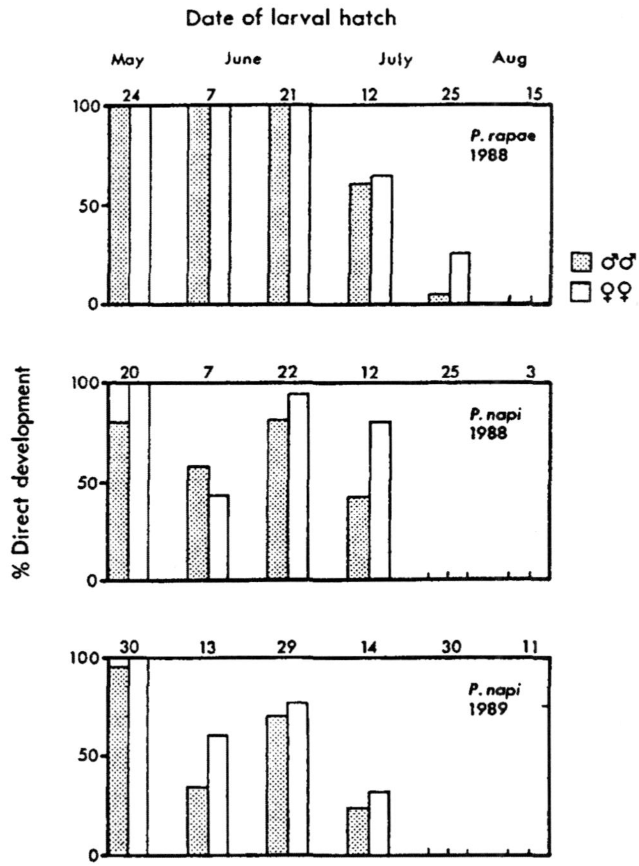

FIGURE 15. Percent directly developing males and females of *Pieris rapae* and *Pieris napi* reared outdoors throughout the season. Cohorts of approximately 40 larvae were reared, and the hatching date of each cohort is shown above the bars. (From Wiklund, C., Wickman, P.-O., and Nylin, S. 1992. *Evolution* 46:519–528. With permission.)

TABLE 3
Number of Males and Females that Enter Direct or Diapause Development under Constant Temperature and Photoperiod Conditions that Produce Mixed Broods in Three Species of Butterfly from Three Different Subfamilies, namely *Pieris napi* (Pieridae; Pierinae), *Polygonia c-album* (Nymphalidae, Nymphalinae) and *Pararge aegeria* (Nymphalidae; Satyrinae)

	Development				
	Direct		Diapause		
Species	Males	Females	Males	Females	P
P. napi	189	295	152	86	<0.001
P. c-album	141	155	91	43	<0.001
P. aegeria	64	95	56	34	<0.001

From Wiklund, C., Wickman, P.-O., and Nylin, S. 1992. A sex difference in the propensity to enter direct/diapause development: a result of selection for protandry. *Evolution* 46:519–528. With permission.

For animals with a short life expectancy, time becomes an essential currency that has a fundamental influence on the life of both males and females. For instance, it is likely to influence the opportunity for female mate choice, an issue that has been debated ever since Darwin raised it in 1871. In butterflies, the degree of female hesitance to accept courting males varies from species like *Anthocharis cardamines*, *P. napi*, and *Coenonympha tullia*, in which some 90% of courtships involving virgin males result in mating to species like *Coenonympha pamphilus*, where the corresponding figure is around 60% (Wiklund and Forsberg;[62] Forsberg and Wiklund;[63] Wickman[64,65]). Lack of strong female mate choice can be rationalized in terms of females being selected to minimize the time spent in unmated condition, conferring that females pay a time cost for every male they reject. In accordance with the concept of life expectancy constraining the female option to be selective in their choice of mates, Wickman[65] recently showed that females of the relatively short-lived satyrid *C. tullia* fly up at, and solicit courtship from, passing males, whereas females of the relatively more long-lived congeneric *C. pamphilus* do not and often resist male mating attempts and wait to mate until they have located a territorial male (which typically can be found close to specific landmarks). Hence, the option for female mate choice is inversely proportional to female longevity (cf. Janetos;[66] Thornhill and Alcock;[1] Hubbel and Johnson;[67] Real[68]).

In systems where females are short-lived and their option for mate choice is limited, male mating success is strongly dependent on their ability to locate receptive females. In species where females mate only once, the timing of male appearance becomes an important factor influencing the number of females that males have the opportunity to mate with. In accordance with the flexible distribution models for protandry, it is easy to understand that the penalty on males that emerge too late is more severe than that on males that emerge too early. This is particularly highlighted in butterflies where males locate females before they emerge, and sit (sometimes in clusters) and wait for the adult female to emerge from the pupa as in some *Heliconius* species (Gilbert[69]) and the lycaenid *Jalmenus evagoras* (Elgar and Pierce[70]).

As a result of the benefit of protandry, males will often be severely pressed in a number of different ways. For instance, since sexual selection seems to favor large relative male size in polyandrous butterflies, these males are pressed to become larger than females in a shorter time period, which can only be achieved by males increasing their growth rate which presumably carries a cost. Moreover, in species where the adults emerge during the summer but do not reproduce until after hibernation, as in *Gonepteryx rhamni*, females can use their energy stores accumulated during summer and autumn for surviving the winter. Although females are able to mate immediately upon becoming active in spring, they seem not to have devoted energy to reproduction prior to mating, as evidenced by the fact that they ususally do not start to lay eggs until about a week after mating (Wiklund and Lindfors, in preperation). Conversely, males not only have to survive the winter, but their being able to mate directly in spring requires that they have their ejaculatory ducts filled with sperm and accessory substances at that time, which means that their energy reserves during the winter have been devoted both to survival and reproduction.

In conclusion, many life history variables are related to male emergence, like variablity in larval growth rate in relation to the acquisition of large size, producing sperm early during hibernation so as to be ready to mate early in spring, and male reluctance to develop directly in partially bivoltine species. This provides evidence that protandry has far-reaching effects on insect life history and represents a truly fundamental aspect of mate acquisition and reproductive success in insects.

ACKNOWLEDGMENT

I am grateful to Peter Abrams for comments on a previous version of this paper.

REFERENCES

1. **Thornhill, R. and Alcock, J.** 1983. *The Evolution of Insect Mating Systems.* Harvard University Press, Cambridge, MA.
2. **Clutton-Brock, T.H. and Harvey, P.H.** 1977. Primate ecology and social organisation. *J. Zool.* 183:1–39.
3. **Clutton-Brock, T.H., Harvey, P.H., and Rudder, B.** 1977. Sexual dimorphism, socionomic sex ratio and body weight in primates. *Nature* (London) 269:797–800.
4. **Clutton-Brock, T.H., Guinness, F.E., and Albon, S.D.** 1985. Parental investment and sex differences in juvenile mortality in birds and mammals. *Nature* (Lond.) 313:131–133.
5. **Shine, R.** 1978. Sexual size dimorphism and male combat in snakes. *Oecologia* (Berlin) 33:269–277.
6. **Shine, R.** 1979. Sexual selection and sexual size dimorphism in the amphibia. *Copeia* 1979:297–306.
7. **Berry, J.F. and Shine, R.** 1980. Sexual size dimorphism and sexual selection in turtles. *Oecologia* (Berlin) 41:185–191.
8. **Clutton-Brock, T.H.** 1985. Selection in relation to sex. In: Bendall, D.S. (Ed.), *Evolution from Molecules to Men.* Cambridge University Press, Cambridge, pp. 457–482.
9. **Clutton-Brock, T.H. (Ed.)** 1988. *Reproductive Success: Studies of Individual Variation in Contrasting Breeding Systems.* University of Chicago Press, Chicago.
10. **Partridge, L., Ewing, A., and Chandler, A.** 1987a. Male size and mating success in *Drosophila melanogaster*: the roles of male and female behaviour. *Anim. Behav.* 35:555–562.
11. **Partridge, L., Hoffman, A., and Jones, J.S.** 1987b. Male size and mating success in *Drosophila melanogaster* and *D. pseudoobscura* under field conditions. *Anim. Behav.* 35:468–476.
12. **Persson, L.** 1985. Asymmetrical competition: are larger animals competitively superior? *Am. Nat.* 126:261–266.
13. **Bengtsson, J.** 1988. Smaller zooplankton species are not superior in exploitative competition: a comment on Persson. *Am. Nat.* 129:928–931.
14. **Darwin, C.** 1871. *The Descent of Man and Selection in Relation to Sex.* J. Murray, London.
15. **Grafen, A.** 1987. Measuring Sexual Selection: Why bother? In: Bradbury, J.W. and Andersson, M.B. (Eds.) *Sexual selection: Testing the alternatives.* John Wiley and Sons, Chichester, pp. 221–233.
16. **Richards, O.W.** 1927. Sexual selection and allied problems in the insects. *Biol. Rev.* 2:298–364.
17. **Wedell, N.** 1993. Protandry and mate assessment in the wartbiter *Decticus verrucivorus. Behav. Ecol. Sociobiol.* 31:301–308.
18. **Simmons, L.W., Llorens, T., Schinzig, M., Hosken, D., and Craig, M.** 1994. Sperm competition selects for male mate choice and protandry in the bushcricket, *Requena verticalis. Anim. Behav.* 47: 117–122.
19. **Demoll, R.** 1908. Die Bedeutung der Proterandrie bei Insekten. *Zool. Jahrb. Abt. Syst. Oekol. Geogr. Tiere* 26:621–628.
20. **Stephen, W.P., Bohart, G.E., and Torchino, P.F.** 1969. The Biology and External Morphology of Bees. Oregon State University Press, Corvallis, OR.
21. **Evans, H.E. and West-Eberhard, M.J.** 1970. *The Wasps.* University of Michigan Press, Ann Arbor.
22. **Petersen, W.** 1892. Über die Ungleichzeitigkeit in der Erscheinung der Geschlechter bei Schmetterlingen. *Zool. Jahrb. Abt. Syst. Oekol. Geogr. Tiere* 6:671–679.
23. **Petersen, B.** 1947. Die geografische Variation einiger Fennoskandischer Lepidopteren. *Zool. Bid. Uppsala* 25:329–531.
24. **Shapiro, A.M.** 1970. The role of sexual behavior in density-related dispersal of pierid butterflies. *Am. Nat.* 104:367–372.
25. **Scott, J.A.** 1986. *The butterflies of North America.* Stanford University Press, Stanford, CA.
26. **Wiklund, C. and Fagerström, T.** 1977. Why do males emerge before females? A hypothesis to explain the incidence of protandry in butterflies. *Oecologia (Berlin)* 31:153–158.
27. **Nielsen, H.T. and Nielsen, E.T.** 1953. Field observations on the habits of *Aedes taeniorhynchus. Ecology* 34:141–156.
28. **Ford, E.B.** 1945. *Butterflies.* Collins, London.
29. **Lundgren, L. and Bergström, G.** 1975. Wing scents and scent-released phases in the courtship behaviour of *Lycaeides argyrognomon. J. Chem. Ecol.* 1:399–412.
30. **Fagerström, T. and Wiklund, C.** 1982. Why do males emerge before females? Protandry as a mating strategy in male and female butterflies. *Oecologia (Berlin)* 52:164–166.
31. **Bulmer, M.G.** 1983. Models for the evolution of protandry in insects. *Theor. Popul. Biol.* 23:314–322.
32. **Iwasa, Y., Odendaal, F.J., Murphy, D.D., Ehrlich, P.R., and Launer, A.E.** 1983. Emergence patterns in male butterflies: A hypothesis and a test. *Theor. Popul. Biol.* 23:363–379.
33. **Parker, G.A. and Courtney, S.P.** 1983. Seasonal incidence: adaptive variation in the timing of life history stages. *J. Theor. Biol.* 105:147–155.

34. **Zonneveld, C. and Metz, J.A.J.** 1991. Models on butterfly protandry: virgin females are at a risk to die. *Theor. Popul. Biol.* 40:308–321.
35. **Zonneveld, C.** 1992. Polyandry and protandry in butterflies. *Bull. Math. Biol.* 54:957-976.
36. **Baughman, J.F., Murphy, D.D., and Ehrlich, P.R.** 1988. Emergence patterns in male checkerspot butterflies: testing theory in the field. *Theor. Popul. Biol.* 33:102–113.
37. **Bulmer, M.G.** 1980. *The Mathematical Theory of Quantitative Genetics.* Oxford University Press, London.
38. **Baughman, J.F.** 1991. Do protandrous males have increased mating success? The case of *Euphydryas editha. Am. Nat.* 138:536–542.
39. **Goddard, M.J.** 1967. Broods of the speckled wood (Pararge aegeria aegerides Stgr.). *The Entomologist* 100:241–254.
40. **Nylin, S., Wiklund, C., Wickman, P.-O., and Garcia-Barros, E.** 1993. Absence of trade-offs in life-history evolution:the case of sexual size dimorphism and protandry in *Pararge aegeria. Ecology.* 74: 1414–1427.
41. **Lederhouse, R.C., Finke, M.D., and Scriber, J.M.** 1982. The contributions of larval growth and pupal duration to protandry in the black swallowtail butterfly, *Papilio polyxenes. Oecologia* (Berlin) 53:296–300.
42. **Wiklund, C. and Forsberg, J.** 1991. Sexual size dimorphism in relation to female polygamy and protandry in butterflies: a comparative study of Swedish Pieridae and Satyridae. *Oikos* 60:373–381.
43. **Wiklund, C. and Solbreck, C.** 1982. Adaptive versus incidental explanations for the occurrence of protandry in a butterfly, *Leptidea sinapis. Evolution* 36:56–62.
44. **Singer, M.C.** 1982. Sexual selection for small size in male butterflies. *Am. Nat.* 119:440–443.
45. **Sibly, R.M. and Calow, P.** 1986. *Physiological Ecology of Animals: An Evolutionary Approach.* Blackwell, Oxford.
46. **Forsberg, J. and Wiklund, C.** 1988. Protandry in the green-veined white butterfly, *Pieris napi. Funct. Ecol.* 2:81–88.
47. **Rutowski, R.L.** 1980. Courtship solicitation by females of the checkered white butterfly *Pieris protodice. Behav. Ecol. Sociobiol.* 7:113–117.
48. **Oberhauser, K.** 1989. Effects of spermatophores on male and female monarch butterfly reproductive success. *Behav. Ecol. Sociobiol.* 25:237–246.
49. **Svärd, L. and Wiklund, C.** 1989. A comparative study on ejaculate mass in relation to female monandry/polyandry. *Behav. Ecol. Sociobiol.* 24: 395–402.
50. **Wiklund, C., Nylin, S., and Forsberg, J.** 1991. Sex-related variation in growth rate as a result of selection for large size and protandry in a bivoltine butterfly, *Pieris napi. Oikos* 60:241–250.
51. **Nylin, S., Wickman, P.-O., and Wiklund, C.** 1989. Seasonal plasticity in growth and development in the speckled wood butterfly, *Pararge aegeria. Biol. J. Linn. Soc.* 38:155–171.
52. **Labine, P.A.** 1966. The population biology of the butterfly Euphydryas editha. VIII. Oviposition and its relation to patterns of oviposition in other butterflies. *Evolution* 22:799–805.
53. **Karlsson, B. and Wiklund, C.** 1984. Egg size variation in satyrid butterflies: adaptive versus historical. "Bauplan", and mechanistic explanations. *Oikos* 43:391–400.
54. **Karlsson, B. and Wiklund, C.** 1985. Egg weight variation in relation to egg mortality and starvation endurance of newly hatched larvae in some satyrid butterflies. *Ecol. Entomol.* 10:205–211.
55. **Svärd, L. and Wiklund, C.** 1988. Fecundity, egg weight and longevity in relation to multiple matings in females of the monarch butterfly. *Behav. Ecol. Sociobiol.* 23:39–43.
56. **Svärd, L. and Wiklund, C.** 1991. The effect of ejaculate mass on female reproductive output in the European swallowtail butterfly, *Papilio machaon. J. Insect. Behav.*
57. **Danilevskii, A.S.** 1965. *Photoperiodism and Seasonal Development of Insects.* Oliver and Boyd, Edinburgh.
58. **Tauber, M.J., Tauber, C.A., and Masaki, S.** 1986. *Seasonal Adaptations of Insects.* Oxford University Press, Oxford.
59. **Danks, H.V.** 1987. *Insect Dormancy: An Ecological Perspective.* Biological Survey of Canada, Ottawa.
60. **Wiklund, C., Wickman, P.-O., and Nylin, S.** 1992. A sex difference in the propensity to enter direct/diapause development: a result of selection for protandry. *Evolution* 46:519–528.
61. **Vogel, S.** 1989. *Life's devices: The Physical World of Animals and Plants.* Princeton University Press, Princeton, NJ.
62. **Wiklund, C. and Forsberg, J.** 1985. Courtship and male discrimination between virgin and mated females in the orange tip butterfly, *Anthocharis cardamines. Anim. Behav.* 34:328–332.
63. **Forsberg, J. and Wiklund, C.** 1989. Mating in the afternoon: time-saving in courtship and remating by females of a polyandrous butterfly, *Pieris napi. Behav. Ecol. Sociobiol.* 25:349–356.
64. **Wickman, P.-O.** 1986. Courtship solicitation by females of the small heath butterfly, *Coenonympha pamphilus*, and their behaviour in relation to male territories before and after copulation. *Anim. Behav.* 34:153–157.
65. **Wickman, P.-O.** 1992. Mating systems of *Coenonympha* butterflies in relation to longevity. *Anim. Behav.* 44:141–148.
66. **Janetos, A.C.** 1980. Strategies of female mate choice: a theoretical analysis. *Behav. Ecol. Sociobiol.* 7:107–112.

67. **Hubbel, S.P. and Johnson, L.K.** 1987. Environmental variance in lifetime mating success, mate choice, and sexual selection. *Am. Nat.* 130:91–112.
68. **Real, L.** 1990. Search theory and mate choice. I. Models of single sex discrimination. *Am. Nat.* 136:376–404.
69. **Gilbert, L.E.** 1975. Ecological consequences of a coevolved mutualism between butterflies and plants. In: Gilbert, L.E. and Raven, P.R. (Eds.) *Coevolution of Animals and Plants.* Texas University Press, Austin, pp. 210–240.
70. **Elgar, M.A. and Pierce, N.E.** 1988. Mating success and fecundity in an ant-tended lycaenid butterfly. In: Clutton-Brock, T.H. (Ed.) *Reproductive Success.* University of Chicago Press, Chicago, pp. 59–75.

Chapter 9

SWARM-BASED MATING SYSTEMS

Athol McLachlan and Rachel Neems

And thick and fast they came at last and more and more and more

Lewis Carroll, Alice's Adventures in Wonderland

CONTENTS

I. INTRODUCTION

Swarms of flying insects, sometimes numbering many millions of individuals (Figure 1), are a common sight. They have been recorded in the following taxa at least: beetles, termites, dragonflies, bees, ants, caddis flies, stone flies, mayflies, and among the Diptera in the Trichoceridae, Empididae, Culicidae, Ceratopogonidae, Simulidae, Chaoboridae, Bibionidae, and Chironomidae. In many of these taxa, such assemblages are composed entirely of male insects, and the obvious question is what function, if any, they serve. Good evidence exists that swarms composed of males are mating assemblages. Typically, these are aggregations of males displaying to attract females which enter the swarm to acquire a mate (reviewed by Thornhill and Alcock[1]). Such displays are necessary properties of a lek,[2] a type of mating system more familiar among birds and mammals. Leks are peculiar mating systems and, despite being widespread, they do not constitute a common way for animals to acquire mates (reviewed by Krebs and Davies[3]). They are peculiar because males seem to have nothing to offer the female but their sperm. Males do not defend a resource sought by females as is the common alternative method of acquiring mates (see Parker[4]). In insect leks, there is the added peculiarity that there is little evidence of females exercising any choice of mate within the lek. Instead, the emphasis shifts to competition between males. In a spectrum of leks,[5] from

0-8493-6695-X/95/$0.00+$.50

FIGURE 1. A swarm of lake flies (Chaoboridae) on Lake George (Uganda). (Photograph by L. MacGowan).

extreme emphasis on female choice as seen in some bird species to extremes of male/male competition, swarming insects considered here occupy a position near the latter end of the spectrum.

In a recent review, Bradbury and Davies,[6] express the opinion that in insects "... the heavier emphasis on male-male conflict when compared to avian leks cries out for detailed study". Over the last 10 years, there have been several studies on insect swarms as arenas for sexual selection. Leks of the lovebug (Bibionidae), mayflies (Ephemeroptera), and the nonbiting midges (Chironomidae) have received attention. The mayflies and chironomids are of particular interest because adults do not feed, or at least are not obligate feeders (Burtt et al.[7]). Adults are, therefore, entirely concerned with reproduction, which should make them easier subjects for studies of sexual selection. Here, we consider leks in these taxa and particularly chironomid leks. Because females do not seem to exercise choice in the matter of partner, we leave aside the question of epigamic selection and consider only intrasexual selection between competing males. This decision diverts the emphasis away from the evolution of male adornment, which is the focus of work on lekking vertebrates. What has become known as the "paradox of the lek"[8] stems from this emphasis. The paradox exists because it is not clear why females should show such a strong preference for males offering little but their sperm. This is a question which does not apply to the insects considered here. The absence of elaborate ornamentation in males compared with females among swarming insects is evidence for the shift away from female choice.

We give prominence to body size as a factor determining mating success in swarms of insects. The emphasis on size has impeccable antecedents (see Darwin,[9] Wallace,[10] and Simpson[11]). There is no sign of size-assortative mating in the taxa we consider (Bibionidae,

Ephemeroptera, and Chironomidae), so sexual selection on male body size is not complicated by considerations of female size and fecundity. This is another reason for leaving females aside in this chapter. We consider swarming in a variety of insects, but focus progressively on the Diptera and finally on the nonbiting midge *Chironomus plumosus*, as a model swarming insect. We intend that much of what is said about *C. plumosus* be extrapolated, not only to other chironomid species, but also to male mating swarms of insects in general. Time will tell whether these generalizations are incautious.

II. HOW ARE SWARMS FORMED?

There is sufficient evidence about sensory physiology in the Diptera at least for a reasonable understanding of how swarms are formed and how they attract females. Swarming evidently depends upon a visual search by males for landmarks followed by a visual station keeping over the landmark. A landmark is simply any visual contrast in the environment. A bush, tree, track, or pond serve equally well. The elaborate antennae of the male, seen in many families of Diptera, are sound detectors that respond to the wing beat sounds of the female and come into play only after the female has entered a swarm. There is neither evidence that the male antenna responds to the higher frequency sounds made by flying males to aid the initial formation of a swarm, nor is there any evidence that the less elaborate antenna of the female is capable of detecting sound at all. She seems to find the swarm entirely on visual cues and is captured there by a male using a sound cue. The different mating behaviors of the sexes are clearly reflected in sexual dimorphism of their sensory equipment (Figure 2). The observations reported here have been reviewed by Downes,[12] but they have a much earlier origin. In 1901, Sir Hiram Maxim, better known for his invention of the machinegun, conducted an elegant series of experiments published in the *Times (London)* newspaper. He clearly demonstrated that the female flight sound was attractive to the male mosquito (reported by Roth et. al.[13]). If the male antenna is necessary to detect the female, it follows that the females lacking such equipment may not be able to detect an individual male at all. This is one reason to suppose that female choice is not a prominent aspect of mating in swarming Diptera.

III. WHY DO INSECTS SWARM?

In this section, we consider questions about how these swarm-based leks have evolved. Such questions hinge on understanding what the individual male stands to gain in terms of mating success, by entering a lek as compared to remaining on his own. There are two main families of hypotheses to explain why it is advantageous for males to lek. These hypotheses center around the "hotspot" and the "hotshot" models (reviewed by Krebs and Davies[3]). "Hotspots" are places where males aggregate because females are likely to be common there. Such places include female emergence sites, as seen in the lovebug, or where female territories overlap as in many bird species. The "hotshot" model concerns the case, again observed among birds, where males aggregate around an especially successful competitor and improve their chances of mating by association with him. Chironomid swarms, and perhaps swarming Ephemeroptera as well (and possibly many others), present a difficulty because, at first sight, neither of these models seems to be applicable. We propose the following testable hypothesis. Females are unable to detect a single male, but are able to see a dancing swarm of males. This is basically the old idea of "stimulus pooling," which Bradbury[2] has criticized on the grounds that joining an aggregation is unlikely to increase the individual males chances of mating. However, we propose that there is a critical minimum swarm size below which the males chances of attracting a mate are zero. There may also be a critical maximum swarm size above which the costs of intrasexual competition become excessive. If we are correct, the idea of stimulus pooling may be applicable to insect swarms. The test requires the determination of

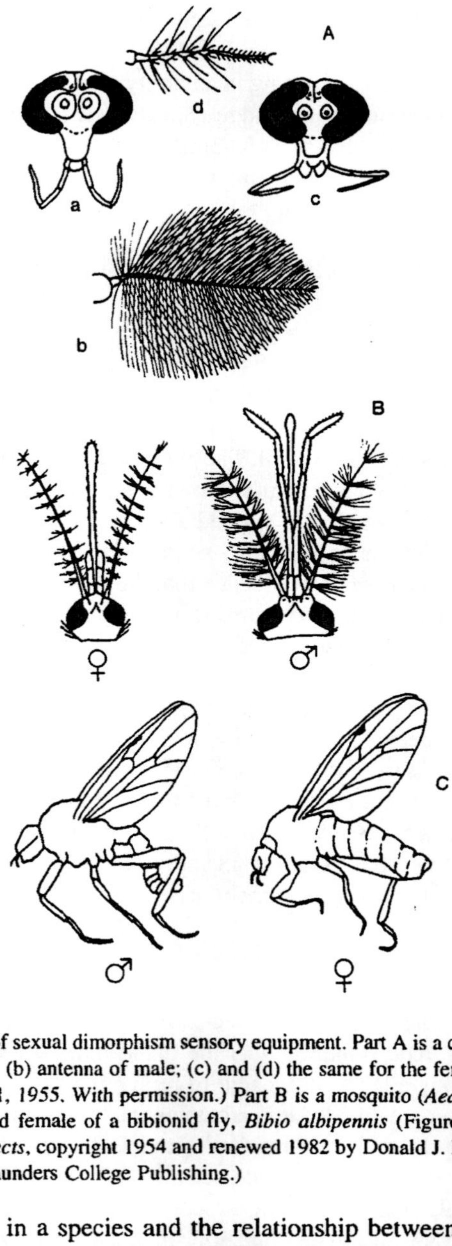

FIGURE 2. Some examples of sexual dimorphism sensory equipment. Part A is a chironomid, *Chironomus bellus* (a) front view of the male head; (b) antenna of male; (c) and (d) the same for the female (Modified from Freeman, P. *Bull. Br. Mus. Nat. Hist.*, 4, 1, 1955. With permission.) Part B is a mosquito (*Aedes*) showing head of male and female. Part C depicts male and female of a bibionid fly, *Bibio albipennis* (Figures B and C modified from *An Introduction to the Study of Insects*, copyright 1954 and renewed 1982 by Donald J. Borror and Dwight M. Delong, reproduced by permission of Saunders College Publishing.)

the range of swarm sizes in a species and the relationship between swarm size and mating success. Some of these matters are taken up in the following section.

There is a second facet to the evolution of swarm-based mating in insects. This concerns the landmark over which swarms form. Landmarks seem central to the formation of such swarms. Indeed, Downes[12] describes experiments on mosquitos where swarm size was manipulated by altering the number of landmarks available. It is widely observed that a swarm will disappear if the landmark is hidden by the observer. So, an understanding of the evolution of such swarms requires information of the origin and use of landmarks. It is intuitively reasonable to suppose that the landmark has evolved from the pond as a lekking site. This idea is feasible both because females emerge from the pond and because the pond represents a resource to the females as an oviposition site. Furthermore, males of most species emerge before females (protandry), so swarming over the home pond to await the arrival of mates makes sense (for review of protandry, see Wiklund and Fagerstom[14]). The landmark, and not the female, thus, becomes the focus of male lekking. The landmark can be seen as symbolic of an ancestral resource. This is the view taken by Thornhill and Alcock.[1] Downes[12] has coined the phrase "landmark swarming" for the type of insect lek we are interested in. We suggest that the term be retained but broadened to include the idea of the lek. What Bradbury[5] calls arbitrary lekking sites may be landmark swarming as we define it here. He suggests that female choice of such sites may have evolved in precisely the same way as for the preference of other male traits, that is, through a closed feedback loop between female choice and male preference. This argument seems not entirely satisfactory because females have become the searchers in lekking species. This point cannot easily be ignored because it raises the difficulty that, in cases of anisogamy, it ought to be the male that has surplus resources to put to searching behaviour (reviewed by Parker[4]). So, there are a number of unresolved problems in understanding the evolution of landmark swarming and of lekking behavior in general.

IV. BODY SIZE AND MATING SUCCESS IN SWARMING INSECTS

The males of most species of animal spend their adult lives fighting between themselves for mates. Since large size carries advantages in such contests, size lies at the center of studies on intrasexual selection in males. In this section, we take up the question of body size in swarming insects and show how new evidence puts a different light on topics such as the evolution of landmark swarming, the role of landmarks and of social interactions in swarm formation. Much of the evidence presented comes from studies of swarming chironomids. Most species produce a cohesive, dancing swarm very amenable to investigation.

A. BODY SIZE

The mating success and fitness of males of all swarming insects thus far studied is found to be strongly influenced by size. In some cases, as in the lovebug *Plecia nearctica*[15] and the mayfly *Epeorus longimanus*,[16] it is the larger males that acquire a disproportionate number of matings. In others, for example among many species of chironomid,[17] the smaller than average male is most successful (Figure 3A and B). Size selective pressures on the male can lead to the sexual size dimorphism so widespread among animals (for review see Adams and Greenwood[18]). Specific differences in what constitutes the successful male are presumably due to differences in the method adopted by males to capture females. Details of mate capture and male/male competition appear to be unknown in the lovebugs and mayflies. For the chironomids, we have some idea of what happens to account for the success of small males. It is well known that small flying animals, like small aircraft, have a better turning moment and greater acceleration than larger ones. An increased aerobatic ability in smaller male insects may therefore help in capturing mates on the wing.[19] A direct test of this idea would be the measurement of the turning moment and acceleration of males in a swarm. Okubo et al.[20] achieved direct measurement using cinefilm of swarms of the cecidomyiid midge *Anarete pritchardii* and found that some individuals had far greater acceleration than others. Unfortunately, they did not relate this finding to body size. An indirect way of relating together size and mating success is to weigh the wing muscles and regress this measure against wing length as a correlate of size. The expected allometric relationship for weight against length (see Calder[21]) is

$$M = ax^3 \tag{1}$$

where

M = mass
x = length
a = constant

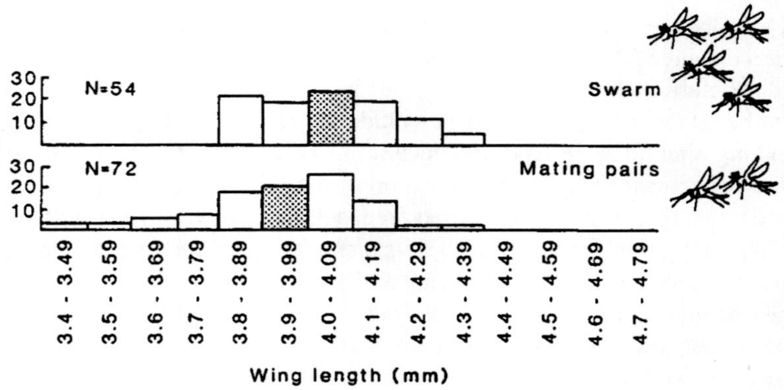

FIGURE 3A.

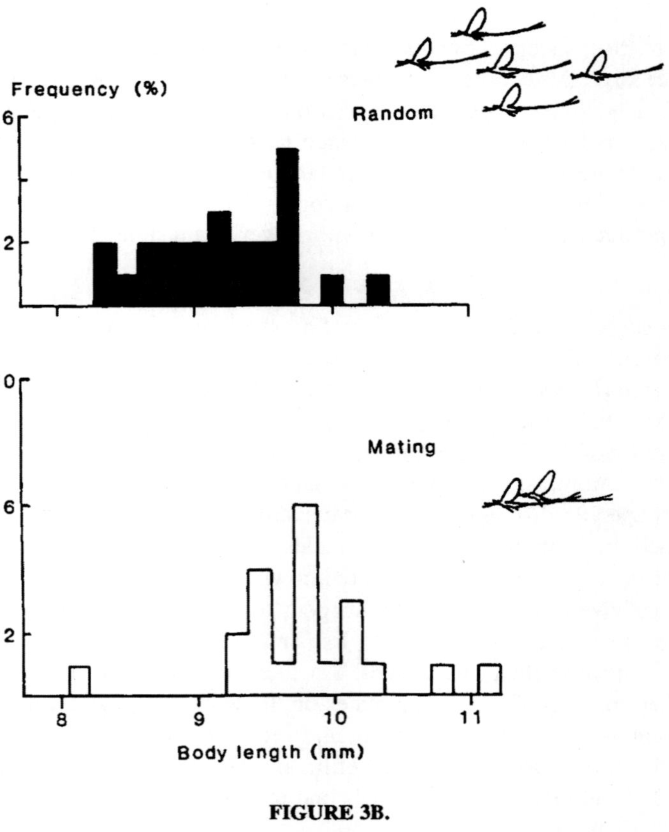

FIGURE 3B.

FIGURES 3A and 3B. Size distribution of males in the unmated male swarm compared with males caught paired with females in (A) the midge *Chironomus plumosus* (Modified from Neems, R.M., McLachlan, A.J., and Chambers, R. *Anim. Behav.* 40, 648, 1990. With permission.) and (B) the mayfly *Epeorus longimanus* (Modified from Flecker, A.S., Allan, J.D., and McClintock, N.L. *Holarctic Ecol.* 11, 280, 1988. With permission.)

This is the reasoning adopted by Convey[22] and by Marden[23] for dragonflies. The relationship between wing length and flight muscle mass in *C. plumosus* is unexpected in that it is a straight line (y = 0.23x – 0.52); that is

$$M = ax \quad (2)$$

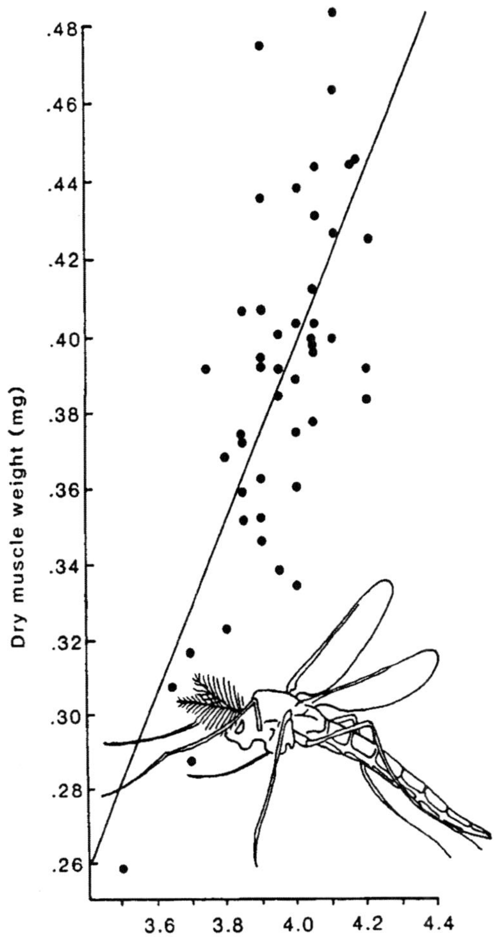

FIGURE 4. The relationship between wing length and dry weight of flight muscles in male *Chironomus plumosus*. Each point represents a single male, with n being the total number of flies measured. (Modified from Neems, R.M., Ph.D. thesis, University of Newcastle upon Tyne, U.K., 1990.)

and not $M = ax^3$ (Figure 4). This means that large male *C. plumosus* do not have the expected increase in muscle mass predicted from the cubic formula. The reciprocal conclusion is that small males are invested with a disproportionate amount of flight muscle. Hence, they may gain in aerobatic performance, not only through being smaller per se, but also in having a proportionately larger flight machinery.

In other words, in cases of size specific mating success it is not the most common size that does best. This is a puzzle, because selection ought to result in males that are most successful also being the most common. Among mammals, the explanation may be that few males ever reach large size, so the advantages to possessors of these genes are probably seldom realized (see Parker[24]). Among damsel flies, small males have a better short-term mating success but a short life span so their lifetime reproductive success (LRS) is reduced in favour of larger competitors.[25] In the case of short-lived insects, longevity is not such an important component of fitness and in chironomids, the LRS of small males is better than that of larger ones. We use LRS as an approximation to fitness (W) so, for chironomids small males are fitter than large. This conclusion takes into account four components of fitness: mating success, stamina, fertility, and longevity (Neems et. al.,[26] Figure 5). The explanation as to why the most fit size of male is not also the most common may lie in a different part of the life cycle. The major

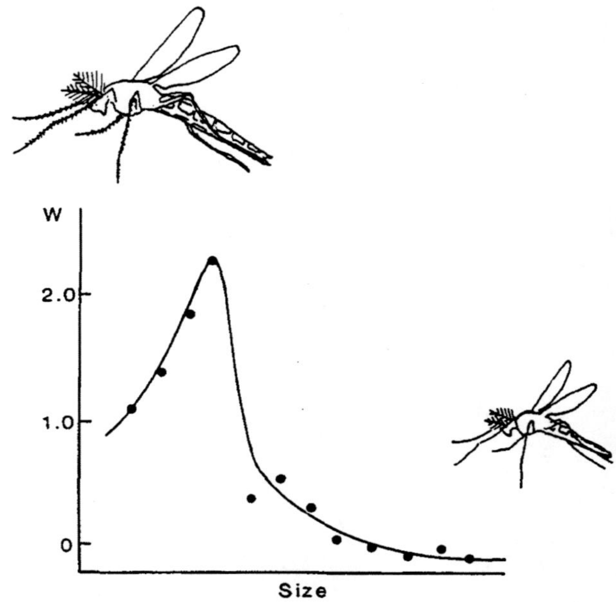

FIGURE 5. Lifetime reproductive success (W) as a function of body size in *Chironomus plumosus* (Modified from Neems, R.M., Ph.D. thesis, University of Newcastle upon Tyne, U.K., 1990.)

part of a chironomids life is spent as a larva and among larvae competing for food and space, large size may be favored. The resulting selection for large size in larvae might account for the discrepancy between the fittest size and the commonest size among adults. We have yet to investigate this possibility.

B. ALTERNATIVE MATING OPTIONS

What has been said about body size and fitness is only part of the story. Among chironomids, males are able to adopt a variety of alternative mating strategies depending upon body size and species. These alternative mating behaviors give evidence of social interactions in the swarm and de-emphasise Downes' view of the landmark as the sole organizing force in cases of landmark swarming (Downes[12]). The most extreme examples are seen in such species as *Bryophenocladius vernalis* and *Procladius crassinervis*. As in most chironomid species, mating involves females entering a swarm singly to emerge, almost immediately, paired with a male. Pairs fall out of the swarm and can be captured when they alight on the ground. In these species, the smaller males in the population are poor at sustaining flight and spend much of their time on the ground near a swarm made up of larger males. Some of the females attracted to the swarm evidently settle on the ground because these small males are not found in samples of the swarm but appear in samples of male/female pairs caught on the ground (Figure 6).

These are small species. In bigger ones, smaller males do not lose their ability to fly, indeed they may be more aerobatic than larger individuals. An example of alternative mating strategies is seen in *Chironomus plumosus,* where males assort themselves into different swarms depending upon body size. Small males can form satellite swarms around a swarm composed of larger males keeping station over a landmark.[27] Males in these satellite swarms appear to intercept females on their way to the main swarm. As in the case of the small males that do not swarm, the existence of satellite swarms opens the possibility of phenotype-determined alternative mating strategies with small individuals avoiding competition without jeopardizing their chances of mating. We must now ask why any of the males undertake the energetically costly flight at all. Why do they all not alight on the ground and await the arrival of a female as do small males of *B. vernalis* and *P. crassinervis*? The answer is not difficult

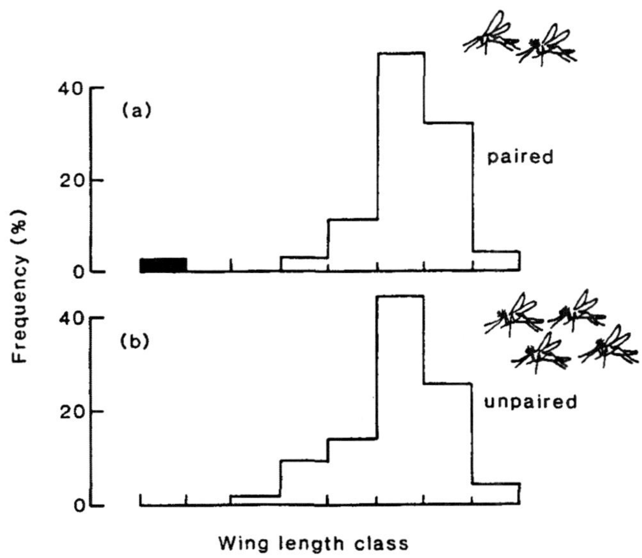

FIGURE 6. The distribition of wing size among males of the midge *Bryophenocladius vernalis* (a) in mating pairs and (b) in the male swarm. (Modified from McLachlan, A.J., *Ecol. Entomol.* 11, 237, 1986. With permission.)

to guess. If too many males join satellite swarms or alight on the ground, there is no signal to attract females. An Evolutionarily Stable Strategy (ESS) occurs where the payoff to each option is equal. In other words, we are seeing a frequency-dependent alternative option strategy.[28]

We have said enough to show that body size is at the center of mating performance in lekking male insects. Since it seems so important, it is desirable to quantify the intensity of sexual selection operating on body size. This has been attempted for dragonflies[29] and more recently for *Drosophila*.[30] We have attempted to measure selection intensity in *C. plumosus* using the same methods. Calculations depend upon the assumption that variance in mating success for males of different sizes is directly related to the intensity of selection. In principle, how much mated males differ in size from those in the male population as a whole is calculated. An example of the procedure for *C. plumosus* is shown in Table 1. The outcome (Table 2) shows that sexual selection does appear to significantly affect body size in this species.

C. THE IMPORTANCE OF SWARM SIZE

We have already shown that an individual's body size is important in determining his success in the swarm and, as with any other group of animals, what everybody else is doing has consequences for the individual as well. We now turn to the question of size of group, or the swarm in this case, on an individual's reproductive success. All the male insects of a particular species in an area do not normally aggregate into a single large swarm, but instead form many discrete aggregations of different size. In an attempt to explain why a range of swarms exist, we model the costs and benefits to the individual of joining swarms of various sizes. On the one hand, the larger the swarm the better the chances of avoiding detection by a predator. This is because, as in Hamilton's "selfish herd", an individual can more effectively hide in a crowd.[31] On the other hand, because of low levels of intrasexual competition, the smaller the swarm the better a males' chances of acquiring a mate. There ought, therefore, to be a balance between the payoff to small vs. large swarm size. We proceed with caution because swarm size may not be stable in the wild. One reason for this, suggested by Sibly,[32] is that group size may become larger than optimal due to excess individuals attempting to join a successful group.

TABLE 1
An Example of the Procedure used for Estimating the Intensity of Sexual Selection of Male Body Size in *Chironomus plumosus*

Mean male wing length (mm)	Paired status	Class frequency
4.2711	1	0.03
4.3220	0	0.97

Note: z is mean size; P, variance around z; W, mean fitness; s, selection differential; i, selection intensity in standard deviation units. Hence, z = 4.3205. P = 0.0148, W = 0.03 and z_1 = 4.2711; s_1 is z_1, −z and i are s_1/P.

From Neems, R.M., Ph.D. thesis, University of Newcastle upon Tyne, U.K., 1990.

TABLE 2
A Summary of the Parameters Governing the Intensity of Sexual Selection. z is the Mean Size, P is the Variance Around z; N, Total Sample Size; z_1, Mean Size After Selection; and i, Selection Intensity

Sample	z	P	N	z_1	i
1	4.3205	0.0148	95	4.2711	−0.4081
2	4.3011	0.0175	68	4.1577	−1.0840
3	4.3044	0.0144	122	4.3394	−0.2917
4	4.0072	0.0187	126	3.9090	−0.7181

To determine whether an optimal swarm size can be demonstrated in the case of chironomids, we gathered data on the rate of predation by *Emphis*.[33] These flies are active predators of the *C. plumosus* swarms we studied in Northumberland. We determined predation rates by counting the number of *Emphis* males leaving the swarm with a chironomid over a 40 minute period. These records were used to calculate a figure for the risk of predation per individual chironomid male (P_{pred}), obtained by dividing the number of attacks by the number of males in a swarm. At the same time, mating pairs leaving the swarm were counted to obtain a figure for the probability of mating per male (P_{mate}). Again, the number of pairs was divided by the number of males in the swarm to generate a figure for the probability of mating.

Combining these data, we arrive at a predicted, combined fitness payoff (W) to individual males to joining swarms of different sizes (Figure 7 and Box 1). The smallest swarms give the highest pay-off in fitness. A swarm size of about 100–500 males is optimal while those of around 4000 individuals give the minimum pay-off. The smallest swarms are not the commonest swarm size observed in the wild which is about ten times larger (1000–2000 males).

This outcome seems to justify Sibly's concern, but thus far we have left considerations of body size aside. As there is a strong covariance between body size and swarm size (Figure 8), this variable must be incorporated into the model. As it stands, the optimal swarm size is calculated from the combined payoff to the average male. We know that males differ widely in competitive ability, and the size of the swarm that is optimal for a small male may not be optimal for a large male. In this way, a range of swarm sizes should be expected rather than just one optimal swarm size (Figure 9). We may, in fact, be dealing with an Ideal Free

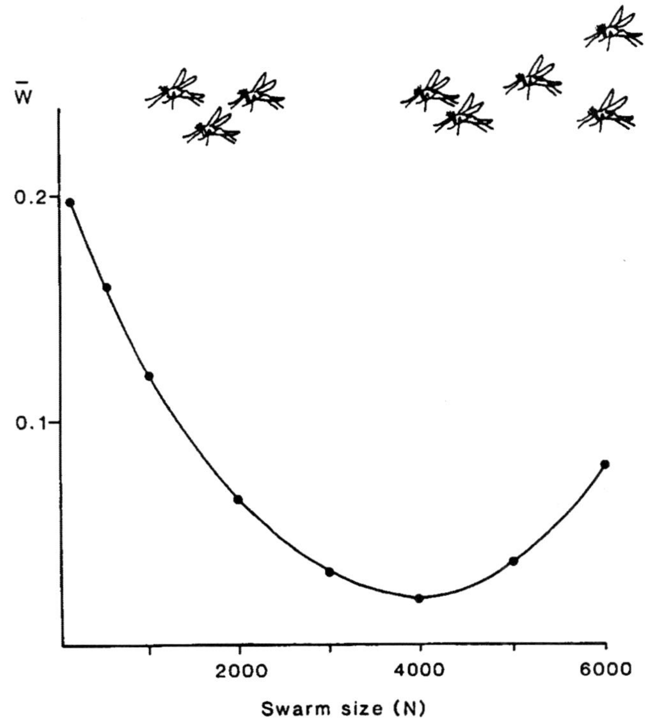

FIGURE 7. The predicted combined payoff (W) to individual males in joining swarms of different sizes.

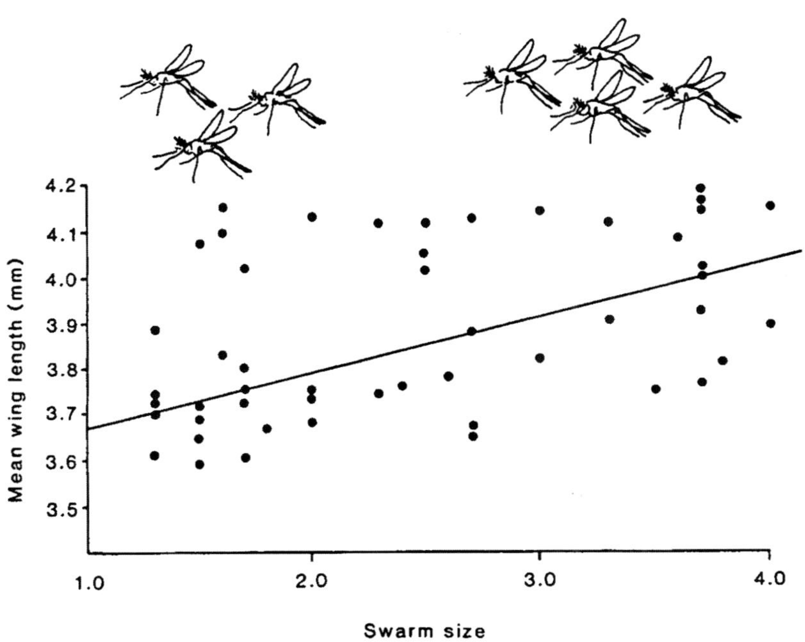

FIGURE 8. The relationship between log swarm size and body size in male *Chironomus plumosus*. Each point represents the average body size in a different swarm. (Modified from Neems, R.M., Ph.D. thesis, University of Newcastle upon Tyne, U.K., 1990.)

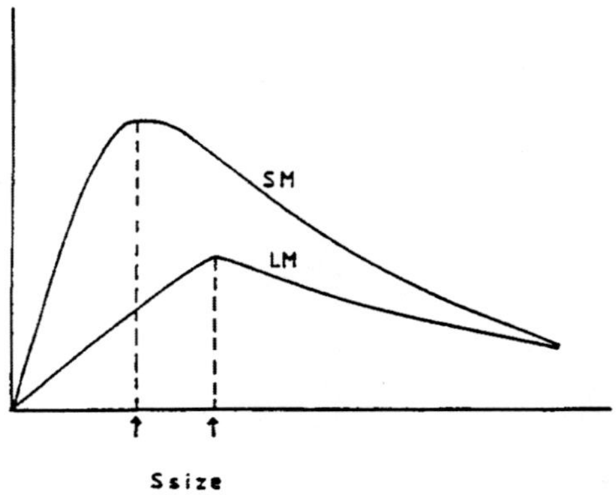

FIGURE 9. The payoff (W) to large males (LM) and small males (SM) in joining swarms of various sizes . The optimal size of swarm for each size of male is indicated by an arrow.

Distribution of unequal competitors,[34] where the best competitors (small males) occupy the best patches (small swarms).

For variance in male body size to be evolutionarily stable, the payoff to males of all sizes must be equal. We started this chapter by showing that the payoff is not equal to all males but that small size is favored. This is an unresolved dilemma. Recall that we believe the discrepancy in fitness between the commonest and the smallest males may be corrected in the larval stage so that over the life cycle as a whole, the two strategies can persist despite an unequal payoff in the adult part of the life cycle. This is not yet tested, but something more can be said here about the immature stages of the life cycle.

Box 1

The number of females attracted to a swarm increases with swarm size ($r = 0.93$, $P < 0.001$). However, since the attraction rate does not increase in proportion to swarm size, the individual males' probability of mating (P_{mate}) decreases with swarm size (but increases again slightly above a size of 4000). The curvilinear nature of this relationship is described by the regression equation

$$P_{mate} = 0.2146 - 0.0001W + 1.3 \times 10^{-8}N^2 \qquad (3)$$

Predatory attack rate also increased with swarm size, but again the probability of predation to the individual(P_{pred}) declines significantly as swarm size increases due to a dilution effect. This relationship is described by the linear equation

$$P_{pred} = 0.0263 - 0.3 \times 10^{-5}N \qquad (4)$$

Equations (3) and (4) can be combined to provide a measure of "mating success" per evening (W) in a common currency that takes into account the fact that the individual might be captured by a predator while in the swarm

$$W = P_{surv} \times P_{mate} \qquad (5)$$

where

$$P_{surv} = 1 - P_{pred} \qquad (6)$$

As Figure 7 in the main text shows, W declines as swarm size increases up to a value of about 4000 males but increases thereafter. In spite of the greater safety of larger swarms, mating success is greatest in the smallest swarms. These relationships are summarized in the figure below.

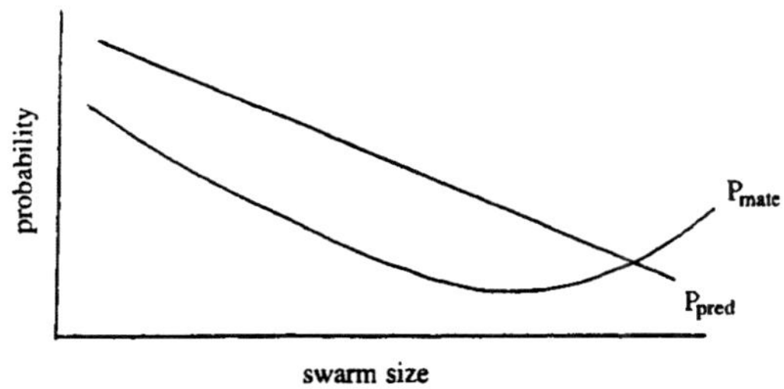

BOX 1 FIGURE. The calculation of optimal swarm size in *C. plumosus*.

D. THE ROLE OF THE LARVAL HABITAT IN DETERMINING BODY SIZE IN THE ADULT

Our conjecture is that under conditions of intense crowding, competition for resources is asymmetric, with larger larvae doing better than smaller ones. If size has a heritable component, which it almost certainly does, and if large larvae always give rise to large adults and small larvae to small adults, then size determined in the larval habitat may have an important bearing on size in the adult. It is often observed in insects that smaller adults emerge from larvae reared under crowded conditions.[35] The immature stages of many insects experience very variable levels of crowding because the habitat is a temporary one and crowding is correlated with progress of the habitat towards extinction. Examples are the conditions experienced by larvae of *Drosophila* in a rotting apple, *Calliphora* in carrion, and larvae of the genus *Chironomus* in rainpools. In some *Chironomus* species, as in some toads (see Duellman and Trueb[36] and McLachlan[37] for reviews), the story is more complex because giant adults as well as dwarfs are produced when crowding is extreme. We do not yet have data on the relationship between size of individual larvae and adults in the chironomids, nor do we have data on differential survival by size among larvae under different levels of competition. However, it is an obvious thing to investigate and seems accessible to study.

V. SUMMARY

The majority of insect swarms are aggregations of flying male animals, maintaining station over a landmark of no value to the female. We consider such swarms to be leks, but they differ from the classical lek seen in vertebrates in at least two important respects. First, there is little evidence of female choice and, as a consequence, the focus of interest moves from the evolution of male ornamentation to questions of competition between males. Second, there are peculiarities about the evolution of swarming in insects which do not easily fit any of the existing models depending upon the distribution of females. Swarming may have evolved from a system originally dependent on female distribution. The occurrence of the lek depends upon male choice of a suitable swarming site or landmark where males aggregate to create a signal to attract dispersed females. The landmark may represent the home pool or stream as a site where emerging females were likely to be encountered. We propose that the term "landmark swarming" be retained for such insect leks.

For our test species *C. plumosus*, we have the data to develop models for the prediction of optimal swarm size. This turns out to depend upon a compromise between costs of predation and benefits of mate acquisition. We also consider the mechanisms of mate acquisition within the swarm. Both swarm size and mating success hinge on body size in the male sex. The larva is the stage of the life cycle where the environmental component to size in the adult is entirely determined. We identify selection on size in the larval stage, the evolution of the insect lek, and the mechanisms of male/male competition as areas for further work.

ACKNOWLEDGMENTS

We owe many thanks to John Lazarus for help in developing the mathematical models considered here and for writing the computer programs used in the modeling section. We thank Nicholas Davies for his helpful comments on earlier versions of this chapter.

REFERENCES

1. **Thornhill, R. and Alcock, J.**, *The Evolution of Insect Mating Systems*, Harvard University Press, London, 1983.
2. **Bradbury, J.W.**, The evolution of leks, in *Natural Selection and Social Behaviour*, Alexander, R.D. and Tinkle, D.W., Eds., Blackwell Scientific Publications, Oxford, 1981, 12.
3. **Krebs, J.R. and Davies, N.B.**, *Behavioural Ecology, An evolutionary approach*, Third ed., Blackwell Scientific Publications, Oxford, 1991.
4. **Parker, G.A.**, Evolution of competitive mate searching, *Annu. Rev. Entomol.*, 23, 173, 1978.
5. **Bradbury, J.W.** Contrasts between insects and vertebrates in the evolution of male display, female choice, and lek mating, in *Experimental Behavioural Evolution, Ecology and Sociobiology*, Holldobler, G. and Lindauer, M., Eds., Fortschr. Zool. Gustav Fischer Verlag, Stuttgart, 31, 273, 1985.
6. **Bradbury, J.W. and Davies, N.B.**, Relative roles of intra- and intersexual selection, in *Sexual Selection: Testing the Alternatives*, Bradbury, J.W. and Andersson, M.B., Eds., Wiley, New York, 1987, 143.
7. **Burtt, E.T., Perry, R.J., and Mclachlan, A.J.**, Feeding and sexual dimorphism in adult midges, *Holarctic Ecol.*, 9, 27, 1986.
8. **Kirkpatrick, M. and Ryan, M.J.**, The evolution of mating preferences and the paradox of the lek. *Nature*, 350, 33, 1991.
9. **Darwin, C.**, *The Descent of Man and Selection in Relation to Sex*, J. Murray, London, 1871.
10. **Wallace, A.R.**, *Darwinism: An Exposition of the Theory of Natural Selection with Some of its Applications*, Macmillan, London, 1889.
11. **Simpson, G.G.**, *The Major Features of Evolution*, Columbia University Press, New York, 1953.
12. **Downes, J.A.**, The swarming and mating flight of Diptera, *Annu. Rev. Entomol.*, 14, 271, 1969.
13. **Roth, M., Roth, L.M., and Eisner, T.E.**, The allure of the female mosquito, *Nat. Hist.*, 75, 27, 1966.
14. **Wiklund, C., and Fagerstom, T.**, Why do males emerge before females? A hypothesis to explain the incidence of protandry in butterflies, *Oecologia*, 31, 153, 1977.
15. **Thornhill, R.**, Sexual selection within mating swarms of the lovebug *Plecia nearctica* (Diptera: Bibionidae), *Anim. Behav.*, 28, 405, 1980.
16. **Flecker, A.S., Allan, J.D., and McClintock, N.L.**, Male body size and mating success in swarms of the mayfly *Epeorus longimanus,Holarctic Ecol.*, 11, 280, 1988.
17. **Mclachlan, A.J. and Allen, D.F.**, Male mating success in Diptera: advantages of small size, *Oikos*, 48, 11, 1987.
18. **Adams, J. and Greenwood, P.J.**, Why are males bigger than females in pre-copula pairs of *Gammarus pulex?*, *Behav. Ecol. Sociobiol.*, 13, 239, 1983.
19. **Mclachlan, A.J.**, Survival of the smallest: advantages and costs of small size in flying animals, *Ecol. Entomol.*, 11, 237, 1986.
20. **Okubo, J., Chiang, J.C., and Ebbesmeyer, C.C.**, Acceleration field of individual midges, *Anarete pritchardi* (Diptera: Cecidomyiidae), within a swarm, *Can. Entomol.*, 109, 149, 1977.
21. **Calder, W.A.**, *Size, Function and Life History*, Harvard University Press, London, 1984.

22. **Convey, P.**, Influences on the choice between territorial and satellite behaviour in male *Libellula quadrimaculata* Linn. (Odonata: Libellulidae), *Behaviour, 109, 125, 1989.*
23. **Marden, J.H.**, Bodybuilding dragon flies; costs and benefits of maximizing flight muscle, *Physiol. Zool.*, 62, 505, 1991.
24. **Parker, G.A.**, Arms race in evolution: an ESS to the opponent- independent costs game, *J. Theor. Biol.*, 101, 619, 1983.
25. **Banks, M.J. and Thompson, D.J.**, Lifetime mating success in the damselfly *Coenagrion puella*, *Anim. Behav.*, 22, 1175, 1985.
26. **Neems, R.M., Lazarus, J., and Mclachlan, A.J.**, Lifetime reproductive success in a swarming midge: stabilising selection for male body size. *Nature.* 1995 (submitted).
27. **Neems, R.M.**, *The Role of Body Size in the Mating Systems of Midges (Chironomidae)*, Ph.D. thesis, University of Newcastle upon Tyne, U.K., 1990.
28. **Maynard Smith, J.**, *Evolution and the Theory of Games*, Cambridge University Press, Cambridge, 1982.
29. **Arnold, S.J. and Wade, M.J.**, On the measure of natural and sexual selection: applications, *Evolution*, 38, 709, 1984.
30. **Wilkinson, G.S.**, Equilibrium analysis of sexual selection in *Drosophila melanogaster*, *Evolution*, 41, 11, 1987.
31. **Hamilton, W.D.**, Geometry for the selfish herd, *J. Theor. Biol.*, 31, 295, 1971.
32. **Sibly, R.M.**, Optimal group size is unstable, *Anim. Behav.*, 31, 947, 1983.
33. **Neems, R.M., Lazarus, J., and McLachlan, A.J.**, Swarming behavior in male chironomid midges: a cost - benefit analysis, *Behav. Ecol.*, 3, 285, 1992.
34. **Fretwell, S.D.**, *Populations in a Seasonal Environment*, Princeton University Press, Princeton, NJ.,1972.
35. **Nicholson, A.J.**, The self-adjustment of populations to change, in *Population Studies: Animal Ecology and Demography*, Warren, K.B., Ed., Cold Spring Harbor Symposium in Quantitative Biology, 22, 1958, 153.
36. **Duellman, W.E. and Trueb, L.**, *Biology of Amphibians*, McGraw-Hill, New York, 1986.
37. **McLachlan, A.J.**, Animal populations at extreme densities: size dimorphism by frequency dependent selection in ephemeral habitats,*Funct. Ecol.*, 3, 633, 1989.

Chapter 10

MALE NUPTIAL GIFTS: PHENOTYPIC CONSEQUENCES AND EVOLUTIONARY IMPLICATIONS

Carol L. Boggs

CONTENTS

0-8493-6695-X/95/$0.00+$.50

I. DESCRIPTION OF MALE NUPTIAL GIFTS

A. DEFINITION

Male insects of many species give a nuptial gift to females. Nuptial gifts, as used here, are potentially nutritious substances given to the female by the male in conjunction with mating and used by the female in her resource budget. Such gifts can be as extreme as allowing the female to eat part of the male,[1-4] as behaviorally complex as presenting a female with a prey item on which she feeds during copulation,[5-7] or as mundane as passing nutritive accessory gland fluid into the female's reproductive tract along with sperm.[8] Thus, nuptial gifts are an investment by the male in a female and/or her prezygotic offspring with potential selective consequences. Activities such as guarding offspring[9] or feeding offspring[10] are a postzygotic investment and are not included in my definition of nuptial gifts. Tallamy[11] reviews postzygotic investment in insects.

Male nuptial gifts are widespread across insect orders, including those with a range of feeding habits and in a diversity of environments. Groups and the type of donation are tabulated in several recent reviews.[6,12-15] Such tables therefore will not be duplicated here.

Nuptial gifts have ramifications throughout the biology of both males and females. Individual-level physiology and behavior, as well as population demography and genetics, may be affected in ecological and evolutionary time. Much of the work to date has focused on only a few aspects of nuptial gifts, in part because nuptial gift giving is an excellent vehicle to test sexual selection theory. However, nuptial gifts function in a much broader biological context. While I review the more well-known aspects of nuptial gift giving, I also point out areas that are currently begging for exploration.

B. PREY ITEMS AND REGURGITANTS

Males of some mecopteran, dipteran, hymenopteran, and heteropteran species present a prey item to the female for her feeding during copulation.[5-6] The male sometimes may feed on the prey item himself before finding a mate. In the most well-described cases in Mecoptera, bittacid males capture an insect prey, dangle in the vegetation, and call females by means of a pheromone; once a female approaches, the prey is presented to the female for feeding during copulation.[5,16-17] If the prey is small or has been largely eaten, females may prematurely terminate copulation.[5,17] Otherwise, the male and female may fight over the prey item once males terminate copulation,[5,17] with the male retaining the prey most of the time. Prey may then be reused by the male in another courtship. In at least one species, *Harpobittacus similis*, males with larger prey did not call as much as males with small prey, but still attracted females.[17]

Dance flies (Diptera: Empididae) exhibit a variant on this theme. Males catch prey, but the two sexes meet either at male[18] or female[7] leks. Prey items in *Empis borealis* can include conspecifics.[7] Other empidids have altered nuptial gift giving in a stereotyped way.[18-19] While male *E. borealis* simply present prey to females to feed on during mating, in more advanced species prey are wrapped in silk (*Hilara* species) or covered with a frothy balloon (*Empis* species). In yet other species, the silk or balloon surround inedible dried prey fragments; finally, in some species the male presents the female only with an empty wrapping. Females do not feed on this package, but only manipulate it during mating. By my definition, neither of these last two types of gifts qualify as nuptial gifts.

D. BODY PARTS AND HEMOLYMPH

The most extreme form of male nuptial gifts is cannibalism of the male by the female. This occurs in three insect orders,[12] including the mantids and ceratopogonine flies. In three ceratopogonine tribes, females pierce the mating male's head or thorax and completely consume him during mating.[1] It is not known if males are always eaten. Numbers of matings obtained by females and effect of the male meal on female foraging and reproduction are not known for these species. However, since cannibalism ends further male reproduction, selection pressures operating on this behavior should be quite large. Buskirk et al.[60] argue that a low expected lifetime number of matings for males (independent of cannibalism) and a significant increase in the male's reproductive success with cannibalism are the two factors selecting for sexual cannibalism.

A similar, but less drastic, form of nutrient donation occurs in some Orthoptera. Within the Haglidae, female *Cyphoderris buckelli* chew on a male's fleshy hind wings, ingesting both tissue and hemolymph.[2,61] Females begin feeding before copulation. The male's "gin trap", or pinching organ which holds the female by the venter, restricts females from leaving before copulation. Feeding is terminated by the male. Females may eat as much as one third of a male's wings at a mating and also consume a sizeable spermatophylax. Other orthopterans feed on hemolymph from wounds inflicted on males. Included in this group are *A. fasciatus* (Gryllidae) females, who open a male's tibial spur with a mandible and feed on the exudate.[4,42] Finally, some orthopteran females feed from secretions from male metanotal glands. Included here are female *Oecanthus* (Gryllidae) and some Eneopterinae and Gryllinae.[3,42]

E. SPERM

Sperm serve as a male nutrient donation to the female in insect species with haemocoelic insemination, including the bedbug families Nabidae, Cimicidae, Plokiophilidae, and Polyctenidae (Hemiptera: Cimicoidea).[62] In primitive groups, the sperm and accessory fluids are injected into the female's haemocoele and some sperm are digested. In other groups, the sperm are injected into a specialized tissue, the mesospermalege, and absorbed either by free phagocytes or by specialized cells in the mesospermalege. Hinton[62] argued that haemocoelic insemination was advantageous because females got a meal thereby, although no data exist on the effects of sperm phagocytosis on female fitness.

Some groups, such as Lepidoptera, produce both apyrene (anucleate) and eupyrene (nucleated) sperm. Apyrene sperm have been proposed to function as nutrient resources for the female or the eupyrene sperm. However, there is no evidence for this, and apyrene sperm do not contain much energy.[63]

II. OVERVIEW OF ADAPTIVE FUNCTIONS OF MALE NUPTIAL GIFTS

As background to examination of the effects of male nuptial gifts on insect physiology, ecology, fitness, and evolution, we need to understand the potential adaptive functions of the nutrient donations. Male nuptial gifts are part of male reproductive effort. This effort can have two potentially adaptive functions: investment in the female's resource budget and/or in obtaining or guarding a mate.[12,64] Both forms of effort can increase male reproductive success. Contributions of resources used by females can increase female fecundity directly through an increase in nutrients available to make eggs, or potentially indirectly through an increase in nutrients available to support female survival. Use of resources as bribes to obtain a mate or to prevent or delay female remating can also increase a male's reproductive success through increasing the number of eggs that he fertilizes. These two forms of effort, contribution to a female's resource budget and control of mating opportunities, are not mutually exclusive. Contributions that influence mating opportunities may also be used by the female to make eggs.

III. PHYSIOLOGICAL COSTS AND BENEFITS OF MALE NUPTIAL GIFTS: EFFECTS ON RESOURCE AND TIME BUDGETS

Nuptial gifts are associated with physiological costs and/or benefits, affecting the resource and time budgets. Changes in resource and time budgets in turn may affect foraging, survival, and reproductive strategies, and thence fitness. I deal with specific impacts on fecundity and survival below under demographic fitness component consequences of nuptial gifts; here, I focus on underlying physiological effects.

A. FEMALE PERSPECTIVE

The benefit gained by females from a nuptial gift will depend on the usable donation size relative to the female's total resource pool. Types of nutrients that are limiting to females should have more effect than those available in abundance.[65-66]

Male-derived nutrients can be allocated as are nutrients derived from other sources: to egg production, to maintenance, to defense, and to foraging activity. The specific allocation pattern of male-derived nutrients should depend on the type of nutrients donated by males, the female's other available reserves, and state of ovarian development.[66] If males donate compounds that can be used directly in egg production with minimal processing and mating occurs during or just prior to a peak of vitellogenic activity in the ovaries, then such compounds should be more likely to be used in egg production. For example, in *X. hamata*, females feed on male-produced urates after mating, and the timing of mating during a reproductive cycle corresponds to the period when uptake of uric acid by the developing eggs is at a peak. As noted earlier, large amounts of male-derived uric acid may be incorporated into eggs in this species.[52]

Male-derived nutrients can reduce the need for females to forage on their own for food; this occurs in *Hylobittacus apicalis*, *E. borealis*, *Heliconius charitonius* and *Heliconius cydno*.[5,7,18,66] I have suggested that reduction in foraging by females with increasing male-derived nutrients is most likely to occur when non-nutritional factors limit egg production such that females cannot increase fecundity by maintaining foraging levels.[66] Such factors may include mortality risk to the female while foraging, time restrictions for oviposition, or body size limitations on the number of oocytes that can be matured at once.

Male nuptial gifts fed on by the female, such as prey items, external spermatophores, and male body parts, will be incorporated into the female resource budget at the same rates as any other similar quality food item to be used, stored or excreted. As noted above, however, rates and targets of allocation of these items could be affected by the male if males have control over the composition of the gift. Internal spermatophores and male accessory gland compounds that are absorbed by the female may be treated differently. The donation is initially present within the female in a form of storage. Donation usage rates should depend on the balance of draw down of other forms of storage such as fat body, use of free nutrients from newly absorbed food, and use of the "stored" male accessory gland products. The dynamics and priority of use of various types of nutrients from these sources in reproduction, survival, and foraging remain to be examined. Timing and amount of reproduction, feeding sources for the adult and juvenile female, effects of storage on risk of predation, etc., may also influence allocation patterns. Such studies, in combination with information on patterns of paternity, will give us an understanding of the effect of the nuptial gift on the female's resource budget and on the reproductive success of each sex. This in turn will provide a mechanistic basis for answers to questions about the evolution of specific forms of nuptial gifts as outlined below.

We have initial information on absorption rates of internal spermatophores and on the usage rates in oogenesis of materials from both internal and external donations. Most of these data come from radiolabel experiments. Males are labeled with radioactive amino acids and mated with females; then eggs and/or females are tested for radioactivity at intervals after mating. In both an orthopteran, *D. verrucivorus*, and two lepidopterans, *D. plexippus* and *H. hecale*, the amount and timing of incorporation of labeled compounds into eggs is independent of the previous female mating history.[8,67] In all Orthoptera and Lepidoptera examined to date, label is incorporated into the next eggs laid by a female although the peak of incorporation may be delayed by several days.[8,64,66-70] Timing of the peak of incorporation varies from immediately after mating in *Colias eurytheme*,[69] to 10 days after mating in *D. verrucivorus*.[67] Elevated levels of label may be found in eggs for only 5 to 6 days as in *C. eurytheme*,[69] for 7 to 10 days as in *Dryas julia*[64] or *D. plexippus*,[8] or for up to 15 to 20 days as in *Heliconius erato*,[8] *H. charitonius*,[64,66] or *D. verrucivorus*.[67] This timing agrees with data on rates of decrease in size of internal spermatophores: spermatophores of *D. plexippus* reach a baseline size within 7 to 10 days,[71] spermatophores of *Pontia protodice*, with a biology similar to *C. eurytheme*, reach a baseline size within 5 days,[72] and spermatophores of *H. charitonius* are absorbed within about 14 days.[64] In *Acanthoscelides obtectus* (Coleoptera), radiolabel declines within 48 hours in the spermatophore, peaks in eggs laid at 36 hours, and declines by 48 hours.[73] In essentially all studies, *all* eggs laid by females after mating with a labeled male contain at least a small amount of radioactivity. This would be expected if nutrients are incorporated into eggs at all stages of development, such that eggs that are in late development stages and incorporating large amounts of nutrients are heavily labeled, whereas eggs in early stages of development, incorporating small amounts of nutrients, are lightly labeled.

The differences among species in temporal pattern of radiolabel incorporation into eggs should depend on: (1) species-specific timing of incorporation of specific nutrient types into eggs; (2) the particular compounds that are labeled; (3) pool sizes of those compounds; (4) the rate at which male compounds are absorbed through the gut or reproductive walls; (5) whether more than one day's batch of eggs is matured at once; (6) the number of eggs laid per day; and (7) the total usable size of the male investment relative to mass of individual eggs or clutches of eggs. These factors have been explored to some extent in *B. germanica*,[53] where timing of uptake of uric acid into developing eggs is known, as is timing of ingestion of male-derived urates. However, for the most part we know little about factors affecting different patterns observed among species. It would also be interesting to expand the studies on physiology of female use of male-derived nutrients to Trichoptera, as Khalifa[74] reports differences among species in the size and timing of absorption of the spermatophore.

The interaction of nuptial gifts with female time budgets is also not well understood. Mating may not interact with the foraging or reproductive time budgets in some species. For example, monarch butterflies (*D. plexippus*) can mate overnight,[75] a time when the female would not be actively engaged in foraging or searching for oviposition sites anyway. However, for many species in which females mate more than once, this lack of impact on the time budget is likely to be the exception rather than the rule. Conflicts over the time spent mating could affect the evolution and maintenance of nuptial gift-giving behavior, particularly if mating is prolonged to allow male nutrient donations. In such cases, the time cost and nutritional rewards of foraging by a female will be balanced against the costs and rewards of male nuptial gifts.

Females may pass up opportunities to use male nuptial gifts altogether. Although *Melanoplus sanguinipes* (Orthoptera: Acrididae) females incorporate protein derived from male accessory gland products into eggs,[26] the amount of material transferred into the female's spermatheca is small, about 5 μg, while the amount of protein in eggs laid between matings is large, on the order of 100 mg.[25] However, the spermatophore, which could represent a larger male investment, was observed to be rubbed off the female's ovipositor and discarded in this species.[25] *Why* females should bypass cheaply available nutrients is unknown. The answer could depend on the composition of the spermatophore, or the conditions under which observations were made.

B. MALE PERSPECTIVE

Nuptial gifts affect the male's resource and time budgets. Resources spent on large ejaculates or capturing prey are not available for other uses; time spent mating, replenishing accessory glands, or capturing nuptial prey is not available for other uses.

Impacts of nuptial gifts on the energy and resource budget may in turn affect ability to attract mates. *Cyphoderris strepitans* males produce an external spermatophore, eaten by the female, and allow females to eat part of their hind wings and hemolymph during mating.[76] Virgin males of this species call for significantly longer than recently mated males, suggesting that energy reserves needed for calling have been reduced by mating and/or that intensity of calling depends on distension of the male's accessory glands. Similarly, *Requena verticalis* males on low-protein diets maintain spermatophore mass, but reduce calling, probably as a result of energy limitation.[77]

Nuptial gifts may result in matings lost because of a refractory period while accessory glands are replenished or fresh prey are obtained. The form of this cost differs among insect orders. As Gwynne[78] points out, Lepidoptera will mate before they have replenished the accessory glands; matings simply take longer, presumably to include time for glands to be refilled and an ejaculate formed. Even so, smaller spermatophores are formed if remating occurs rapidly,[30,64,78-79] which could affect male reproductive success. Further, in at least one lepidopteran, recently mated males' courtship persistence time was an order of magnitude lower than males that had not mated recently,[80] which will also affect male success. Orthoptera, Megaloptera, and Coleoptera, however, have a male refractory period during which time males will not attempt mating and accessory glands are replenished. The length of this refractory period depends on the relative size of the spermatophore: in two species without spermatophylaxes, *Gryllus bimaculatus*[37] and *Gryllus veletis*,[81] male refractory periods are 1 hour and 30 minutes, respectively; in a species with a relatively small spermatophylax, *Gryllodes sigillatus*,[81-82] male refractory period is 3 hours; and in a species with a relatively large spermatophylax, *R. verticalis*,[78] the male refractory period is about 3 days. Megaloptera exhibit a similar pattern. *Protohermes immaculatus* males have smaller spermatophores relative to male body weight than do *P. grandis* males; *P. immaculatus* males can remate daily, whereas *P. grandis* males have a refractory period of about 2 days.[38] Differences among species in the length of the refractory period may also result from differences in the normal diet. Within Meloidae (Coleoptera), seed feeding species can mate every 4 hours, whereas species feeding on relatively protein-poor flower petals have a refractory period of 1 to 2 days.[55]

Nuptial gifts may also affect survival through impacts on the energy budget. This is discussed in detail below.

Nuptial gift size should affect the relative costs to the male's resource and time budgets. Size of the nutrient donation is expected to vary within species with the male resource budget, female ability to receive and process nutrients, and sperm precedence patterns. First, ejaculate or spermatophore size is positively correlated with male size (and hence, presumably nutrient reserves) within a species in many groups, including *Lytta magister* and *Tegrodera alogra* (Coleoptera: Meloidae),[55] *H. charitonius* and *D. julia* (Lepidoptera: Nymphalidae),[64] *C. eurytheme* (Lepidoptera: Pieridae),[30] *Papilio machaon* (Lepidoptera: Papilionidae),[83] *Plodia interpuctella* (Lepidoptera: Pyralidae),[68] *Ostrinia nubilalis* (Lepidoptera: Pyralidae),[84] *G. sigillatus* (Orthoptera: Gryllidae),[85-86] *R. verticalis* (Orthoptera: Tettigonidae),[87] *Conocephalus nigropleurum* (Orthoptera: Tettigonidae),[88] and *D. verrucivorus* (Orthoptera: Tettigonidae).[67,89] In *G. sigillatus* and *D. verrucivorus*, the correlation occurs because the male's investment in the spermatophylax or whole spermatophore, respectively, is a constant proportion of his body weight;[85,89] male investment as a proportion of body weight actually declines with male size in *R. verticalis*.[87] The data suggest that absolute spermatophore size is more constrained in *R. verticalis* than in the other two species. These differences could be related to differences among species in diet, in spermatophore size relative to male body size, and/or in spermatophore function.

In contrast, no relationship was found between male size and ejaculate mass or volume in *P. protodice* (Lepidoptera: Pieridae)[72] or *Pararge aegeria* (Lepidoptera: Satyridae).[90] Rutowski[72] suggests that this may be due to differences in the primary function of the spermatophore among Lepidoptera, with *P. protodice* donations functioning more to prevent female remating than as a nutrient investment. This idea is explored in more detail below.

Among species, Reiss[91] argues that the investment *per unit time* in reproduction by either sex should scale allometrically with body weight, with a coefficient between 0.5 and 0.9; the precise predictions for each group will depend on the allometries of energy intake and nonreproductive expenditure with body weight. Extensive data to test this hypothesis do not exist for male nutrient donations.

Male ejaculate or spermatophore size increases with male age in virgin male *D. plexippus*,[92] but not in *O. nubilalis*.[84] This suggests that males of some species may need time after eclosion to reach complete sexual maturity (as measured by filling of the accessory glands) and/or that adult feeding contributes to the formation of spermatophores.

Spermatophore size decreases with number of previous matings by a male in some Orthoptera,[89] Lepidoptera[30,72,84,93] and Trichoptera.[74] Change in size of spermatophores with previous mating history is likely to be determined by the opportunity for males to replenish reserves from feeding, the average and range in number of matings by males, and the intensity of selective pressure for maintaining a minimum size to guarantee sperm transfer.

Quality of the male's diet and presence of parasites affect the male's resource budget and hence may affect the size of the nutrient donation. Zuk[94] showed that the number of spermatophores produced in 24 hours decreased with number of gregarine parasitic cysts in the feceae of *G. veletis* and *G. pennsylvanicus* (Orthoptera: Gryllidae), but no relationship existed for shorter term spermatophore production measures. Further, *R. verticalis* males infected with a protozoan gut parasite had lower mating frequencies than uninfected males when fed a poor diet. However, the effect of parasitic infection disappeared when males were maintained on a rich diet, indicating that parasites have the effect of lowering experienced diet quality as far as impact on mating success is concerned.[95]

Female ability to receive or process the nutrient donation could constrain the size of the nuptial gift, placing upper limits on size variation. In species with an internal spermatophore, female size and previous number of matings (if the spermatophore is not completely absorbed) may constrain the size of the male's nutrient investment due to space available in the bursa copulatrix or appendix bursa. Some evidence for this comes from *D. julia*, where female winglength interacted with other parameters to affect spermatophore size.[64] Finally, female mating status has an effect on spermatophylax and ampulla size in *D. verrucivorus*; virgin females obtain larger spermatophores.[89]

IV. EFFECTS OF NUPTIAL GIFTS ON DEMOGRAPHIC FITNESS COMPONENTS

Fecundity and survival are components of individual fitness and, as noted earlier, observed patterns of birth and death may be direct consequences of male nutrient donations because of the impact of these donations on the time and resource budgets of each sex. In this section, I explore the translation of effects on time and resource budgets into effects on fecundity and survival. Explicitly evolutionary relationships between nuptial gifts and fitness components are dealt with later.

A. AGE-SPECIFIC FECUNDITY PATTERNS

Mating can affect age-specific fecundity patterns either through hormonal mechanisms stimulating oogenesis and/or oviposition[22-23] or through alteration of the female's resource budget via nuptial gifts. These two causes can be difficult to separate, especially in species with internal spermatophores. Here, I am not concerned with oogenesis or oviposition stimulating factors, but rather with effects on fecundity via effects on the female's resource budget.

Male nutrient donations may affect fecundity independently of whether nuptial gifts arose and/or are maintained specifically in the selective context of increasing female reproductive success (and hence, male success). Rather, whether fecundity is affected by nuptial gifts should depend on the size and composition of the donation relative to the female's overall nutrient stores and the timing of the donation relative to oogenesis.[66] Fecundity enhancement is expected if: (1) at least some oogenesis occurs after mating; and (2) the type of nutrient donated is a limiting factor in egg production. A nutrient may be limiting either due to lack of foraging opportunities either in the adult or juvenile stage or due to an evolutionary history during which females have become dependent on male donations to replace foraging. In these cases, we expect that egg numbers will increase over the short term after mating, consistent with timing of incorporation of male nutrients into eggs outlined above. Alternatively, lifetime fecundity may also be enhanced if male donations allow females to reduce foraging and by so doing, decrease death rates from predation or increase time available to lay eggs. This means that species which live either in harsh or dangerous environments or which use ephemeral but nutritious resources, are likely to show an effect of male donations on fecundity.[13,62]

Some support exists for these ideas, as outlined by Boggs.[66] Within Orthoptera, spermatophylax consumption has an effect on the next few days' fecundity in several species, including *R. verticalis*[65] and *Chorthippus brunneus*.[96] This impact is affected by the quality of the diet fed to experimental females. More directly, spermatophylax consumption and a seasonally available high quality diet had equivalent effects on female fecundity in an unnamed zaprochiline katydid.[97] Not all tettigonids show an effect of male-donated nutrients on fecundity, however. Even on a restricted diet, no effect was detected for *D. verrucivorus*.[98] The spermatophylax in this species is just large enough to ensure that all sperm are transferred to the female after mating,[67,99] and has a relatively low protein content,[67] suggesting little opportunity for male-donated nutrients to be important in the female's resource budget, and, hence, to affect fecundity — unless mating were to occur frequently. Within Lepidoptera, larger spermatophores increased post-mating fecundity in *C. eurytheme*.[100] Females in this experiment were fed a relatively poor diet consisting of a 10% sucrose solution; we do not know how often females are food stressed in the field. Within Diptera, semi-starved *D. subobscura* females showed an increase in fecundity the first 2 days after mating if they were fed by males as compared to not fed.[21] A comparison of crop sizes in wild and lab-fed or lab-starved flies showed no difference in crop size between wild and starved flies, but wild flies had significantly smaller crops than fed flies.[21] This suggests that females are frequently nutrient stressed in the field. Hence, male-donated nutrients likely play an important role in egg production in this species. In *Drosophila mojavensis*, receipt of a large male donation increases early fecundity only if females are held without access to yeast.[101] Further, females of *Panorpa* spp. feeding on dead arthropods during mating showed increased fecundity compared to those not feeding.[20] Thornhill[20] found evidence for interspecific competition for food in this group, and for significant mortality when individuals forage from spider webs, suggesting that food was scarce and costly to obtain. Finally, within Coleoptera, unfed *Caryedon serratus* females show higher fecundity if allowed multiple matings instead of only one mating, although the authors are not convinced that their results were due to a nutritional rather than hormonal stimulus effect.[34]

B. SURVIVAL PATTERNS

In cases of sexual cannibalism, male nutrient donation ipso facto reduces male survival rate. The effect of donations on male survival is seldom this severe for other types of male donations. Nonetheless, we lack a detailed understanding of the impact of nutrient donations and mating numbers on male survival. Interesting questions include: Do males whose nutrient donation is a sizeable portion of their body mass suffer a relatively greater survival cost than males with a relatively smaller donation? Does the survival rate differ depending on whether specialized male structures are eaten or accessory gland products are donated? If accessory gland products are involved, does the effect on survival depend on the match between the composition of the male-donated nutrients and the composition of the adult male diet and hence the ability to replenish the nutrient pool? Understanding the answers to these questions will allow us to understand physiological costs of mating to the males and how such costs translate into fitness-related survival patterns.

Refractory periods found in Orthoptera and Megaloptera, or the decrease in nutrient donations with frequent matings seen in Lepidoptera, may act to buffer survival rates against effects from reproductive expenditures by limiting reproduction. Within Lepidoptera, *D. plexippus* males given the opportunity to mate every day had the same life span as males with no opportunity to mate;[102] a similar result was obtained for *P. aegeria*[90], the orthopterans *G. sigillatus* and *G. veletis*,[81] and the megalopterans *Prothermes grandis* and *P. immaculatus*.[38] This lack of effect could stem from a limitation on reproductive expenditure, from compensatory feeding by males with more matings[103] as occurs in *H. charitonius*,[104] or from a lack of impact of mating expenditures on types of nutrients important in survival.

In an interesting twist, a *lack* of opportunity to pass nutrients to females can impact male survival under some conditions in at least one roach. *X. hamata* males fed on foods with a high nitrogen content (as are chosen in the lab and field) can die from uric acid toxicity if not allowed to mate and give some urates to females.[52] How important this is under field conditions as a source of mortality is unknown, but if the variance in distribution of mating numbers among males is at all high, significant numbers of males could be exposed to this mortality source under appropriate nutritional environments.

Effects of male donations on female survival are somewhat more variable than those seen to date on male survival. No survival effect of increasing the quantity of male nutrient donations has been found for *D. plexippus*[75,79] or *O. nubilalis*[84] within Lepidoptera, or *D. verrucivorus*[98] within Orthoptera. However, increasing male nutrient donations increased female survival in *C. eurytheme*[100] and *Psuedaletia unipuncta*[105] within Lepidoptera, and *G. sigillatus* and *G. veletis*[81] within Orthoptera. As for males, the observed results may be affected by possibilities (in the field or lab) for compensatory feeding if females receive small donations, or by the type of nutrients donated by males at mating and their importance to survival processes.[66] In the latter case, it is also possible that a shortage of male-donated nutrients normally used for survival could cause a reallocation of female-derived resources away from reproduction, maintaining female survival rates at the expense of reproduction. Such reallocation driven by decreases in male-derived nutrients has not been explored, but *Speyeria mormonia* females (Lepidoptera: Nymphalidae) reallocate resources from reproduction to survival in the face of adult female food shortages.[106]

Effects on survival may occur not just through impacts on the resource budget, but also through increases or decreases in exposure to predators because of changes in foraging activity by either males or females. Through feeding on nuptial gifts rather than foraging on their own, the incidence of spider predation is apparently reduced in the hanging fly *H. apicalis*, as females are found significantly less frequently than males in webs, whereas there is no difference by sex in a similar species without nuptial prey gifts.[16] Conversely, in mormon crickets and conocephaline katydids, the sex competing for mates (and hence, with a relatively smaller investment) is more active and suffers more wasp predation.[107]

Finally, nuptial gifts can affect offspring survival rates. Increasing the female's supply of defensive compounds, such as pyrrolizidine alkaloids in ithomiine butterflies[57] or moths,[58] could increase the survival rates of offspring. Offspring survival can also be increased if male donations increase the female's supply of trace nutrients, including sodium, as seen in *Thymelicus lineola*.[46] The same effect may explain data for *R. verticalis*:[65] females eating more or larger spermatophylaxes laid larger eggs; larger eggs have higher over-winter survival rates, and, if male, differ in developmental rate compared to smaller eggs. Improved female diet alone did not increase egg size in this species.

V. EFFECTS OF NUPTIAL GIFTS ON POPULATION STRUCTURE

Individual fecundity and survival influence fitness relative to the fecundity and survival of other members of a population. The composition of the population is affected by its age structure, individual dispersal habits, and effective population size. Male nuptial gifts can have effects on these three parameters, and hence on the population context within which selection may occur.

A. AGE STRUCTURE

Adult age structure is affected by development time of juveniles, as well as age-specific death rates of adults. Male nuptial gifts may interact with juvenile development time through impacts on the resource budget of each sex. If nutrients gathered during the juvenile stage are used by males in procuring or making nuptial gifts, then longer development times resulting in larger male resource stores and consequently larger donations should be favored, all else being equal.[108] Conversely, females receiving substantial nutrient input from males could have shorter development times if resources normally obtainable only in the juvenile stage are provided by males and juvenile mortality rates are high. These ideas could be tested using selection experiments in a species with a suitably short life cycle. Alternatively, development times could be compared for closely related species from the same habitat with similar food sources and sperm precedence patterns, but which differ in male nutrient donations.

Further, sperm precedence patterns favoring protandry, or adult emergence of males before females, may also affect the size of male nutrient donations.[89] If sperm from multiple matings mix, then a male's net fitness increment from mating with a female decreases with increasing number of previous matings by the female. Males could offset this decline by increasing the number of sperm transferred, which in Orthoptera can entail a larger nutrient donation to ensure sperm transfer. However, the net gain from a given increase in number of sperm transferred diminishes with increasing numbers of previous matings by the female. As predicted in a protandrous system with sperm mixing and in which the male nutrient donation is just large enough to ensure complete sperm transfer, *D. verrucivorus* males make their largest contribution to virgin females, independent of male mating status.[89]

B. DISPERSAL

Male nuptial gifts may interact with dispersal in two ways: nutrient donations may provide necessary resources for successful dispersal by females, or females may view males as a resource and refrain from leaving areas with males.[109]

The first idea, that nuptial gifts allow females to disperse or to migrate, has been explored in monarch butterflies (*D. plexippus*).[110-112] Overwintering monarchs in Mexico and coastal California have different mating regimes before the spring migration.[110] In Mexico, nearly a third of migrating females are virgins, and males and females leave the colony at about the same time. For virgin females, mating presumably occurs on the trip north, since females oviposit on milkweeds during the migration. Prior to migration, mating males are more worn and smaller than the average for the population, whereas mating females are less worn and larger, and males attempting matings appear to discriminate against mated females.[111] In coastal California, 95% of females mated between 1 and 7 times before moving north, and many males may not leave the colony at all. Thus, there are different patterns in these two populations for the timing of entry of male-derived nutrients into the female resource budget with respect to the timing of migration. These differences could be affected by sperm precedence patterns, physical environmental parameters, activity levels of over-wintering individuals, and distance to the nearest milkweed.[110]

Wells et al.[112] have shown that female lipid content increases once mating begins in the California monarch aggregations, while male lipid content continues to decline. They attributed this difference to a shift of nutrients to the female from males at mating. Using measured energy expenditures by monarchs, the amount of energy estimated to be obtained at each mating by the female, and the timing of first access to milkweed plants for oviposition in the spring, they did Monte Carlo simulations examining expected long-term fate of monarch populations. Their results indicate that multiple mating upon leaving the winter roosting site is necessary for long-term persistence of the population.

Experimental measurements of effects of male nutrients on either dispersal ability or on reproductive ability once a new habitat has been reached have not yet been published. However, several authors have turned the prediction around, arguing that species which are long-lived migrants should be expected to have large nutrient donations from the male to the female.[105] *P. unipuncta* is one such case, as female fecundity and life span are increased by multiple mating in this migrant.[105]

An alternative effect of significant male nutrient donations is for females to refrain from dispersing and to remain in areas with males, using males as a nutrient resource. Monarch butterflies in Australia, which do not form overwintering aggregations but breed year-round, may provide an example of this effect.[109] Female density is lower inside dense milkweed patches than on the edges, and male density shows the reverse pattern, suggesting that male harrassment may drive females away from the center of the patches. However, if males are removed from an area, females tend to disperse, whereas if population density is simply reduced, the effect is not seen. Use of males as a resource, and hence females remaining in an area with males, need not be the only explanation for this result, but it is certainly a possible cause. It could be instructive to test this hypothesis by contrasting the female dispersal patterns in *Hylobittacus apicalis*, whose females depend on male-provided prey items,[5,16] with related species with no prey gifts and whose females hunt.

C. EFFECTIVE POPULATION SIZE

The effective population size (N_e) is a measure of the number of individuals actually contributing offspring to the next generation, as opposed to the total number of individuals in a population. Among other things, effective population size governs the power of genetic drift as a significant evolutionary factor. As the effective population size becomes smaller, drift may play a more powerful role affecting evolutionary change.

A given population will exhibit a characteristic mean number of matings per individual, but some individuals will mate more often than others. Male nuptial gifts may affect N_e by affecting the male refractory period after a mating, as noted for Orthoptera and Megaloptera.[37-38,78,81-82] The male refractory period in turn affects the size of the pool of males available to mate by removing males from the pool for a length of time equal to the refractory period. Thus, the identity of the males in the pool of potential mates is constantly changing, with some turnover rate dependent on the size of the donation. The larger the pool of males available for mating relative to the absolute number of males, the greater the chances that the pool contains males that have recently mated, and thus the greater the chances that a male that has already mated will secure the next mate, or that a male that has not yet mated will not secure the next mate. Note that this effect only occurs if the operational sex ratio is male biased. Otherwise, all males in the pool at any point in time should be able to obtain matings, since there are more females willing to mate than males available, resulting in all males in the population obtaining at least one mating.

The practice of males donating nutrients at mating may also affect N_e through resource-based effects on the operational sex ratio (OSR). When the OSR is biased towards one sex, the possibility exists that not all members of the over-represented sex will obtain matings and contribute to the effective population size. Hence, the component of N_e determined by mating success (a first step in representation in the next generation) may be significantly lower than the actual observed population size for the over-represented sex, but close to the actual observed population size for the under-represented sex. The OSR can be affected by resource availability in populations with large nutrient donations at mating. If females are in a resource-poor environment, they attempt to replace nutrients unavailable from the environment with nutrients from males at mating; conversely, males may experience longer refractory periods since a given donation is more expensive relative to available resources. The OSR may become female biased under such conditions. In a rich resource environment, females are less willing to spend time mating, since nutrients are not scarce and males experience shorter refractory times. The OSR may become male biased under such conditions, even in the same population.

Male refractory times are not the only item that may constrain the pool of available males in a resource-poor environment. In species with nuptial prey, the abundance of prey items in the environment may affect hunting times, which could in turn affect the number of males with prey available as mates, affecting male N_e. In environments with low prey availability, males which are better hunters or more able to steal prey items may get more matings, increasing the number of males with zero lifetime reproductive success and decreasing N_e.

Species for which diet affects the male refractory period or the size of the male nuptial gift, or for which prey availability is important, then, may experience fluctuations in N_e as a function of variation in the nutrient environment, even without fluctuations in observed numbers of individuals. These fluctuations may have important consequences for the population genetic structure, and the role of genetic drift.

Data on effects of donation size on N_e do not currently exist; for that matter, population data on insect lifetime reproductive success in reasonably natural environments are scarce.[113] Species which alter the size of donations as a function of the food environment would be good candidates for study.

VI. PATTERNS OF THE EVOLUTION OF NUPTIAL GIFTS

A. ORIGINS: PHYLOGENETIC HISTORY

Nuptial gifts may be a very old trait within Insecta. Khalifa[74] and Davey[29,43] argue that presence of spermatophores is a primitive trait within this class and has been lost multiple times during radiation of various groups. Thysanura possess spermatophores, as do the orthopteroids and neuropterans. Davey[29] points out that other groups without spermatophores often retain reduced accessory glands. The antiquity of spermatophores within Insecta means that the *potential* for male nutrient donations via the spermatophore or similar accessory gland secretions is probably at least as old as the class. Although spermatophores are believed to have arisen in the context of facilitating sperm transfer in a terrestrial environment,[43] the possibility of using the structure as a means to provide females with extra nutrients was present. Thus, spermatophores may have been exaptations for male nutrient donations.

Within orders, or genera, phylogenetic history can be an important determinant of both the presence and size of male nutrient donations. Pitnick et al.[114] found that monophyletic species groups in *Drosophila* have similar sized ejaculates and similar levels of incorporation into ovaries and female soma of radiolabeled male nutrient donations. This is in spite of the fact that species compared within a species group differed in nutritional ecology and habitat. Likewise, comparison of *D. mojavensis* and *D. pachea*, both cactophilic species endemic to the Sonoran desert but members of different species groups, showed that these two species differ dramatically in ejaculate size and incorporation by the female, but were similar to other members of their species group.

Phylogenetic history undoubtably plays an important role in determining the degree of male nutrient donation within Orthoptera as well. As noted above, families differ in the presence or absence of a spermatophylax, which increases the size of the male donation. Boldyrev[115] argues that the ancestral spermatophore was probably similar to a simple ampulla without the spermatophylax; in that case, the spermatophylax could have evolved as an elaboration in circumstances where increased male nutrient donation was favored.[65] There is some uniformity of spermatophore structure within families,[116] so the elaboration of the spermatophore occurred relatively early. Many of the relationships between male nutrient donations and other aspects of the biology of a given orthopteran species may be constrained by the phylogenetic history associated with spermatophore production.

Phylogenetic history may also play a role in the development of presentation of prey items to females at mating or other behaviors associated with male nuptial gifts. Certainly the variations seen in *Empis* or *Hilara* species, ranging from presentation of a prey item in primitive species to wrapping the prey item in silk or a balloon to presentation of an empty wrapping in the most derived species, indicate that phylogeny can have significant effects.[18-19] At a more basic level, a survey of mecopteran species which do and do not have male nuptial gifts, examining both phylogenetic relationships and habitat or ecological specialization, could be instructive as to the relative roles of phylogeny and ecology or habitat in influencing the evolution of presentation of nuptial prey.

B. CURRENT VARIATION AND FUTURE PROSPECTS: HERITABILITY

Male nuptial gifts range from secretions, whose evolution must be based on metabolic control of resource allocation, to presentation of prey items, whose evolution must be based on behavioral adaptations. The discussion so far assumes that variation in nuptial gifts has a genetic basis, and hence selection and evolutionary change is possible. However, the genetics underlying nuptial gift giving is not yet understood for any species. In fact, in only one case has heritability of any aspect of the nuptial gift been measured. Based on father-son regressions, Sakaluk demonstrated a ~47% heritability for spermatophylax mass/male body mass in *G. sigillatus*.[85-86] None of ampulla mass/male body mass, ampullas mass alone, or spermatophylax mass alone showed significant heritabilities. Sakaluk points out that this result suggests that the size of the ampulla and spermatophylax are genetically uncoupled. This makes sense given that the composition of the two are not the same, and hence the two are likely produced through different metabolic pathways with the potential for differing control. Further, the fact that percent body mass allocated to spermatophylaxes showed heritable variation while absolute spermatophylax mass did not suggests that the controlling factor is an enzyme regulating allocation rate from some finite pool to male accessory glands and/or gene(s) affecting the relative size of the accessory glands.

Sakaluk suggests that variation for relative spermatophylax size may be maintained by fluctuating food environments. In a high food environment, a large spermatophylax may not increase the female's fecundity, but may decrease the number of matings a male obtains due to a longer male refractory period. The disadvantage of a longer refractory period still holds in low food environments, but female fecundity may be increased by a large spermatophylax. Similarly, lack of heritable variation for ampulla mass may have resulted from lack of variation in selection pressures affecting ampulla mass since the costs or benefits of number of sperm transferred do not vary with environmental food availability.

C. INTERACTION OF NUPTIAL GIFTS WITH MATE COMPETITION AND MATE CHOICE

Given the possible adaptive function of nuptial gifts in enhancing female fecundity and securing matings, we can ask whether or not aspects of the reproductive biology of insects are coadapted with male nuptial gift giving. That is, are types, compositions, sizes, or timing of nuptial gifts coadapted with female fecundity, security of paternity, mate competition, or choice, or mating system? Are there patterns of association among traits which have high fitness value and could be selected for in natural environments over evolutionary time? I first consider interactions with sex roles, then with mating systems, then with age-specific fecundity, and then I conclude by pointing out that these interactions themselves are not independent.

Coadaptation of nuptial gifts occurs not just with mating systems, but with the combination of mating systems and age-specific fecundity.

1. Sex Roles and Sexual Selection: Theory

Much of the work on insect nuptial gifts has been done to explore the effects of relative reproductive investment by each sex on sex roles and the operation of sexual selection. This work was stimulated by a series of authors, beginning with Darwin. Before examining the results of experiments relating male nuptial gifts to sexual selection, I will explore the conceptual context for those experiments.

Trivers[117] built on arguments by Darwin,[118] Fisher,[119] Bateman,[120] and Williams,[121] and argued that differences between the sexes in amount of investment in offspring should control the operation of sexual selection. Trivers defined this critical parameter, parental investment, as "any investment by the parent in an individual offspring that increases the offspring's chance of surviving (and hence reproductive success) at the cost of the parent's ability to invest in other offspring."[117] Thus, individual fitness is the currency in which parental investment is measured; investments must increase the fitness through one offspring's survival at the cost of ability to realize fitness increases through other offspring. Trivers goes on to argue that the sex with the larger parental investment will limit the realized reproductive success of the sex with the smaller investment. Thus, the sex with the smaller investment should experience intrasexual competition for mates. Further, the sex with the larger investment has more to lose from a bad choice of mates than does the sex with the smaller investment. Thus, the sex with the larger investment should exhibit mate choice. Presumably to avoid circularity, Trivers defined parental investment explicitly to exclude investment in obtaining mates through "sexual competition for mates".[117]

Trivers' definition has lead to a proliferation of papers on male nuptial gifts as parental investment vs. mating investment.[99,102,122-127] This distinction is critical to testing Trivers' hypothesis, since only shifts in relative parental investment are predicted to lead to shifts in sex roles.

Taking a different approach to the problem of sex roles, Emlen and Oring[128] examined environmental and ecological effects on mating systems. They argued that which sex experiences intrasexual competition for mates depends on the Operational Sex Ratio (OSR), defined as the sex ratio of individuals currently ready to mate. That is, if receptive females are present in smaller numbers than reproductively active males, females represent a scarce resource for males, and males should compete among themselves for access to females and *vice versa*. As Gwynne[78,129] pointed out, the OSR may be driven by parental investment patterns if the size of the parental investment affects the amount of time spent out of the mating pool by either sex.

Recently, these arguments have been substantially expanded by Clutton-Brock and Parker.[130] They include time in their analysis, build explicit connections between investment and the OSR, and acknowledge a variety of other factors that may influence the OSR or intrasexual mate competition. They begin by stating that the OSR determines which sex will compete for mates, using arguments similar to Emlen and Oring's.[128] However, rather than parental investment, Clutton-Brock and Parker focus on potential reproductive rates of each sex as the major factor determining the OSR, along with the local adult sex ratio and sexual differences in survival rates. Potential reproductive rates are in turn affected by investment in offspring, courtship, and mating, as well as physiological and environmental constraints. In effect, Clutton-Brock and Parker have interposed the OSR and potential reproductive rates as intermediary steps between parental investment and mate competition, and have acknowledged other factors that can affect the outcome as to which sex is competitive.

Clutton-Brock and Parker's[130] theory does not require the distinction between mating and parental investment required by Trivers. Each type of investment is important in so far as it affects the potential reproductive rate. Since a nuptial gift may serve both functions, as for example nuptial prey in Mecoptera or spermatophores in some Lepidoptera, this theory is more amenable to experimental test.

There is a remaining problem yet unaddressed. Many authors explicitly or implicitly assume that where intrasexual mate competition occurs, mate choice must be occurring in the opposite sex or that conditions leading to mate competition produce mate choice by the opposite sex. Clutton-Brock and Parker[130] are obvious exceptions to this pattern, as they only address mate competition in their analysis. Trivers,[117] however, considers factors affecting the operation of sexual selection in general, and does not examine potential differences in causes of intrasexual competition and intersexual mate choice.

There is no expectation for mate choice to automatically accompany mate competition or vice versa, however. Males may compete with each other for access to a female's feeding territory, or, in some *Heliconius*, males defend female pupae from each other.[131] In these cases, the female is not exerting mate choice. Alternatively, in *Colias*, females choose mates,[132-134] but males are not competing with each other for access to females.

Three factors are important prerequisites for the mate choice to be expected. First, more than one potential mate must be reliably available, to allow a choice to be made. This does not mean that the operational sex ratio must be skewed towards the nonchoosy sex, although that certainly helps. Second, fertility (for females) or paternity (for males) must be assured, unless choice is to be made among potential mates on the basis of which one will assure genetic representation of the chooser in offspring. Third, one sex must be able to control resources needed by the other,[6,128,135] **AND** there must be differences among individuals in quality or quantity of the resources controlled, in order for a basis for choice to exist. Such resources may be genetic, nutritional, or portions of habitat. Thus, choice is likely to be seen in populations with skewed OSRs due to density or lekking behavior, in populations whose individuals mate multiply and/or show last male sperm precedence, and in species in which variation in male nutrient donations or female fecundity are discernible due to age, size, or behavioral differences among differing individuals.

2. Experimental Evidence: Nuptial Gifts, OSR, and Mate Competition

Mate competition can take at least two forms in insects with nuptial gifts. The first is the standard one of fighting between members of one sex for access to a receptive individual of the opposite sex. Fighting for access to mates has been documented within Orthoptera[36,129,136] and some Lepidoptera.[131] Within Lepidoptera, some *Heliconius* species' males find a female pupa and sit on it about 24 hours before female emergence. Two males will fit on a pupa, one on each side. Males may attempt to take over pupae by dislodging resident males in much the same way they attempt to take over matings in other species. One defending male then mates with the female either just before or at female eclosion.

The second form of mate competition is theft of nuptial prey items, as occurs in *H. bittacus*.[16] Males can act as "transvestites", approaching a calling male as if for mating; in one study, calling males actually offered the prey item to the thief 67% of the time.[16] Males which gain a prey item may also thereby be able to gain a mate at the expense of delayed mating possibilities by the male losing the prey item.

Clutton-Brock and Parker's[130] theory predicts that changes in the potential reproductive rate of each sex, which may be driven by the relative value of the nuptial gift, affect the operational sex ratio and hence, determine which sex is the competitive sex. Evidence from Orthoptera is consistent with the theory. For example, at high food availability, mating frequency by a zaprochiline katydid female is reduced; at low food availability, female fecundity is maintained only through spermatophore consumption.[97] Thus, the fraction of females in the mating pool at any one time is lower at high food availability such that the operational sex ratio should be more likely to be male biased at high food availability and female biased at low food availability. As expected, males compete for mates when adult food is plentiful, but females compete for mates when adult food is scarce.[136] In another case,

Anabrus simplex occurs in either high or low density in Colorado, U.S.[137] Males in the high density site have lighter accessory glands than males in the low density site. Given the refractory period after a mating while glands are refilled, these data are consistent with the observation that few males were calling at the high density site. Average number of matings by females did not differ between sites. If high density led to longer male refractory periods at the high density site, the OSR may have been biased towards females at that site, and towards males at the lower density site. This would produce the observed sex role reversal between the two sites, with competitive females at the high density site only.

The fact that sex role reversals tend to be observed in Orthoptera with relatively large male investments and multiple matings by females is also consistent with the theory. Larger male investments lead to increased refractory times (see above), decreasing the number of sexually active males at any point in time. Further, the possibility of female dependence on male nutrient donations when other adult food resources are scarce means that the proportion of sexually active females in the population can increase under some circumstances. Thus, the OSR can swing from male to female biased depending on food availability, accompanied by changes in which sex competes for mates.

The interaction between energy needed for male nutrient donations and for mate calling can also produce a biased OSR, but only if noncalling males are not part of the mating pool. For example, *C. strepitans* males' wing pads are fed on by females during mating, impairing the ability of nonvirgin males to call for mates.[76] Likewise, *R. verticalis* males held on low diets maintain spermatophore size, but reduce calling for mates due to energetic limitations.[77] Since female competition has not been observed in the latter species, the theory predicts that male refractory periods should be relatively unchanged by low quality diets and that noncalling males really are part of the mating pool — that females are able to find them for mating purposes.

In species without direct mate competition, sex role reversals may take the form of shifts in the sex initiating courtship. This has been documented for the pierid butterflies *P. protodice*[138] *Colias philodice*, and *C. eurytheme*,[139] whose females approached males and elicited courtship chases once the bursal spermatophore contents were largely depleted. This system would be interesting to examine in the context of interactions among resource availability, potential reproductive rates, and the operational sex ratio.

3. Mate Choice Based on Nuptial Gifts

Mate choice can be accomplished in several ways. The first way is primary choice through rejection or acceptance of mating with a prospective mate. Secondary choice by females occurs if the female rapidly remates with another male (in systems without first male sperm precedence) or if females control the length of copulation and hence, the amount of sperm transferred, as occurs in some bittacids.[5] In the orthopteran *Gryllus bimaculatus*, females use both approaches: females may remove some males' spermatophores before all sperm are transferred, but remain near large males, remating several times.[140]

Above, I predicted that presence of choosiness should depend on reliable availability of mates, assurance of fertility or paternity, and ability of one sex to control resources needed by the other combined with distinguishable differences among individuals in ability to provide those resources. Available data for choice involving nuptial gift donations supports these predictions. First, food availability can influence choosiness in each sex through effects on mate availability. For example, in a zaprochiline tettigoniid, calling males were frequent under conditions of high food availability and females were choosy, rejecting males. However, females were eager to mate and receive nutrients under low food availability, the OSR was female biased, and males were choosy.[136]

Second, in *D. julia* (Lepidoptera), females tend to mate on their second and subsequent matings with males who have had large numbers of matings, but have not mated recently; these males provide larger spermatophores than males that have mated more recently. This pattern is not observed for the first matings by females.[141] Females in this species can fertilize all eggs from a single mating if access to males is restricted.[142] A similar increase in female discrimination is observed in *Colias* butterflies.[143] Thus, female choosiness can increase after a first mating if sufficient sperm are obtained from one mating to fertilize all of a female's eggs. In a twist on this idea, *D. verrucivorus* males contribute a larger spermatophore to virgin females than to nonvirgins, which constitutes a form of investment choice; males probably father more offspring of virgins than nonvirgins.[89]

Both sexes can exercise mate choice simultaneously. For example, in *P. protodice*, male courtship duration depends on the type of female: larger, younger females are courted longer than others.[144] In the same species, females discriminate among males based on the duration and intensity of courtship.[80] Since choice can be exercised by both sexes simultaneously, then the identity of the choosy sex(es) cannot be dependent only on the operational sex ratio or on the relative amount of parental investment by each sex. An examination is needed of threshold conditions for choice, considering mate availability, fertility/paternity assurance, and discernible differences among individuals in resource control.

D. INTERACTION OF NUPTIAL GIFTS WITH MATING SYSTEMS AND SECURITY OF PATERNITY

Nuptial gifts can function as investments in obtaining a mate or in ensuring complete sperm transfer. A male either presents a prey or is expected to donate a spermatophore,[80] which gives the female sufficient net benefit that she is willing to mate with that male. Nuptial gifts may thus function as bribes, as an index of male genetic quality, or both. For species which have external spermatophores or which present arthropod prey for the female to feed on during mating, the size and quality of the male donation can determine whether the maximum number of sperm are transferred. For example, in *G. sigillatus*, the length of time the ampulla remains attached, allowing sperm to be transferred and stored in the female, depended on the size of the spermatophylax and how long it takes females to eat it; once they had eaten the spermatophylax, they removed the ampulla and ate it as well.[86,122] Copulation duration in *H. apicalis* depended on size of the offered nuptial prey item, with a threshold size above which copulations lasted long enough for complete sperm transfer.[5] In these cases, the function of the donation may differ, depending on the vantage point: for females, it is still an investment in their resource budget; for males, it is expenditure in obtaining a mate and not necessarily a direct investment in offspring.

A prediction following from this function is that a threshold investment in the female's resource budget exists, and males that cross that threshold obtain mates and/or complete sperm transfer. The threshold might shift, depending on the state of the female's resource pool, the relative time cost of mating, and the current risks associated with foraging herself. Quantitative shifts in thresholds have not been explored in detail yet, except in the extreme case where females are on very poor diets, which can lead to sex role reversals where females are willing to accept any male as they compete for males, rather than choosing among males.[97,136-137]

Male nuptial gifts can also function to prevent females from remating. This can be accomplished in several ways. First, female refractory periods may be induced through compounds in the ejaculate.[22-23] Second, mating plugs (which might also function as nuptial gifts if they are absorbed by the female) may be formed in the female's genital tract, which physically prevent other males from mating with the female.[28] Third, for some species with internal spermatophores, the spermatophore triggers a stretch receptor in the bursa, initiating nonreceptivity; this has been shown for *Pieris rapae*.[145] The decay of female nonreceptivity in another Lepidoptera, *D. plexippus*, is correlated with the initial size of the spermatophore;[71] whether resumption of receptivity is cued by a threshold size of the absorbed spermatophore, by decay of some factor transferred by the male whose volume is correlated with spermatophore size, or by changes in the female's nutritional status as a result of decreasing input of spermatophore nutrients remains to be determined.

Males make a nutrient donation to the female prior to the transfer of sperm in all known cases, even if female utilization of that donation is delayed.[30] In Lepidoptera and other groups with internal spermatophores, the spermatophore is passed to the female prior to sperm movement; in Orthoptera, the sperm are often deposited in an external ampulla, and the ampulla, spermatophylax, and/or male body parts are available to the female prior to sperm movement into the female's body; in Mecoptera and Diptera with nuptial prey or salivary gifts, prey are presented to females prior to the onset of copulation. This uniformity among cases where the nutrient value of the donation to the female varies suggests that selective pressures associated with mate acquisition, complete insemination and mate guarding were crucial to the evolution of nuptial gifts. That is, males that were able to present gifts first obtained matings from females and also obtained complete inseminations. In some of these groups, then, female choice of mates with an initial gift may have played an important role in the evolution of the form and sequence of gift giving.

While mate acquisition, mate guarding, and security of paternity may have played a role in the evolution of male nutrient donations, male nutrient donations may also function as the selective context for changes in mating systems and sperm precedence patterns. In short, the two sets of traits, mating systems and donation quality/quantity, should be coadapted. In species whose females mate several times, the timing of remating can be closely linked to the timing of the decay of use of the previous male's nutrients in egg production. Female *C. eurytheme* remate after 4 to 6 days in a field population;[30] male nutrients are primarily found in eggs laid during the first 3 to 4 days after a mating.[69] However, both *R. verticalis* and *D. verrucivorus* females show a peak of incorporation of male nutrients into laid eggs much later than the end of the female refractory period.[70,85,146] Actual timing of female remating in the field will depend on female access to mates as well as the length of the refractory period.[146] Intermating intervals may thus be longer than the refractory period which was observed under experimental conditions.

Differences not just in nutrient use as outlined above, but also in donation quality or quantity, are expected for related species that differ in expected number of female matings, sperm precedence patterns, and/or alternative mechanisms of mate guarding not involving the male nutrient donation. Extensive data to test this do not yet exist. We do know that spermatophore size is correlated with the duration of the intermating interval within some species but not others, and that the species-specific time required to regain the ability to make a normal sized spermatophore after a mating is correlated with mean number of matings in some groups.

Time since the last mating by the male is positively correlated with spermatophore size in several Lepidoptera.[30,64,72,83-84,90,92] Further, male lepidopterans remate rapidly, increasing the time in copula if mating has occured recently, whereas male orthopterans generally have a post-mating refractory period, with subsequent spermatophore sizes unaffected. This suggests differences between the two groups in mate acquisition strategies, with lepidopterans remating whenever possible even if a small spermatophore results, whereas orthopteran males appear to need the ability to make a minimum investment before mating will occur.[78]

Differences in the mean number of lifetime matings among lepidopteran species are correlated with the rate of recovery of ability to produce a "normal" sized spermatophore.[35] Four species from a diversity of families with an average number of matings greater than 2.0 were able to make a spermatophore equal in mass to that of the first mating within 2 to 3 days after the first mating; individual males from three species with an average number of matings less than 2.0 had not recovered the ability to make normal sized spermatophores even 5 days post-mating. Further, male mating success is highly skewed in at least three species including one with mean number of matings less than 2.0; in *H. cydno*, *H. charitonius*, and *D. julia*, less than 15% of the males obtain half the matings.[147] This difference in recovery time, then, means that a significant number of females which only mate once may be receiving small spermatophores without the opportunity to obtain more nutrients from males at a later time.

E. INTERACTION OF NUPTIAL GIFTS WITH AGE-SPECIFIC FECUNDITY: FECUNDITY ENHANCEMENT

Female intake of male-donated food, in whatever form, is a contribution by the male to the female's resource budget.[66] Depending on subsequent allocation of that food by the female, it may also be an investment in offspring; depending on paternity patterns, it may be an investment in the donating male's offspring. Males can theoretically influence allocation patterns by the type of nutrients they donate; nutrients which are in short supply for egg production and which are usable with little modification should be allocated to offspring by the female. Although the fitness of the male may also be enhanced if "his" nutrients are allocated to female survival rather than invested in his offspring, this effect has never been experimentally tested.

From a female perspective, the nuptial gift may be regarded as another food source, with attendant risks and benefits which differ in detail from those of "ordinary" food sources. Females should be expected to allocate male-donated resources in a manner consistent with maximizing their own lifetime reproductive success. Thus, the possibility exists for conflict in evolutionary time between the sexes over female allocation of male-derived nutrients.[15] Such conflicts may have molded the present-day size, composition and timing of male nutrient donations, and the consequent role donations have in enhancing female fecundity.

Most work has focused on the end product — number of eggs produced — as the trait affected by changes in available resources due to male nuptial gifts. However, other aspects of reproductive resource allocation patterns may be coadapted with nuptial gift giving. That is, the sources of resources and dynamics of their use in reproduction may be shifted in expectation of a given level of nutritional intake by the female from the male. This can have two manifestations: the first is allocation to reproductive reserves, and the second is ovarian dynamics.

Holometabolous insects in particular have the luxury of reallocating resources during metamorphosis to suit adult needs. Boggs[108] predicted that allocation of larval resources to body vs. reproductive reserves during metamorphosis should vary among species and sexes as a function of the expected adult intake of nutrients, including nuptial gifts, and output of nutrients in reproduction. Three species of heliconiine butterflies fit the predictions. These species differ in the quality of the spermatophore received by the female and in the number of matings. Thus, allocation of larval reserves to reproduction accomodates expected nuptial gifts.

Boggs[66] further suggests that the nutritional function of the male nuptial gift should be matched with the female's ovarian dynamics. Females whose eggs are matured after adult emergence have the opportunity to use male donations in egg production, rather than just for somatic maintenance. Species with large nutrient donations thus are expected to yolk eggs after adult eclosion.

F. NONINDEPENDENCE OF ADAPTIVE FUNCTION: INTERACTIONS BETWEEN FECUNDITY ENHANCEMENT AND MATING SYSTEMS EFFECTS OF NUPTIAL GIFTS

Nuptial gifts' interactions with mating systems can be affected by the interaction between nuptial gifts and female reproduction, and vice versa. For example, once male nutrient donations of whatever form exist, the size and quality of the donation should be affected by selective pressures related to female and male resource budgets for reproduction, in the context of security of paternity.[15,148] The effectiveness of mechanisms delaying remating in species with sperm precedence will constrain the possible size/quality of the nutrient donation through pressure on males not to provide nutrients to make another male's offspring. For example, complete sperm precedence coupled with rapid female remating in particular should select for males to donate only small quantities of rapidly utilizable nutrients to females. The state of male and female resource budgets will depend on foraging efficiencies, food availability, and demands of survival expenses on the resource pool. The fecundity-enhancement function of male nutrient donations thus should have been most strongly selected for in environments which are resource poor, because of low prey or high predator densities or low quality diets, and in species whose males have high confidence of paternity. Note that a species feeding on a highly dispersed but high quality prey could still be considered to reside in a resource-poor environment when compared with species feeding on less dispersed prey items. By this criteria, many parasites, including bedbugs, could be considered to reside in resource-poor environments. Quality of the resource environment should also put a constraint on the evolution of the fecundity-enhancement function of male nutrient donations, however. Nutritional environments which are very poor or risky should not be able to support the resource pools needed by males to make large donations. Tallamy[11] presents a variant of this hypothesis, including both prezygotic and postzygotic paternal reproductive expenditures. He provides extensive supporting evidence from a wide variety of arthropods.

Male nuptial gifts may thus have mixed functions, both as resource and mating investments — but only from the male's perspective. From the female's perspective, the sole role of the nutrient donation is as investment in her resource budget; what differs is the size of the investment, its possible uses, and the possibility of exercising choice among males with different quality gifts.

The observed relative importance of the two functions should depend on the following factors. Timing of oogenesis relative to mating and certainty of paternity should affect elaboration of the role of gifts as investment in the female's resource budget (from a male viewpoint). The ability of the female to terminate mating before sperm transfer is complete, the alternative behavioral or hormonal methods of mate guarding, the ability of the male to mate with unwilling females, and the availability of other female mates should all affect elaboration of the role of gifts as mating investment.

Within Orthoptera, we now know that spermatophores can function primarily as a mating investment guaranteeing complete sperm transfer, or have an additional primary function as an investment in the female's resource budget. Conclusions are based on size of the spermatophylax relative to that needed to ensure complete sperm transfer, protein content of the spermatophylax, and the sensitivity of its size to male diet.[99,122-124] Similar studies have not yet been done in other groups.

VII. FUTURE PROSPECTS

Male nuptial gifts have primarily served as a case example to study theories of sexual selection. The focus has been testing hypotheses concerning the effect of the balance of investment between the two sexes on sex roles and the operational sex ratio. Work in this area should continue to expand, with a better understanding of causes of mate competition and mate choice and inclusion of study of the physiological and genetic underpinnings of observed behaviors.

Study of male nuptial gifts, however, has broader potential than just illuminating sexual selection theory. Like many other reproductive traits, male donations have effects at organizational levels ranging from individual physiology through behavior to population demography. As discussed above, effects of size or frequency of gift donation, for example, can impact the male's resource budget, affecting courting intensity, the operational sex ratio, sex roles, survival, and effective population size. Nuptial gifts also have the potential to affect the evolutionary pathway of traits within a population or of whole phylogenetic groups of organisms through direct effects on fitness or by interacting with other traits to affect fitness. Thus, we may expect to see suites of traits associated with a particular pattern of male nuptial gift giving in the context of a particular resource environment and constraints imposed by phylogenetic history. The degree of variability in the resource environment, the relative harshness of the environment, and the level of survival risk associated with obtaining resources should be important environmental descriptors in this case. As a corollary, we should only see some combinations of traits as a transient state in evolutionary time, if at all. Traits at issue here include the size and composition of the nuptial gift, the timing of oogenesis, the mean and variance in number of matings by each sex, mate competition, mate choice, allocation of male-derived resources by females, time and allocation costs to males of mating, and age at maturity for each sex. Some of these traits will determine other traits considered above. For example, the mean and variance in number of matings, combined with total population size, determines the population effective population size.

In particular, male nuptial gifts provide a nearly unique opportunity to examine emerging ideas concerning resource allocation. Since male nutrient donations are an allocation to reproduction by the male, but a meal to the female, they are part of the life history strategy of one sex, but the foraging strategy, broad sense, of the other. Resource allocation links foraging and life history allocation; the study of male nutrient donations will allow us to link not only foraging and allocation, but the life strategies (*sensu* Gatto et al.[149]) of both sexes. Other possibilities abound as well. Internal spermatophores are a form of "stored" nutrients for the female. Questions concerning the use of available vs. stored nutrients as a function of environmental food availability can be addressed in this system. Questions concerning the effects of variation in availability of different nutrient types at different times in the life cycle can be addressed as well. With the expansion of groups in which male nutrient donations have been studied, we can begin to examine the role of phylogenetic history in constraining allocations related to male nutrient donations, thus limiting suites of traits observed in nature. We should also be able to explore circumstances under which convergent evolution is more or less likely to occur.

ACKNOWLEDGMENTS

I thank P. Grandcolas, D. Gwynne, J. McNeil, D. Mullins, C. Nalepa, K. Oberhauser, S. Sakaluk, L. Simmons, L. Svärd, D. Tallamy, and N. Wedell for preprints or discussion, and R. Rutowski and W. Watt for comments on the manuscript.

REFERENCES

1. **Downes, J. A.,** Feeding and mating in the insectivorous Ceratopogoninae (Diptera), *Mem. Entomol. Soc. Can.*, 104, 1978.
2. **Morris, G. K.,** Mating systems, paternal investment and aggressive behavior of acoustic Orthoptera, *Fla. Entomol.*, 62, 9, 1979.
3. **Bell, P. D.,** Opportunistic feeding by the female tree cricket, *Oecanthus nigricornis* (Orthoptera: Gryllidae), *Can. Entomol.*, 112, 432, 1980.
4. **Bidochka, B. J. and Snedden, W. A.,** Effect of nuptial feeding on the mating behaviour of female ground crickets, *Can. J. Zool.*, 63, 207, 1985.
5. **Thornhill, R.,** Sexual selection and nuptial feeding behavior in *Bittacus apicalis* (Insecta: Mecoptera), *Am. Nat.*, 110, 529, 1976.
6. **Thornhill, R. and Alcock, J.,** *The Evolution of Insect Mating Systems*, Harvard University Press, Cambridge, 1983.
7. **Svensson, B. G. and Petersson, E.,** Sex-role reversed courtship behaviour, sexual dimorphism and nuptial gifts in the dance fly, *Empis borealis* (L.), *Ann. Zool. Fennici*, 24, 323, 1987.
8. **Boggs, C. L. and Gilbert, L. E.,** Male contribution to egg production in butterflies: evidence for transfer of nutrients at mating, *Science*, 206, 83, 1979.
9. **Smith, R. L.,** Evolution of exclusive postcopulatory paternal care in the insects, *Fla. Entomol.*, 63, 65, 1980.
10. **Nalepa, C. A. and Jones, S. C.,** Evolution of monogamy in termites, *Biol. Rev.*, 66, 83, 1991.
11. **Tallamy, D. W.,** Nourishment and the evolution of paternal investment in subsocial arthropods, in *Nourishment and Evolution in Insect Societies*, Hunt, J. and Nalepa, C., Eds., Westview Press, Boulder, 1994.

12. **Thornhill, R.,** Sexual selection and paternal investment in insects, *Am. Nat.*, 110, 153, 1976.
13. **Zeh, D. W. and Smith, R. L.,** Paternal investment by terrestrial arthropods, *Am. Zool.*, 25, 785, 1985.
14. **Ridley, M.,** Mating frequency and fecundity in insects, *Biol. Rev.*, 63, 509, 1988.
15. **Parker, G. A. and Simmons, L. W.,** Nuptial feeding in insects: theoretical models of male and female interests, *Ethology*, 82, 3, 1989.
16. **Thornhill, R.,** Adaptive female-mimicking behavior in a scorpionfly, *Science*, 205, 412, 1979.
17. **Gwynne, D. T.,** Nuptial feeding behaviour and female choice of mates in *Harpobittacus similis* (Mecoptera: Bittacidae), *J. Aust. Entomol. Soc.*, 23, 271, 1984.
18. **Downes, J. A.,** The feeding and mating behaviour of the specialized Empidinae (Diptera): observations on four species of *Rhamphomyia* in the high arctic and general discussion, *Can. Entomol.*, 102, 769, 1970.
19. **Kessel, E. L.,** The mating activities of balloon flies, *Syst. Zool.*, 4, 97, 1955.
20. **Thornhill, R.,** *Panorpa* (Mecoptera: Panorpidae) scorpionflies: systems for understanding resource-defense polygyny and alternative male reproductive efforts, *Ann. Rev. Ecol. Syst.*, 12, 355, 1981.
21. **Steele, R. H.,** Courtship feeding in *Drosophila subobscura*. I. The nutritional significance of courtship feeding, *Anim. Behav.*, 34, 1087, 1986.
22. **Leopold, R. A.,** The role of male accessory glands in insect reproduction, *Annu. Rev. Entomol.*, 199, 1976.
23. **Chen, P. S.,** The functional morphology and biochemistry of insect male accessory glands and their secretions, *Annu. Rev. Entomol.*, 29, 233, 1984.
24. **Kasuga, H., Aigaki, T. and Osani, M.,** System for supply of free arginine in the spermatophore of *Bombyx mori*: arginine liberating activities of contents of male reproductive glands, *Insect Biochem.*, 17, 317, 1987.
25. **Cheeseman, M. T. and Gillott, C.,** Long hyaline gland discharge and multiple spermatophore formation by the male grasshopper, *Melanoplus sanguinipes*, *Physiol. Entomol.*, 14, 257, 1989.
26. **Friedel, T. and Gillott, C.,** Contribution of male-produced proteins to vitellogenesis in *Melanoplus sanguinipes*, *J. Insect Physiol.*, 23, 145, 1977.
27. **Osani, M., Aigaki, T., Kasuga, H. and Yonezawa, Y.,** Role of arginase transferred from the vesicula seminalis during mating and changes in amino acid pools of the spermatophore after ejaculation in the silkworm, *Bombyx mori*, *Insect Biochem.*, 16, 879, 1986.
28. **Dickinson, J. L. and Rutowski, R. L. 1989.** The function of the mating plug in the chalcedon checkerspot butterfly, *Anim. Behav.*, 38, 154, 1989.
29. **Davey, K. G.,** *Reproduction in the Insects*, Freeman, San Francisco, 1965.
30. **Rutowski, R. L. and Gilchrist, G. W.,** Copulation in *Colias eurytheme* (Lepidoptera: Pieridae): patterns and frequency, *J. Zool. London (A)*, 209, 115, 1986.
31. **Markow, T. A.,** *Drosophila* males provide a material contribution to offspring sired by other males, *Funct. Ecol.*, 2, 77, 1988.
32. **Gerber, G. H.,** Evolution of the methods of spermatophore formation in pterygotan insects, *Can. Entomol.*, 102, 358, 1970.
33. **Rutowski, R. L., Newton, M., and Schaefer, J.,** Interspecific variation in the size of the nutrient investment made by male butterflies during copulation, *Evolution*, 37, 708, 1983.
34. **Boucher, L. and Huignard, J.,** Transfer of male secretions from the spermatophore to the female insect in *Caryedon serratus* (Ol.): analysis of the possible trophic role of these secretions. *J. Insect Physiol.*, 33, 949, 1987.
35. **Svärd, L. and Wiklund, C.,** Mass and production rate of ejaculates in relation to monandry/polyandry in butterflies, *Behav. Ecol. Sociobiol.*, 24, 395, 1989.
36. **Gwynne, D. T.,** Sexual difference theory: mormon crickets show role reversal in mate choice, *Science*, 213, 779, 1981.
37. **Simmons, L. W.,** Male size, mating potential and lifetime reproductive succes in the field cricket, *Gryllus bimaculatus* (De Geer), *Anim. Behav.*, 36, 372, 1988.
38. **Hayashi, F.,** Male mating costs in two insect species (*Protohermes*, Megaloptera) that produce large spermatophores, *Anim. Behav.*, 45, 343, 1993.
39. **Alexander, R. D. and Otte, D.,** The evolution of genitalia and mating behavior in crickets (Gryllidae) and other Orthoptera, *Misc. Publ. Mus. Zool. Univ. Mich.*, 133, 1967.
40. **Sakaluk, S. K. and Cade, W. H.,** Female mating frequency and progeny production in singly and doubly mated house and field crickets, *Can. J. Zool.*, 58, 404, 1980.
41. **Evans, A. R.,** Mating systems and reproductive strategies in three australian gryllid crickets: *Bobilla victoriae* Otte, *Balamara gidya* Otte and *Teleogryllus commodus* (Walker) (Orthoptera: Gryllidae: Nemobiinae; Trigonidiinae; Gryllinae), *Ethology*, 78, 21, 1988.
42. **Mays, D. L.,** Mating behaviour of Nemobiinae crickets *Hygronemobius*, *Nemobius* and *Pteronemobius* (Orthoptera: Gryllidae), *Fla. Entomol.*, 54, 13, 1971.
43. **Davey, K. G.,** The evolution of the spermatophore in insects, *Proc. R. Entomol. Soc. London Ser. A*, 35, 107, 1960.
44. **Marshall, L. D.,** Male nutrient investment in the Lepidoptera: what nutrients should males invest?, *Am. Nat.*, 120, 273, 1980.

45. **Engebretson, J. A. and Mason, W. H.,** Transfer of ^{65}Zn at mating in *Heliothis virescens, Environ. Entomol.*, 9, 119, 1980.
46. **Pivnick, K. A. and McNeil, J. N.** Puddling in butterflies: sodium affects reproductive success in *Thymelicus lineola, Physiol. Entomol.*, 12, 461, 1987.
47. **Adler, P. H. and Pearson, D. L.,** Why do male butterflies visit mud puddles? *Can. J. Zool.*, 60, 322, 1982.
48. **Arms, K., Feeny, P., and Lederhouse, R. C.,** Sodium: stimulus for puddling behavior by tiger swallowtail butterflies, *Science*, 185, 372, 1974.
49. **Boggs, C. L. and Jackson, L. A.,** Mud puddling by butterflies is not a simple matter, *Ecol. Entomol.*, 16, 123, 1991.
50. **Sculley, C. and Boggs, C.L.** unpublished manuscript.
51. **Mullins, D. E. and Keil, C. B.,** Paternal investment of urates in cockroaches, *Nature*, 283, 567, 1980.
52. **Schal, C. and Bell, W. J.,** Ecological correlates of paternal investment of urates in a tropical cockroach, *Science*, 218, 170, 1982.
53. **Mullins, D. E., Keil, C. B., and White, R. H.,** Maternal and paternal nitrogen investment in *Blattella germanica* (L.) (Dictyoptera; Blattellidae), *J. Exp. Biol.*, 162, 55, 1992.
54. **Sierra, J. R., Woogon, W.-D., and Schmid, H.,** Transfer of cantharidin during copulation from the adult male to the female *Lytta vesicatoria* ("Spanish flies"), *Experientia*, 32, 142, 1976.
55. **Alcock, J. and Hadley, N. F.,** Tests of alternative hypotheses for assortative mating by size: a comparison of two meloid beetles (Coleoptera: Meloidae), *J. Kans. Entomol. Soc.*, 60, 41, 1987.
56. **Achey unpublished, cited in Brower, L. P.,** New perspective on the migration biology of the monarch butterfly, *Danaus plexippus* L. in *Migration: Mechanisms and Adaptive Significance*, Rankin, M. A., Ed., Univ. of Texas Contributions in Marine Science, Austin, TX, 1985, 748.
57. **Brown, K. S.,** Adult-obtained pyrrolizidine alkaloids defend ithomiine butterflies against a spider predator, *Nature*, 309, 707, 1984.
58. **Dussourd, D. E., Ubik, K., Harris, C., Resch, J., Meinwald, J., and Eisner, T.,** Biparental defensive endowment of eggs with acquired plant alkaloid in the moth *Utethesia ornatrix, Proc. Natl. Acad. Sci. U.S.A.*, 85, 992, 1988.
59. **Marshall, L. D. and McNeil, J. N.,** Spermatophore mass as an estimate of male nutrient investment: a closer look in *Pseudaletia unipuncta* (Haworth) (Lepidoptera: Noctuidae), *Funct. Ecol.*, 3, 605, 1989.
60. **Buskirk, R. E., Frohlich, C., and Ross, K. G.,** The natural selection of sexual cannibalism, *Am. Nat.*, 123, 612, 1984.
61. **Morris, G. K., Gwynne, D. T., Klimas, D. E., and Sakaluk, S. K.,** Virgin male mating advantage in a primitive acoustic insect (Orthoptera: Haglidae), *J. Insect Behav.*, 2, 173, 1989.
62. **Hinton, H. E.,** Sperm transfer in insects and the evolution of haemocoelic insemination, in *Insect Reproduction, Symposia of the Royal Entomological Society, London*, Highnam, K. C., Ed., Royal Ent. Soc. London, 2, 95, 1964.
63. **Silbergleid, R. E., Shepherd, J. G., and Dickinson, J. L.,** Eunuchs: the role of apyrene sperm in Lepidoptera?, *Am. Nat.*, 123, 255, 1984.
64. **Boggs, C. L.,** Selection pressures affecting male nutrient investment at mating in heliconiine butterflies, *Evolution*, 35, 931, 1981.
65. **Gwynne, D. T.,** Courtship feeding and the fitness of female katydids (Orthoptera: Tettigoniidae), *Evolution*, 42, 545, 1988.
66. **Boggs, C. L.,** A general model of the role of male-donated nutrients in female insects' reproduction, *Am. Nat.*, 136, 598, 1990.
67. **Wedell, N.,** Mating effort or paternal investment: incorporation rate and cost of male donations in the wartbiter. *Behav. Ecol. Sociobiol.*, 32, 239, 1993.
68. **Greenfield, M. D.,** The question of paternal investment in Lepidoptera: male-contributed proteins in *Plodia interpunctella, Int. J. Invert. Repro.*, 5, 323, 1982.
69. **Boggs, C. L. and Watt, W. B.,** Population structure of pierid butterflies. IV. Genetic and physiological investment in offspring by male *Colias, Oecologia*, 50, 320, 1981.
70. **Bowen, B. J., Codd, C. G., and Gwynne, D. T.,** The katydid spermatophore (Orthoptera: Tettigoniidae): male nutritional investment and its fate in the mated female, *Aust. J. Zool.*, 32, 23, 1984.
71. **Oberhauser, K. S.,** Rate of ejaculate breakdown and intermating intervals in monarch butterflies, *Behav. Ecol. Sociobiol.*, 31, 367, 1992.
72. **Rutowski, R. L.,** Production and use of secretions passed by males at copulation in *Pieris protodice* (Lepidoptera, Pieridae), *Psyche*, 91, 141, 1984.
73. **Huignard, J.,** Transfer and fate of male secretions deposited in the spermatophore of females of *Acanthoscelides obtectus* Say (Coleoptera Bruchidae), *J. Insect Physiol.*, 29, 55, 1983.
74. **Khalifa, A.,** Spermatophore production in Trichoptera and some other insects, *Trans. Entomol. Soc. Lond.*, 100, 18, 1949.
75. **Svärd, L. and Wiklund, C.,** Fecundity, egg weight and longevity in relation to multiple matings in females of the monarch butterfly, *Behav. Ecol. Sociobiol.*, 23, 39, 1988.

76. **Sakaluk, S. K. and Snedden, W. A.**, Nightly calling durations of male sagebrush crickets, *Cyphoderris strepitans*: size, mating and seasonal effects, *Oikos*, 57, 153, 1990.
77. **Simmons, L. W., Teale, R. J., Maier, M., Standish, R. J., Bailey, W. J., and Wither, P. C.**, Some costs of reproduction for male bushcrickets, *Requena verticalis* (Orthoptera: Tettigoniidae): allocating resources to mate attraction and nuptial feeding, *Behav. Ecol. Sociobiol.*, 31, 57, 1992.
78. **Gwynne, D. T.**, Testing parental investment and the control of sexual selection in katydids: the operational sex ratio, *Am. Nat.*, 136, 474, 1990.
79. **Oberhauser, K. S.**, Effects of spermatophores on male and female monarch butterfly reproductive success, *Behav. Ecol. Sociobiol.*, 25, 237, 1989.
80. **Rutowski, R. L.**, The butterfly as an honest salesman, *Anim. Behav.*, 27, 1269, 1979.
81. **Burpee, D. M. and Sakaluk, S. K.**, Repeated matings offset costs of reproduction in female crickets, *Evol. Ecol.*, 7, 1, 1993.
82. **Sakaluk, S. K.**, Spermatophore size and its role in the reproductive behaviour of the cricket, *Gryllodes supplicans* (Orthoptera: Gryllidae), *Can. J. Zool.*, 63, 1652, 1985.
83. **Svärd, L. and Wiklund, C.**, Different ejaculate delivery strategies at first versus subsequent matings in the swallowtail butterfly, *Papilio machaon* L., *Behav. Ecol. Sociobiol.*, 18, 325, 1986.
84. **Royer, L. and McNeil, J. N.**, Male investment in the European corn borer, *Ostrinia nubilalis* (Hübner) (Lepidoptera: Pyralidae): Impact on female longevity and reproductive performance, *Funct. Ecol.*, 7, 209, 1993.
85. **Sakaluk, S. K. and Smith, R. L.**, Inheritance of male parental investment in an insect, *Am. Nat.*, 132, 594, 1988.
86. **Sakaluk, S. K., Burpee, D. M., and Smith, R. L.**, Phenotypic and genetic variation in the stridulatory organs of male decorated crickets, *Gryllodes sigillatus* (Orthoptera: Gryllidae), *Can. J. Zool.*, 70, 453, 1992.
87. **Gwynne, D. T., Bowen, B. J., and Codd, C. G.**, The function of the katydid spermatophore and its role in fecundity and insemination (Orthoptera: Tettigoniidae), *Aust. J. Zool.*, 32, 15, 1984.
88. **Gwynne, D. T.**, Mate selection by female katydids (Orthoptera: Tettigoniidae, *Conocephalus nigropleurum*), *Anim. Behav.*, 30, 734, 1982.
89. **Wedell, N.**, Protandry and mate assessment in the wartbiter *Decticus verrucivorus* (Orthoptera: Tettigoniidae), *Behav. Ecol. Sociobiol.*, 31, 301, 1992.
90. **Svärd, L.**, Parental investment in a monandrous butterfly, *Parargae aegeria*, *Oikos*, 45, 66, 1985.
91. **Reiss, M. J.**, The allometry of reproduction: why larger species invest relatively less in their offspring, *J. Theor. Biol.*, 113, 529, 1985.
92. **Oberhauser, K. S.**, Male monarch butterfly spermatophore mass and mating strategies, *Anim. Behav.*, 36, 1384, 1988.
93. **George, J. A. and Howard, M. G.**, Insemination without spermatophores in the oriental fruit moth, *Grapholitha molesta* (Lepidoptera: Tortricidae), *Can. Entomol.*, 100, 190, 1964.
94. **Zuk, M.**, The effects of gregarine parasites, body size, and time of day on spermatophore production and sexual selection in field crickets, *Behav. Ecol. Sociobiol.*, 21, 65, 1987.
95. **Simmons, L. W.**, Some constraints on reproduction for male bushcrickets, *Requena verticalis* (Orthoptera: Tettigoniidae): diet, size and parasite load, *Behav. Ecol. Sociobiol.*, 32, 135, 1993.
96. **Butlin, R. K., Woodhatch, C. W., and Hewitt, G. M.**, Male spermatophore investment increases female fecundity in a grasshopper, *Evolution*, 41, 221, 1987.
97. **Simmons, L. W. and Bailey, W. J.**, Resource influenced sex roles of zaprochiline tettigoniids (Orthoptera: Tettigoniidae), *Evolution*, 44, 1853, 1990.
98. **Wedell, N. and Arak, A.**, The wartbiter spermatophore and its effect on female reproductive output (Orthoptera: Tettigoniidae, *Decticus verrucivorus*), *Behav. Ecol. Sociobiol.*, 24, 117, 1989.
99. **Wedell, N.**, Sperm competition selects for nuptial feeding in a bushcricket, *Evolution*, 45, 1975, 1991.
100. **Rutowski, R. L., Gilchrist, G. W., and Terkanian, B.**, Female butterflies mated with recently mated males show reduced reproductive output, *Behav. Ecol. Sociobiol.*, 20, 319, 1987.
101. **Markow, T. A., Gallagher, P. D., and Krebs, R. A.**, Ejaculate-derived nutritional contribution and female reproductive success in *Drosophila mojavensis* (Patterson & Crow), *Funct. Ecol.*, 4, 67, 1990.
102. **Gwynne, D. T.**, Male mating effort, confidence of paternity and insect sperm competition, in *Sperm Competition and the Evolution of Animal Mating Systems*, Smith, R. L., Ed., Academic Press, New York, 117, 1984.
103. **Tuomi, J., Hakala, T., and Haukioja, E.**, Alternative concepts of reproductive effort, costs of reproduction and selection in life-history evolution, *Am. Zool.*, 23, 25, 1983.
104. **Woods, H. A., Boggs, C. L., and Karlsson, B.**, unpublished manuscript.
105. **Svärd, L. and McNeil, J. N.**, Female benefit, male risk: Polyandry in the true armyworm *Pseudaletia unipuncta*, *Behav. Ecol. Sociobiol.* 35, 319, 1994.
106. **Boggs, C. L. and Ross, C. L.**, The effect of adult food limitation on life history traits in *Speyeria mormonia* (Lepidoptera: Nymphalidae), *Ecology*, 74, 433, 1993.

107. **Gwynne, D. T. and Dodson, G. N.,** Nonrandom provisioning by the digger wasp, *Palmodes laeviventris* (Hymenoptera: Sphecidae), *Ann. Entomol. Soc. Am.*, 76, 434, 1983.
108. **Boggs, C. L.,** Nutritional and life-history determinants of resource allocation in holometabolous insects, *Am. Nat.*, 117, 692, 1981.
109. **Zalucki, M. P.,** Sex around the milkweed patch — the significance of patches of host plants in monarch reproduction, in *Biology and Conservation of the Monarch Butterfly*, Malcolm, S. B. and Zalucki, M. P., Eds., Natural History Museum of Los Angeles County, Los Angeles, 69, 1993.
110. **Brower, L. P.,** New perspective on the migration biology of the monarch butterfly, *Danaus plexippus* L. in *Migration: Mechanisms and Adaptive Significance*, Rankin, M. A., Ed., University of Texas Contributions in Marine Science, Austin, TX, 748, 1985.
111. **van Hook, T.,** Non-random mating in monarch butterflies overwintering in Mexico, in *Biology and Conservation of the Monarch Butterfly*, Malcolm, S. B. and Zalucki, M. P., Eds., Natural History Museum of Los Angeles County, Los Angeles, 49, 1993.
112. **Wells, H., Wells, P. H., and Rogers, S. H.,** Is multiple mating an adaptive feature of monarch butterfly winter aggregation?, in *Biology and Conservation of the Monarch Butterfly*, Malcolm, S. B. and Zalucki, M. P., Eds., Natural History Museum of Los Angeles County, Los Angeles, 61, 1993.
113. **Clutton-Brock, T. H.,** *Reproductive Success: Studies of Individual Variation in Contrasting Breeding Systems*, University of Chicago Press, Chicago, 1988.
114. **Pitnick, S., Markow, T. A., and Riedy, M. F.,** Transfer of ejaculate and incorporation of male-derived substances by females in the Nannoptera species group (Diptera: Drosophilidae), *Evolution*, 45, 774, 1991.
115. **Boldyrev, B. T.,** Contributions à l'étude de las structure des spermatophores et des particularités de la copulation chez Locustodea et Gryllodea, *Horae Soc. Entomol. Ross.*, 41, 1, 1915.
116. **Gwynne, D. T.,** Male nutritional investment and the evolution of sexual differences in the Tettigoniidae and other Orthoptera, in *Orthopteran Mating Systems: Sexual Competition in a Diverse Group of Insects*, Gwynne, D. T. and Morris, G., Eds., Westview Press, Boulder, 337, 1983.
117. **Trivers, R. L.,** Parental investment and sexual selection, in *Sexual Selection and the Descent of Man*, Campbell, B., Ed., Aldine-Atherton, Chicago, 1972.
118. **Darwin, C.,** *The Descent of Man and Selection in Relation to Sex*, John Murray, London, 1871.
119. **Fisher, R. A.,** *The Genetical Theory of Natural Selection*, Oxford University Press, Oxford, 1930.
120. **Bateman, A. J.,** Intra-sexual selection in *Drosophila*, *Heredity*, 2, 349, 1948.
121. **Williams, G. C.,** *Adaptation and Natural Selection*, Princeton University Press, Princeton, NJ, 1966.
122. **Sakaluk, S. K.,** Male crickets feed females to ensure complete sperm transfer, *Science*, 223, 609, 1984.
123. **Gwynne, D. T.,** Courtship feeding in katydids (Orthoptera: Tettigoniidae): investment in offspring or in obtaining fertilizations?, *Am. Nat.*, 128, 342, 1986.
124. **Reinhold, K. and Heller, K.-G.,** The ultimate function of nuptial feeding in the bushcricket *Piecilimon veluchianus* (Orthoptera: Tettigoniidae: Phaneropterinae), *Behav. Ecol. Sociobiol.*, 32, 55, 1993.
125. **Alexander, R. D. and Borgia, G.,** On the origin and basis of the male-female phenomenon, in *Sexual Selection and Reproductive Competition in Insects*, Blum, M. S. and Blum, N. A., Eds., Academic Press, New York, 1979.
126. **Low, B. S.,** Environmental uncertainties and the parental strategies of marsupials and placentals, *Am. Nat.*, 112, 197, 1978.
127. **Quinn, J. S. and Sakaluk, S. K.,** Prezygotic male reproductive effort in insects: why do males provide more than sperm?, *Fla. Entomol.*, 69, 84, 1986.
128. **Emlen, S. T. and Oring, L. W.,** Ecology, sexual selection, and the evolution of mating systems, *Science*, 197, 215, 1977.
129. **Gwynne, D. T.,** Role-reversal in katydids: habitat influences reproductive behaviour (Orthoptera: Tettigoniidae, *Metaballus* sp.), *Behav. Ecol. Sociobiol.*, 16, 355, 1985.
130. **Clutton-Brock, T. H. and Parker, G. A.,** Potential reproductive rates and the operation of sexual selection, *Q. Rev. Biol.*, 67, 437, 1992.
131. **Gilbert, L. E.,** Postmating female odor in *Heliconius* butterflies: a male-contributed antiaphrodisiac?, *Science*, 193, 419, 1976.
132. **Rutowski, R. L.,** The form and function of ascending flights in *Colias* butterflies, *Behav. Ecol. Sociobiol.*, 3, 163, 1978.
133. **Rutowski, R. L.,** Evidence for mate choice in a sulphur butterfly (*Colias eurytheme*), *Z. Tierpsychol.*, 70, 103, 1985.
134. **Watt, W. B., Carter, P. A., and Donohue, K.,** Females' choice of "good genotypes" as mates is promoted by an insect mating system, *Science*, 233, 1187, 1986.
135. **Borgia, G.,** Sexual selection and the evolution of mating systems, in *Sexual Selection and Reproductive Competition in Insects*, Blum, M. S. and Blum, N. A., Eds., Academic Press, New York, 19, 1979.
136. **Gwynne, D. T. and Simmons, L. W.,** Experimental reversal of courtship roles in an insect, *Nature*, 346, 172, 1990.

137. **Gwynne, D. T.,** Sexual selection and sexual differences in mormon crickets (Orthoptera: Tettigoniidae, *Anabrus simplex*), *Evolution*, 38, 1011, 1984.
138. **Rutowski, R. L.,** Courtship solicitation by females of the checkered white butterfly, *Pieris protodice*, *Behav. Ecol. Sociobiol.*, 7, 113, 1980.
139. **Rutowski, R. L., Long, C. E., Marshall, L. D., and Vetter, R. S.,** Courtship solicitation by *Colias* females (Lepidoptera: Pieridae), *Am. Midl. Nat.*, 105, 334, 1981.
140. **Simmons, L. W.,** Female choice in the field cricket *Gryllus bimaculatus* (De Geer), *Anim. Behav.*, 34, 1463, 1986.
141. **Boggs, C. L.,** Resource allocation and reproductive strategies in several heliconiine species, Ph.D. dissertation, University of Texas at Austin, 1979.
142. **Boggs, C. L.,** personal observation.
143. **Watt, W. B.,** Eggs, enzymes, and evolution: natural genetic variants change insect fecundity, *Proc. Natl. Acad. Sci. U.S.A.*, 89, 10608, 1992.
144. **Rutowski, R. L.,** Epigamic selection by males as evidenced by courtship partner preferences in the checkered white butterfly (*Pieris protodice*), *Anim. Behav.*, 30, 108, 1982.
145. **Sugawara, T.,** Stretch reception in the bursa copulatrix of the butterfly, *Pieris rapae crucivora*, and its role in behaviour, *J. Comp. Physiol.*, 130, 191, 1979.
146. **Gwynne, D. T.,** Courtship feeding in katydids benefits the mating male's offspring, *Behav. Ecol. Sociobiol.*, 23, 373, 1988.
147. **Boggs, C. L.,** unpublished data.
148. **Markow, T. A. and Ankney, P. F.,** *Drosophila* males contribute to oogenesis in a multiple mating species, *Science*, 224, 302, 1984.
149. **Gatto, M., Matessi, C., and Slobodkin, L. B.,** Physiological profiles and demographic rates in relation to food quantity and predictability: an optimization approach, *Evol. Ecol.*, 3, 1, 1989.

INDEX

A

B

H

I

T

For Product Safety Concerns and Information please contact our EU representative GPSR@taylorandfrancis.com Taylor & Francis Verlag GmbH, Kaufingerstraße 24, 80331 München, Germany